Collins

Revision

NEW GCSE SCIENCE

Science and Additional Science

Foundation and Higher

for Edexcel

Authors: **Alison Dennis**
Sarah Mansel
Caroline Reynolds
Gemma Young

Revision Guide +
Exam Practice Workbook

Contents

Contents

Classification and naming species

Grouping living things

- We can classify (group) living organisms according to their shared characteristics.
- Some characteristics (such as feathers) are unique to a group, while others (such as a **backbone**) are common to several groups.
- Animals with backbones are called **vertebrates**. There are five vertebrate groups:
  - *Fish* exchange oxygen and carbon dioxide across gills, and lay eggs (**oviparous**) that are fertilised externally.
  - *Amphibians* exchange gases through their moist, permeable skin and lay externally fertilised eggs.
  - *Reptiles* and *birds* exchange gases via their lungs and lay internally fertilised eggs.
  - *Mammals* give birth to live young, which grow inside the body of the mother. Eggs are fertilised internally.
- Vertebrates are either cold-blooded (**poikilotherms**), which means they can't control their internal body temperature, or warm-blooded (**homeotherms**), which means they can.
- Fish, amphibians and reptiles are poikilotherms. Birds and mammals are homeotherms.

- All living things belong to one of five groups called **kingdoms**:
  - **Animalia** are **multicellular**; cells do not have chlorophyll or a cell wall; they feed heterotrophically (find food from their environment).
  - **Plantae** are multicellular; cells have chlorophyll and a cellulose cell wall; they obtain food autotrophically (make food by photosynthesis).
  - **Fungi** are multicellular; cells do not have chlorophyll and are surrounded by a cell wall not made of cellulose; they feed **saprophytically** (on dead organic matter).
  - **Protoctista** are **unicellular** (except seaweed), with a distinct nucleus.
  - **Prokaryotes** are unicellular, but without a distinct nucleus.
- Within each kingdom, living things are further divided into smaller and smaller groups: phylum; **class; order; family; genus; species**.
- An example of a shared characteristic among the phylum **chordata** is a rod that supports the body, such as the backbone in vertebrates.

> **Remember!**
> Not all living things fit neatly into categories based on anatomy and reproductive methods. Also, new organisms with unusual characteristics are always being discovered. This makes it difficult to place them into distinct groups.

- The more characteristics organisms have in common, the more closely they are related – they share a **common ancestor**.
- Viruses do not belong to a kingdom because scientists cannot decide whether they are living organisms.

What is a species?

- Most biologists define species as organisms that are capable of breeding together to produce **fertile** offspring. This means their offspring can reproduce.
- The offspring of two different species are called **hybrids** and they are usually infertile.
- However, there are complications with this definition of a species because:
  - not all hybrids are sterile (many plant hybrids are fertile)
  - not all organisms reproduce sexually (some produce asexually).

Binomial classification

- All organisms have a two-part (**binomial**) name. For example, the species name for human is *Homo sapiens*. *Homo* is our genus name, *sapiens* is our species name.
- The binomial naming system is used because it prevents confusion over having many different names for the same species.
- Binomial classification is important because it enables scientists to:
  - communicate information about the thousands of different species
  - recognise areas of great **biodiversity** that should be targets for conservation efforts.

Improve your grade

Naming new species

Higher: Scientists exploring an area of the South American rainforests discovered several species of frog. They gave each species a new binomial name.

Explain why classifying organisms in this way is important.

AO1 [3 marks]

Identification, variation and adaptation

Variation within a species

- If you come across a species but you don't know what it is, a **key** will help you to identify it.
- The descriptions in a key come as opposite statements.
- Choosing one statement that fits the species will lead you to the next.

Using a key. You find an animal with eight legs. How do you identify what kind of animal it is?

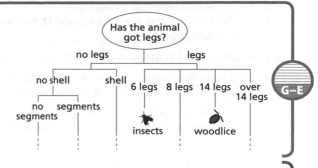

G–E

- Members of the same species have different characteristics. For example, humans vary in their hair colour and blood groups. This is called **variation**.
- Variation can be either continuous or discontinuous:
  - In **continuous variation**, characteristics are spread over a range of values. Height in humans is an example – a few people are either very short or very tall, and there is a full range of 'in-betweens' (intermediates).
  - Tongue rolling is a characteristic that shows **discontinuous variation**. There are no intermediates, but rather distinctly different groups: people who can roll their tongue and people who cannot.

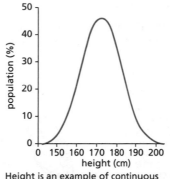

Height is an example of continuous variation.

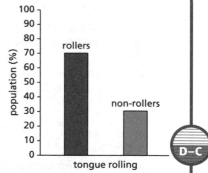

Tongue rolling is an example of discontinuous variation.

D–C

EXAM TIP

You may be shown graphs of variation data and asked to identify which type is being shown. Continuous variation is displayed as a line graph; discontinuous variation is displayed as a bar chart.

Variations complicate identification

- Variations such as those seen in hybrid ducks – which have characteristics of both parent species – can make classification complicated.
- **Ring species** refers to a chain of related species that are closely connected geographically.
- Species within the chain show variation, but they can still interbreed and produce hybrid offspring.
- However, the variations between the species at each end of the chain are so great that they cannot interbreed. They are distinct species.

B–A*

Adaptations

- **Adaptations** are the characteristics of an organism that enable it to survive in its environment.
  - The desert fox is adapted to survive in hot, dry conditions. It has large ears that lose heat and help to cool it. Its light-coloured fur reflects heat and provides camouflage.
  - The polar bear is adapted to survive in very cold conditions. Thick, white fur insulates and camouflages it.

G–E

- Animals living in cold regions are usually larger than those living in warmer environments.
- Large bodies have a smaller surface area relative to their mass than smaller ones. Heat is lost from the body at its surface, so if a body is larger, proportionately less heat is lost from it.

D–C

- Some species of bacteria have adapted in order to live in the extreme environment around deep-sea **hydrothermal vents**, where temperatures can reach more than 90 °C.
- Adaptations of these bacteria allow them to use a chemical produced by the vents (hydrogen sulfide) to produce energy.

B–A*

Improve your grade

Polar bear adaptations
Foundation: Explain how the fur of the polar bear is a good adaptation for surviving in the Arctic.

AO2 [4 marks]

Evolution

G–E

Darwin and his theory

- Charles Darwin was the first person to propose how **evolution** takes place through **natural selection.**

- He described his ideas in a book titled *On the Origin of Species*, published in 1859.

- His theory was a great shock, as people at the time believed in the special place of humans in the natural world and in a benevolent creator or God.

- Darwin incorporated the ideas and observations of other people, including Alfred Russel Wallace, into his theory of natural selection.

Evolution in action

- These ideas are central to Darwin's theory:
  - There is variation in characteristics between individuals of the same species.
  - Overproduction means that organisms produce more offspring than will survive to adulthood.
  - There is a struggle for existence, or competition, between individuals for resources.
  - Individuals best suited to compete for resources (better **adapted**) will survive to reproduce and pass on the gene controlling their advantageous traits to their offspring.
  - It is possible that individuals less suited for competition may become extinct.

- The result of individuals becoming more suited to a particular way of life means that, over time, new species emerge.

- You can see this in the many different species of birds that developed from a common ancestor. The change is called evolution.

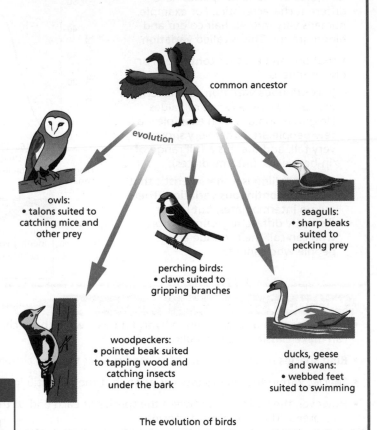

common ancestor

evolution

owls:
• talons suited to catching mice and other prey

perching birds:
• claws suited to gripping branches

seagulls:
• sharp beaks suited to pecking prey

woodpeckers:
• pointed beak suited to tapping wood and catching insects under the bark

ducks, geese and swans:
• webbed feet suited to swimming

The evolution of birds

EXAM TIP

You may be asked to explain how certain species evolved. Do this by applying the stages of natural selection: variation, overproduction, struggle for existence, survival and reproduction.

Speciation

- **Speciation** (the formation of new species) provides evidence for the natural selection of organisms.

- Geographical isolation is one way speciation occurs. For example, on one side of a newly formed mountain range the climate might be wetter and colder than on the other side.

- Over many generations, the part of the population living on the colder, wetter mountainside might change through natural selection to enable it to better withstand the harsher climate.

- Eventually, the difference in characteristics is so great that the population living in the harsher climate becomes a new species.

Improve your grade

Human evolution

Higher: Scientists believe that humans evolved from ape-like animals. Millions of years ago the Earth became drier and forests were replaced by grasslands. Before this change all apes walked on four feet. Afterwards, populations of apes emerged that walked upright on two feet and were able to see further across the grassland.

Use your knowledge of natural selection to explain why apes evolved in this way. *AO1* [3 marks]

Genes and variation

Genetic variation

- There are two types of variation:
  - Environmental variations are **acquired characteristics** caused by environmental factors.
  - Genetic variations are inherited characteristics caused by **mutation** or reproduction.
- Inherited characteristics are controlled by **genes**, which are sections of **chromosomes**. Chromosome are long molecules of **DNA** tightly wound around proteins.
- Genes enable cells to make proteins, which result in characteristics such as eye colour and leg length.

- Each sperm and egg contains only a half-set of chromosomes – in the case of humans this means 23 (human cells have 46 chromosomes).
- When the sperm and the egg fuse during **fertilisation**, the half-set of chromosomes from the male pairs up with the half-set of chromosomes from the female, to form a full set.
- Our parents pass on their genes to us (50% from each parent) during fertilisation. We therefore inherit their characteristics.
- Genetic variation between parents and offspring may also be caused by:
  - Crossing over: the exchange of bits of chromosomes and their genes between chromosome pairs.
  - Gene mutations: changes in the number of chromosomes or in the genes that chromosomes carry.

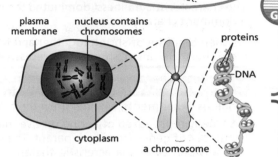

Remember!
Some characteristics can be caused by both inherited and environmental factors. For example, the genes that control tallness can be inherited, but a child will not reach their potential height without a healthy diet.

G–E

plasma membrane · nucleus contains chromosomes · proteins · DNA · cytoplasm · a chromosome

Chromosomes exist in the nucleus of most types of cell. Each gene is a section of a DNA molecule.

D–C

Supporting Darwin's theory

- There is a 99% similarity between human DNA and chimpanzee DNA, but only 88% between human and horse DNA. This illustrates how organisms have evolved over time, but share a common ancestry.
- Bacterial resistance to antibiotics is an example of evolution in action (see page 17 for how this happens).
- Darwin's *On the Origin of Species* is an example of how scientific knowledge was disseminated. Through correspondence and presentations, Darwin discussed his theory both before and after his book's publication.
- Today, the scientific community plays an important role in **validating** evidence:
  - Research results are checked anonymously by other scientists (**peer review**).
  - If validated, the work is published in **scientific journals** available to the wider community.
  - **Conferences** are held to communicate new ideas in a particular area of research.

B–A*

Genetic terms

- Each chromosome in a pair carries the same genes in exactly the same place (**locus**). However, chromosome pairs may have different **alleles** on each chromosome.
- An allele is one form of a gene, e.g. the gene for eye colour has many different alleles, causing blue, green or brown eyes.

G–E

- If the alleles of a pair are the same, the individual is **homozygous** for the characteristic.
- If the alleles are different, the individual is **heterozygous**.
- An allele may be **dominant** or **recessive**. Dominant alleles mask the effects of a recessive allele.
- All of the characteristics that make up an individual are their **phenotype**.
- All of the genes are the individual's **genotype**.

D–C

Family pedigrees

- **Pedigree** charts show the way characteristics of related individuals pass from one generation to the next.
- They provide helpful information about the purity of **lineage** of plants and animals, which is important to farmers.

B–A*

Improve your grade

Hair-colour inheritance
Foundation: Katy has red hair. Both her parents have brown hair.

Explain how Katy inherited red hair, even though her parents do not have it.

AO2 [3 marks]

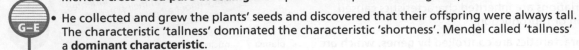
Monohybrid inheritance

Mendel's experiments

G–E

- Gregor Mendel (1822–84) was a priest who studied how single characteristics were inherited in pea plants. This type of inheritance is called **monohybrid**.

- Mendel **cross-bred pure-breeding** short plants with pure-breeding tall plants.

- He collected and grew the plants' seeds and discovered that their offspring were always tall. The characteristic 'tallness' dominated the characteristic 'shortness'. Mendel called 'tallness' a **dominant characteristic**.

- He then bred the offspring together and found that some of these plants were tall and some were short. He called 'shortness' a **recessive characteristic**.

How characteristics are inherited

D–C

- From his experiments, Mendel drew the following conclusions:
  - Sexually reproduced offspring receive the same number of genes from each parent. The development of any particular characteristic, therefore, must be controlled by a pair of genes (one from each parent).
  - Alleles must split when **gametes** (sex cells) form.

- Mendel used letters to symbolise alleles. For example, he used 'T' to show the allele which controls 'tallness' in pea plants and 't' to show the allele which controls 'shortness'.

- The **Punnett squares** opposite summarise the results of Mendel's crosses between pure-breeding parent plants and first-generation plants.

EXAM TIP

You may be asked to create a Punnett square. The question may tell you which letters to use to represent the alleles. If not, use any upper-case letter for the dominant allele and its corresponding lower-case letter for the recessive allele. Be sure to choose letters that look different, e.g. Aa not Zz.

	Cross: TT × tt		Parent plants
Parental gametes	t	t	Pure-breeding recessive parent
Pure-breeding dominant parent — T	Tt	Tt	First-generation plants
T	Tt	Tt	

	Cross: Tt × Tt		First-generation plants
First-generation gametes	T	t	
T	TT	Tt	Second-generation plants
t	Tt	tt	

The results of Mendel's experiments.

Probabilities

B–A*

- Punnett squares are used to predict the probabilities of outcomes from crosses.

- First-generation plants each have two different alleles (heterozygous). However, all of the plants are tall because the T allele is dominant, and this masks the effect of the t allele, which is recessive.

- Not all of the second-generation plants have the same combination of alleles:
  - 50% are heterozygous (Tt)
  - 25% are pure-breeding tall (TT) (homozygous, dominant)
  - 25% are pure-breeding short (tt) (homozygous, recessive).

- The characteristic height of plants separates in the second generation in a ratio of three tall plants to one short plant.

- The phenotype 'tall' is expressed in either homozygous or heterozygous plants because the allele T is dominant.

- The phenotype 'short' is expressed only in homozygous plants because the allele t is recessive.

⦿ How science works

You should be able to:

- format data into diagrams such as Punnett squares and pedigree charts

- interpret the data and predict outcomes based on it.

⦿ Improve your grade

Mendel's experiments

Foundation: Mendel discovered that the colour of pea flowers was controlled by a single gene. The red allele (R) was dominant, and the white allele (r) was recessive. He crossed a pure-breeding red-flowered plant with a pure-breeding white-flowered plant. What would be the genotype and phenotype of the pea plants that were produced? Use a genetic diagram to help you. *AO2* [4 marks]

Genetic disorders

Gene mutations

- A change in a gene's DNA is called a **mutation**.
- As genes carry instructions for building a protein, a mutation may alter a type of protein, or cause no protein to be made at all.
- If sperm and eggs carry a mutation, then any offspring will inherit the mutated gene.
- **Genetic disorders** are the result of inheriting gene mutations.

G–E

When things go wrong

- **Cystic fibrosis** and **sickle cell disease** are examples of inherited recessive genetic disorders.
- Cystic fibrosis affects the movement of fluid in and out of cells, causing a thick, sticky **mucus** to form, particularly in the lungs and digestive tract.
- Symptoms of cystic fibrosis include:
  - mucus blocking the airways of the lungs, causing breathing difficulties
  - lung infections because of bacteria becoming trapped in the mucus
  - problems digesting food, which can lead to malnutrition
  - bone disease.
- Treatments for cystic fibrosis include physiotherapy and massage, medication, and a nourishing diet.
- Sickle cell disease is caused by a mutation that alters **haemoglobin** molecules and causes them to absorb less oxygen. This also results in the red blood cells becoming sickle-shaped.
- Symptoms of sickle cell disease include:
  - feeling weak and tired
  - sudden pain, known as a **sickle cell crisis**, caused by sickled red blood cells forming clumps in the bloodstream, blocking blood flow to organs and causing organ damage.
- Treatments for sickle cell disease include medication, lots of fluids and blood transfusions.

D–C

Inheriting genetic disorders

- If a mutated allele is dominant, or a person inherits two copies of a mutated recessive allele, then the individual in question will be affected if the mutated allele is the cause of a genetic disorder.
- The risk of someone inheriting a particular disorder can be predicted using pedigree analysis. This looks at a disorder's inheritance pattern and predicts the risk for future generations.
- In pedigree charts, generations are indicated from the top using Roman numerals, and members of each generation are numbered from the left.
- In the pedigree chart opposite I-1 and I-2 are carriers because II-3 (one of their children) has the disorder.
- II-4 and II-5 have a 1 in 4 probability of being a carrier.
- III-5 has a 1 in 4 probability of being a carrier because they have a sibling with the disorder.

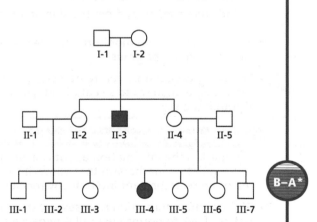

A sample pedigree analysis for cystic fibrosis.
The squares are males and the circles females.
A coloured shape indicates an affected individual.

B–A*

EXAM TIP

You may be asked to analyse the outcomes of genetic crosses using probabilities, ratios or percentages. In the example above, a child having cystic fibrosis is 1 in 4 (probability), 1: 4 (ratio) or 25% (percentage).

Improve your grade

Sickle cell disease inheritance

Higher: Ben carries the mutated gene for sickle cell disease in his sperm. Janet is neither affected by the disease nor is she a carrier of it.

What is the probability of their children having sickle cell disease? Show how you worked out your answer.

AO2 [3 marks]

Homeostasis and body temperature

Constant conditions

G–E

- Living things keep their body's internal environment stable, or constant (no changes). This is called **homeostasis**.

- Regulating body water content and body temperature are examples of homeostasis.

- The process by which the body regulates water content is called **osmoregulation**.
  - Signals to the brain send information about the water content of the blood.
  - The brain sends signals – **hormones** – to the kidneys, to regulate the amount of water they remove in urine.

D–C

- All examples of homeostasis are self-regulating – that is, the body adjusts automatically to keep the internal environment stable.

- The triggers for these responses are called stimuli.

- For example, we need to keep our body temperature at 37 °C. The stimuli that trigger responses are the body becoming either too warm or too cold.

> **Remember!**
> Human body temperature needs to remain at around 37 °C because the **enzymes** that control the chemical activity of our cells work best at this temperature.

Thermoregulation

G–E

- The body regulates its temperature through **thermoregulation**.

- Responses to *increased* body temperature include:
  - Increased blood flow through the blood vessels in the skin, making you look flushed. More blood causes more heat to be lost from the skin's surface.
  - Sweat is produced. Heat from the body evaporates the water in sweat, transferring heat away from the skin.
  - Body hair lies flat against the skin, preventing air becoming trapped next to it. Air is a poor heat conductor, so the absence of an air layer allows more heat to escape.

- Responses to *decreased* body temperature include:
  - Less blood flows through the blood vessels, which means less heat is lost from the skin's surface.
  - Shivering is caused by tiny muscles under the skin contracting and relaxing very quickly. This causes the muscle cells to release heat.
  - Body hair rises away from the skin, trapping a layer of air next to it to insulate the body.

D–C

- Body temperature is monitored by the thermoregulatory centre, in the part of the brain called the **hypothalamus**.

- The thermoregulatory centre detects the temperature of the blood, processes the information and sends nerve impulses to sweat glands and hair erector muscles, which control shivering and blood flow through the skin.

Vasoconstriction and vasodilation

B–A*

- **Feedback** is the information about changes in a self-regulating mechanism that allows it to adjust, maintaining a constant internal environment.

- Because the information reverses any change away from normal back to normal, we call it negative feedback.

- **Vasoconstriction** refers to the narrowing of blood vessels in the skin. Blood flow through the skin is reduced, therefore heat loss is decreased and body temperature rises.

- **Vasodilation** refers to the widening of blood vessels. Blood flow is increased and more heat is lost through the skin.

- Nerve impulses pass along the nerves from the thermoregulatory centre to the muscles in the walls of blood cells, stimulating contraction and causing them to narrow. When the muscles relax the blood vessels widen.

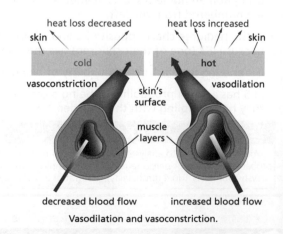

Vasodilation and vasoconstriction.

Improve your grade

Cooling down

Foundation: Sandeep is running a marathon. When he gets too hot he begins to sweat. Explain how sweating cools the body down.

AO2 [2 marks]

Senses and the nervous system

Nerve cells and the nervous system

- A **neurone** is a nerve cell that consists of a cell body with thin fibres stretching out from it. The fibres carry electrical impulses.

- Bundles of neurones form nerves and these nerves form the nervous system:
  - The central nervous system (CNS) consists of the brain and spinal cord.
  - The peripheral nervous system consists of the nerves connecting the sense organs with the central nervous system.

- Nerves consist of different types of neurone, which send electrical impulses in particular directions:
  - Sensory neurones send impulses from **receptors** in the sense organs to the central nervous system.
  - Motor neurones send impulses from the central nervous system to muscles and glands.

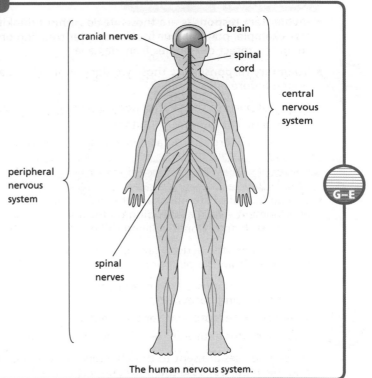

The human nervous system.

Receptors

- Receptors in the sensory neurone fibres detect stimuli. They convert them into electrical impulses (**nerve impulses**) and send them along neurones to muscles and glands.

- Different types of receptor detect different types of stimuli:
  - photoreceptors detect light
  - thermoreceptors detect changes in body temperature.

- Some types of receptor can be found all over the body. For example, thermoreceptors are in the skin.

- Other types of receptor are concentrated in sense organs. For example, photoreceptors form a layer of cells in the eye called the **retina**.

- Sensitivity to touch depends on:
  - the force bearing down on an area of skin
  - the number of touch receptors in the area.

- Fingertips are very sensitive to touch because they contain many touch receptors.

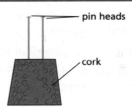

A simple touch-tester using a cork and pins.

EXAM TIP

You can test skin sensitivity by using a touch-tester. If your blindfolded volunteer can feel two pins, the chosen area has many touch receptors and shows high sensitivity. If they can only feel one pin, the area is less sensitive.

Dendrons and axons

- Sensory and motor neurones have similarities and differences.

- They both have a cell body and fine branches called **dendrites**.

- The fibre that carries electrical impulses from the cell body to the dendrites is called the **axon**. Motor neurones have very long axons, whereas sensory neurones have short ones.

- The fibre that carries electrical impulses to the cell body is called the **dendron**.

Improve your grade

Paralysis

Foundation: An injury that results in a broken spine may cause a person to be paralysed. Explain why.

AO1 [3 marks]

Responses and coordination

Responding to stimuli

- **Involuntary responses** are those we do without thinking – they are automatic (for example, pulling your foot away after treading on a sharp pin). We have them to protect our bodies from damage.

- **Voluntary responses** are those we think about (for example, checking your mobile phone before answering it).

- Different parts of the central nervous system coordinate different **responses**:
  - The brain coordinates voluntary responses.
  - The spinal cord coordinates involuntary responses.

Stimuli and response pathway

- Stimuli (plural of **stimulus**) are changes in the environment – such as stepping on the pin – which cause an action in a living thing.

- Responses are the actions taken, such as jerking the foot away from the pin's point.

- Stimuli and responses are linked by the nervous system in the following pathway:

  stimulus → receptor → sensory neurone →

  CNS → motor neurone → effector → response

- Nerve impulses are sent to the effectors. Muscles are effectors and the impulses cause them to contract (shorten).

- Receptors are linked to effectors by a chain of neurones.

- The fibres at the end of one neurone are separated from the beginning of the next neurone by tiny gaps called **synapses**.

- **Neurotransmitters** are chemical messengers that carry information across the synapse.

- The myelin sheath is a fatty substance surrounding dendrons and axons that speeds up the nerve impulses along neurons.

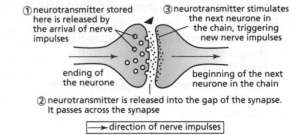

① neurotransmitter stored here is released by the arrival of nerve impulses

③ neurotransmitter stimulates the next neurone in the chain, triggering new nerve impulses

ending of the neurone

beginning of the next neurone in the chain

② neurotransmitter is released into the gap of the synapse. It passes across the synapse

→ direction of nerve impulses

How a nerve impulse travels across a synapse.

The reflex arc

- The involuntary behaviour of the person pulling their foot away after stepping on a pin is known as a **reflex response**.

- Reflex responses are automatic and usually fast. They help to protect the body from damage.

- A reflex response is brought about by a chain of nerves called a **reflex arc**.

- The diagram below shows how the reflex arc works. The numbers indicate the path of the nerve impulses.

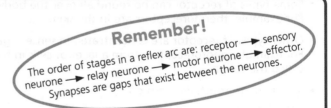

Remember!

The order of stages in a reflex arc are: receptor → sensory neurone → relay neurone → motor neurone → effector. Synapses are gaps that exist between the neurones.

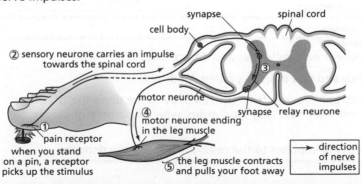

synapse   spinal cord

cell body

② sensory neurone carries an impulse towards the spinal cord

③

motor neurone

synapse   relay neurone

④ motor neurone ending in the leg muscle

① pain receptor

when you stand on a pin, a receptor picks up the stimulus

⑤ the leg muscle contracts and pulls your foot away

→ direction of nerve impulses

Nerve impulses passing from receptors to effectors along the reflex arc.

Improve your grade

Synapses

Higher: Explain how an impulse travels over a synapse.

AO1 [3 marks]

Hormones and diabetes

Hormones

- **Hormones** are chemicals produced by **endocrine glands** in the body and released into the blood. They circulate in the bloodstream and affect different tissues and organs.

- Hormones help to regulate the body's activities and maintain **homeostasis**.

- The pancreas is an organ that produces two hormones: **insulin** and **glucagon**. These help regulate blood glucose levels.

G–E

Regulating blood glucose

- Blood glucose level is normally about 90 mg per 100 cm³ of blood, but it rises or falls depending on circumstances:
  - It rises following a meal, as digested food is absorbed from the intestine into the bloodstream.
  - It falls during exercise, as vigorously contracting muscles use extra glucose for energy.

D–C

- The pancreas monitors the level of blood glucose and triggers a response to return the levels to normal if they change:
  - Insulin is released from the pancreas when the level of blood glucose is high. It helps convert glucose into another type of carbohydrate called **glycogen**, which is stored in liver and muscle tissue.
  - Glucagon is released from the pancreas when the level of blood glucose is low. It promotes the conversion of glycogen into glucose, which is released into the bloodstream.

- Hormones are **specific** to their **target tissue** – they can only bind to the membrane of their target tissue's cells and not to any others.

B–A*

Diabetes

- **Diabetes** is a condition in which the body cannot properly regulate its blood glucose level so it becomes dangerously high, increasing the risk of serious health problems.

- There are two types of diabetes:
  - Type 1 occurs when the pancreas does not produce enough insulin.
  - Type 2 occurs when the pancreas still produces insulin but the target tissues – liver and muscles – become **insensitive** to it.

G–E

- Type 1 diabetes usually occurs in younger people. Type 2 diabetes can develop as people get older.

- People with Type 1 diabetes usually need daily injections of insulin, into the body's **subcutaneous fat** (such as the thigh), to reduce blood glucose levels.

- The amount of insulin required for Type 1 diabetes depends on factors such as diet and the amount of exercise a person does.

D–C

- People with Type 2 diabetes can regulate their blood glucose level by careful eating, regular exercise and losing weight. However, drugs are sometimes needed to help control blood glucose levels.

- Type 1 diabetes is the result of either:
  - an **auto-immune disease** (when a person's immune system destroys the pancreatic cells that produce insulin)
  - a genetic disorder (a **mutation** of the gene encoding the production of insulin).

- Recent research has indicated a strong correlation between obesity and the development of Type 2 diabetes.

- One way of determining if someone is overweight is to measure their **body mass index (BMI)**:

$$\text{BMI}\left(\frac{\text{kg}}{\text{m}^2}\right) = \frac{\text{mass (kg)}}{\text{height (m)}^2}$$

B–A*

> **Remember!**
> BMI is a measure of body fat that applies to normal-sized adult men and women. A body-builder's BMI will be high, but this does not mean he is overweight. His body will have a lot of muscle, which has a higher density than fat.

Improve your grade

Calculating BMI

Higher: Susan is a normal-sized adult woman. She is 1.6 m tall and has a mass of 82 kg.

Work out her BMI by using this equation: $\text{BMI}\left(\frac{\text{kg}}{\text{m}^2}\right) = \frac{\text{mass (kg)}}{\text{height (m)}^2}$

Use the graph opposite to explain how much of a risk Susan has of developing Type 2 diabetes.

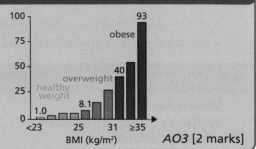

AO3 [2 marks]

Plant hormones

Plant movements

G–E

- Plants can move very slowly as a result of growth.
- When the movements are the result of stimuli coming mainly from one direction, then the response is called a **tropism**.
- The tropism is **positive** if plants grow towards the stimulus and **negative** if they grow away from it.

Auxin

D–C

- Plant **hormones** such as **auxin** are responsible for tropisms.
- **Phototropism** refers to the response of plants to the stimulus of light.
- **Gravitropism** (or **geotropism**) refers to the response of plants to the stimulus of gravity.
- Phototropism can be investigated using cress seeds. For example, by putting cress seeds on a sunny windowsill and watching as the shoots bend towards the light (positive phototropism).
- The discovery of auxin as the substance responsible for regulating shoot growth was made in the 1920s by Dutch biologist Frits Went.
- Went investigated the development of cereal **seedlings** and carried out many experiments to investigate the effects of auxin on shoot growth in response to light.

> ### EXAM TIP
> You may be asked why a control is important in experiments such as this. The control for this experiment would be some cress seeds placed in a dark area. If the shoots in the dark also bend, you know that it is not due to the light.

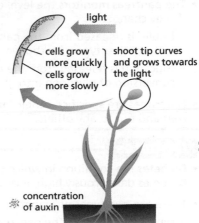

light

cells grow more quickly
cells grow more slowly

shoot tip curves and grows towards the light

concentration of auxin

The shoot grows more quickly where auxin is concentrated.

B–A*

- In shoots, auxin is more concentrated in tissues on the side where light is least intense. The cells here grow more quickly than those on the brightly lit side. In this way, the shoot grows towards the brightest light.
- In roots where the concentration of auxin is high, the cells of the tissues grow more slowly. Auxin is more concentrated on the underside of roots, so the cells of the tissues of the underside grow more slowly than those on the upperside, causing the roots to grow down.

Plant hormones in food production

B–A*

- Copies of plants can be made by taking **cuttings** and dipping the end in rooting powder. This contains plant hormones that encourages cut stems to develop roots.
- Fruit can be picked unripe and transported or stored in warehouses. A plant hormone called **ethene** gas is released into the air around the fruit to ripen it just before it is delivered to shops.
- Some types of **herbicide** contain plant hormones that stimulate the growth of plant stems. Because the rate of root growth does not keep pace with the stem, the roots are not able to absorb enough water to support the growing plant and it dies.
- Herbicides only affect the weeds because they are broad-leaved and absorb more herbicide than narrow-leaved crop plants.
- Growers can produce seedless fruits by smearing the plants' female sex organs with auxin paste, to stimulate the development of the fruit. However, the egg cells within the female sex organs have not been fertilised so seeds are not produced.

◉ How science works

You should be able to:
- plan and execute an experiment to show how light affects the growth of seedlings
- analyse and interpret the data gathered from the experiment
- explain the difference between data and data interpretation.

◉ Improve your grade

Weedkiller

Higher: Explain how plant hormones can be used as weedkillers that kill weeds only, without affecting the crops.

AO1 [3 marks]

Drugs, smoking and alcohol abuse

What are drugs?

- A **drug** is a substance from outside the body that affects the central nervous system and brings about changes in the body that can lead to **addiction**.
- There are different types of drugs:
  - **Painkillers**, such as morphine, deaden pain or affect the way we think about it.
  - **Hallucinogens**, like cannabis, LSD and solvents, produce sensations of false reality called hallucinations.
  - **Stimulants**, including caffeine, increase the speed of our reactions.
  - **Depressants**, such as alcohol, slow the activity of the brain and reaction times.

G–E

- A person can become addicted to a drug because:
  - It can give someone a false sense of well-being, which the person then craves when it goes away.
  - The body gets used to the changes taking place within its tissues.
- Addiction is expensive, because many drugs cannot be obtained without a prescription, so can only be bought from drug dealers.
- **Abuse** is a word often used to refer to the non-medical use of drugs.

D–C

Painkillers and their actions

- Painkillers block the release of neurotransmitter into the synapses which separate the neurones.
- Stimulants enhance the release of neurotransmitter.
- **Reaction time** is how long it takes for a person to respond to a stimulus. This depends on how quickly and how much neurotransmitter is released.

Remember!

Scientists can see how drugs affect reaction times by doing reaction-time tests on people before and after they take a drug, for example, by measuring how long it takes to press a button in response to a flashing light.

B–A*

Legal drugs

- Cigarette smoke contains many chemicals that are harmful:
  - Carbon monoxide reduces the amount of oxygen that red blood cells can carry.
  - Nicotine is an addictive stimulant that raises blood pressure and increases the risk of heart disease.
  - Tar contains substances that can cause cancer (carcinogens). It collects in the lungs and can cause emphysema.
- Alcoholic drinks contain ethanol, which can cause lowered inhibitions, slowed reaction times, blurred vision, and difficulty controlling the arms and legs.

G–E

Alcohol abuse

- Alcohol abuse usually refers to heavy drinking over a long period, or drinking large volumes of alcohol in a short time (**binge drinking**).
- It can cause liver disease, brain damage, heart disease, cancer and raised blood pressure.
- Sex, age, body mass and how quickly the body's cells break down alcohol affect how much an individual can drink safely.
- Four units of alcohol a day for men and three units for women is reckoned to be safe.
- A pregnant woman who drinks alcohol regularly increases the risk of her baby developing abnormally.

D–C

Smoking and disease

- Data collected between 1940 and 1980 established a link between smoking and disease.
- Deaths from lung cancer increased sharply, whereas deaths from other types of lung disease fell.
- A positive correlation was later found between the number of cigarettes someone smoked and their risk of dying from bronchitis and lung cancer.

B–A*

Improve your grade

Drink driving
Foundation: Explain why it is dangerous to drink alcohol and drive.

AO1 [3 marks]

Ethics of transplants

Transplanting organs

- Transplant surgery replaces a diseased organ with a healthy one.
- The person supplying the healthy organ is known as the **donor**.
- The person receiving the organ is the **recipient**.
- Organ donors can be:
  - Living: live donors can donate tissues such as bone marrow or an organ such as a kidney, where there are two doing the same job.
  - Dead: people can give permission for their organs to be used if they are killed in an accident. Most organ donors are victims of accidents.
- **Rejection** of the organ is less likely to happen if the donor is related to the recipient.

Supply and demand

- There are far more people in need of organs for transplantation than there are donors.
- This has resulted in a search for alternative sources of organs. These include:
  - Animal donors (**xenotransplantation**). This was first tried in 1963 using a kidney from a chimpanzee, but the organ was rejected by the recipient.
  - **Genetically engineering** the organs of animals to stop them being rejected by the recipient. This practice is not yet widely used, but further research may enable it to become a powerful treatment for disease in the future.
  - **Transplantation tourism**, in which wealthy people needing a transplant pay for organs. Poor people in developing countries are often prepared to donate their organs for money, even if it is illegal to do so.
- There are many **moral** and **ethical** issues concerned with all of these options, including:
  - whether it is right to use animals as a source of organs
  - whether money should be an incentive in a decision to donate organs
  - wider human rights concerns relating to transplant tourism, such as the exploitation of the poor people in developing countries
  - the increased risk of transplanting diseased organs into recipients because poor donors do not receive regular health care.

Choosing who

- Modern medicine can have amazing effects, but it is expensive and resources are limited, including organs for transplant. This raises ethical concerns about how we **prioritise** transplants.
- A transplant might be needed for a problem that the sufferer has brought on themselves. Should a heart transplant for someone who is obese have the same priority as a life-saving transplant for someone suddenly taken ill?
- Who most 'deserves' the chance to live? Should 'deserve' even come into the equation? There are no easy answers to these difficult questions.

> ### EXAM TIP
> You may be asked to comment on 'moral' and 'ethical' issues. Morals are what people think is right or wrong. Ethics are the actions people take as a result of their moral judgement.

How science works

You should be able to:
- explain how and why scientists continue to research transplantation in areas that are considered controversial
- understand the social effects of such research.

Improve your grade

Ethical concerns

Higher: Jon and Margaret are both on the liver transplant waiting list. Jon is a 36-year-old father of three. He has cirrhosis caused by alcohol abuse. The cause of Margaret's liver failure is unknown. She is 83 years old. Doctors have to decide who receives the next available liver.

Choose who the liver recipient should be and argue your case. *AO2* [5 marks]

Infectious diseases

Causes and spread of disease

- Organisms that cause infectious disease are called **pathogens**. We often call pathogens microbes (microorganisms), because most of them are only visible under a microscope.

- The human body is warm and moist, providing a perfect environment for pathogens to multiply.

- **Infectious** diseases are those caused by pathogens that spread from person to person.

- Cholera **bacteria** are found in water contaminated with sewage.

- **Viruses** that cause colds and flu are airborne – passed from person to person via moisture droplets from coughs and sneezes.

- *Salmonella* is a type of food poisoning that comes from chickens. Cooking food thoroughly and good hygiene decreases the risk of getting *Salmonella*.

- **HIV** is a virus that causes **AIDS**. It destroys some of the white blood cells that help defend the body against pathogens. HIV is spread by the exchange of body fluids.

- Athlete's foot is a fungal infection that thrives in the warm, moist environment of sweaty feet. Athlete's foot can be spread through skin contact and contact with contaminated surfaces.

- The bacteria that cause diarrhoea produce **endotoxins** (poisons). These stimulate the small intestine wall to contract violently and more frequently than normal.

- Diarrhoea prevents digested food and water being absorbed, which can cause severe dehydration that can lead to death if not treated.

Insect-borne disease

- The term animal **vector** refers to animals that spread pathogens.

- Flies can transfer the bacteria which cause dysentery on to food via their feet.

- *Anopheles* mosquitoes are vectors for malaria. They feed on the blood of a victim and suck in the pathogen (a protozoan called *Plasmodium*) before biting another person and passing it on.

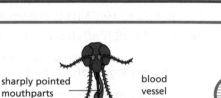

sharply pointed mouthparts   blood vessel

The head of a female *Anopheles* mosquito.

Stopping the spread of infection

- **Antiseptics** are chemicals that stop microbes from multiplying. We use them to swab a wound or clean skin to prevent infection.

- **Antibacterials** are substances that interfere with the growth of bacteria, e.g. antiseptics and antibiotic drugs. However, the term is more often used to describe cleaning products.

- We use **antibiotic** drugs to control infection caused by bacteria. Some kill the bacteria while others prevent them from multiplying.

- **Antifungal** drugs are used to treat fungal infections by killing the fungal cells but not the human cells.

> **Remember!**
> You can test how effective antibiotics are by inoculating agar plates with bacteria and adding small paper discs that have been soaked in antibiotics. An effective antibiotic will stop the growth of the bacteria.

Resistance in bacteria and antibiotic misuse

- Widespread use of antibiotics has led to strains of bacteria that cannot be treated with most antibiotics, e.g. **MRSA**.

- **Resistance** arises because of the high mutation rates of bacterial genes and the ongoing exposure of bacteria to antibiotics.

- Populations of bacteria always contain individuals with resistance genes. These individuals survive antibiotic treatment, reproduce and spread quickly.

Improve your grade

Malaria prevention

Higher: The spread of malaria can be prevented cheaply and easily by covering beds with mosquito nets. Explain how this technique works.

AO2 [3 marks]

Defences and interdependency

Physical barriers

- Human skin forms a physical barrier to protect the body's interior from pathogens outside.
- The lungs contain mucus, which traps bacteria and other particles, and is swept away by hair-like cilia until it reaches the throat, when it is swallowed.
- Plant surfaces are also a physical barrier to pathogens.
- Plants may also have thorns as a defence against plant-chewing animals or hair-like structures that secrete sticky substances to trap insects.

G–E

Chemical defences

- Glands in the skin produce an oily substance called sebum, which kills bacteria and fungi. Glands in the stomach wall produce hydrochloric acid, which kills bacteria on food.
- Tears contain the enzyme lysozyme, which destroys bacteria.
- Some types of white blood cells bind to substances on the surface of pathogens and destroy them. Other types produce proteins called **antibodies**, which also bind to the pathogens.
- Platelets in the blood help form scabs over cuts in the skin, which stops blood loss and prevents pathogens entering.

D–C

New medicines

- Plants also use chemicals as a defence. Animals avoid bitter-tasting plants and prey, as the chemicals they contain can trigger vomiting.
- Bacterial attack is often a signal to the plants to produce antibacterial chemicals.
- Plant antibacterials such as lemon balm, garlic and tea tree are also effective against bacterial pathogens that infect humans.
- Scientists are searching for plants that could be new medicines. This is important, as bacterial pathogens are developing resistance to current antibiotic drugs.

B–A*

Mutualism and parasitism

- All living things are interdependent – they have a dynamic relationship that ensures survival.
- **Mutualism** refers to a relationship where both species benefit. For example, cleaner fish feed on the dead skin and external parasites of other fish species. This keeps the **host** free of health-threatening parasites.

> **EXAM TIP**
>
> Make sure you can describe the difference between mutualism and parasitism and give an example of each.

- **Parasitism** is a one-sided relationship between two species. A parasite obtains food at the other species' expense.
- Parasites can live on the outside (an **ectoparasite** such as a flea) or on the inside (an **endoparasite** such as a tapeworm) of its host.
- Mistletoe is a plant parasite, taking water and mineral salts from the tree on which it grows.

D–C

Useful bacteria

- **Leguminous** plants such as peas, beans and clover have swellings on their roots called nodules. These contain **nitrogen-fixing bacteria**. The bacteria convert nitrogen from the air into compounds which the plants use to make proteins. The bacteria obtain sugars from the plant's roots.
- In hydrothermal deep-sea vents, the substances released by the underwater volcanic activity are the raw materials for another process carried out by bacteria, called **chemosynthesis**:

carbon dioxide + hydrogen sulfide ⟶ sugars + sulfur

- Chemosynthetic bacteria live in the bodies of giant tube worms. The worms supply the bacteria with oxygen, enabling the bacteria to make food that the worm needs.

B–A*

Improve your grade

Interdependence
Foundation: Plants and humans are interdependent.
Explain what this statement means, giving an example.

AO1 [2 marks]

Energy, biomass and population pressures

Food chains and energy flow

- A food chain begins with a **producer** (a plant, algae or some bacteria which produce food by photosynthesis).

- The next in a food chain is always an animal. Animals are **consumers**: they eat food – herbivores eat plants, carnivores eat other animals, omnivores eat both.

- Most carnivores are **predators**. They catch and eat other animals (**prey**) – often herbivores.

- **Scavengers** are carnivores that feed on the dead bodies of animals.

- A food chain represents one pathway of food energy through a community of organisms. A **food web** represents many pathways and is usually a more accurate description of feeding relationships.

- At each link in a food chain, energy is lost in waste products and as a result of **metabolism** in the form of heat. This means that there can be only a limited number of links in the chain.

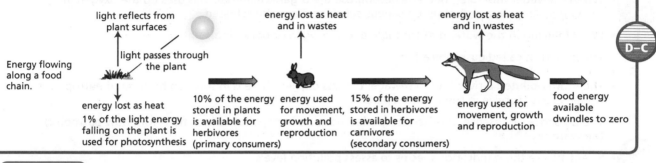

Energy flowing along a food chain.

light reflects from plant surfaces

light passes through the plant

energy lost as heat
1% of the light energy falling on the plant is used for photosynthesis

10% of the energy stored in plants is available for herbivores (primary consumers)

energy used for movement, growth and reproduction

energy lost as heat and in wastes

15% of the energy stored in herbivores is available for carnivores (secondary consumers)

energy used for movement, growth and reproduction

energy lost as heat and in wastes

food energy available dwindles to zero

Biomass

- We refer to the amount of tissue in an organism as **biomass**.

- **Pyramids of biomass** tell us about the biomass of producers and consumers.

- Different **trophic** (feeding) levels make up the pyramid. The amount of biomass decreases along the food chain (up the pyramid). This is because energy is lost at each link. Each trophic level has less biomass than the one below it.

Population and resources

- The human population of the world is increasing as populations of developing nations like India are growing, even though populations of developed nations like France are levelling off.

- Population growth of a country **depends** on the birth rate, life expectancy and health care.

- Population pyramids reveal why growth may be high or low, as they show how many people of each age live in a country.

EXAM TIP

You may be asked to analyse, interpret and evaluate data on population growth, so make sure you are able to comment on what population pyramids show, e.g. one that shows a high percentage of children indicates an increasing population.

- Human population growth means that **resources** are in constant demand.

- Resources are the raw materials we take from the environment to run industry and our homes.

- These include the ores we use to extract metals and the oil we use to produce plastics. Both ores and oil are **non-renewable resources**. They cannot be replaced.

- Even renewable resources, like the wood for making paper, can be used at an unsustainable rate.

Waste: recycle or dump?

- Waste that isn't recycled needs to be incinerated or put into landfill.

- We are running out of landfill sites and they need careful management to stop them polluting water supplies or releasing methane.

- When paper and plastics are incinerated, some of the energy produced can be used as heat for industrial processes and to warm homes. However, the process can release pollutant gases.

- Recycling paper, plastic and metals saves the energy used to make them, avoids the need for raw materials and solves the problem of disposal.

Improve your grade

Metal recycling
Higher: Evaluate the use of metal recycling as an alternative to landfill.   *AO3* [4 marks]

Water and air pollution

Water pollutants

- Chemicals that are harmful to our health and to wildlife are called **pollutants**. These can be released into our environment, causing **pollution**.
- Nitrates and phosphates are pollutants in fertiliser, which is used to increase crop yield.
- These chemicals can pollute groundwater, ponds and rivers when they are not absorbed by crops.
- Groundwater provides one-third of Britain's drinking water. High nitrate and phosphate concentrations in drinking water are a health hazard.

Eutrophication and indicator species

- Nitrate- and phosphate-rich water in ponds and rivers (from fertilisers or sewage) stimulates the growth of algae and water plants, which clog the water.
- When the vegetation dies, bacteria decompose the organic material. This uses up the oxygen in the water. Ammonia and other poisonous substances are also released.
- Wildlife living in the water dies through lack of oxygen or poisoning.
- The process is called **eutrophication**.

- The more polluted a river is with sewage, the less oxygen there is in solution because of eutrophication. Fewer species survive in the polluted parts.
- The presence or absence of different species indicates how polluted (and therefore how eutrophic) the water is.
- Scientists use these **indicator species** to assess pollution levels.
- Stonefly and shrimps are clean-water indicators because they cannot survive in water with low oxygen concentrations.
- Bloodworms and sludgeworms are polluted-water indicators. They can survive at low levels of oxygen because they are tolerant to pollution.

Sulfur dioxide

- Burning fossil fuels releases gases such as sulfur dioxide.

  | sulfur | + | oxygen | → | sulfur dioxide |
  | (from fossil fuel) | | (from air) | | (gas) |

- Pollution with sulfur dioxide reduces air quality and dissolves in water vapour to form **acid rain** or acid snow, which passes into lakes and rivers.

- Acidic water causes fish to produce too much mucus. This clogs the gills and kills the fish through oxygen deprivation.
- Acid rain also washes important substances out of the soil. Poisons are released from the soil and trees die.
- **Lichens** are an indicator species for air pollution. There are different types: shrubby, leafy and slightly leafy.
- The number of lichen species varies according to the level of sulfur dioxide in the air. A greater variety is seen where sulfur dioxide pollution is low.
- Black spot is a fungal disease that covers rose leaves. Its presence is an indicator of good air quality.

Remember!
You can investigate the effects of sulfur dioxide on plant growth by growing germinated cress seeds in a box containing a chemical that gives off the gas. You will also need a control in which the seeds grow with no pollutants, so you can compare them.

- For much of the year the prevailing wind blows from the west/south-west to the east/north-east across Europe.
- These prevailing winds carry acid water vapour from the UK, France and other countries, causing acid rain or acid snow to fall on Norway and Sweden, hundreds of kilometres from the source of the pollution.

Improve your grade

Pond pollution

Foundation: A farmer grows wheat on his field. He notices that the water in a small pond next to the field has turned green and the fish have all died.
Explain what has happened.

AO2 [5 marks]

Topic 3: 3.21, 3.22, 3.23, 3.24

Recycling carbon and nitrogen

Decomposition

- Decay and **decomposition** are caused by fungi and bacteria called **decomposers**.
- **Nutrients** such as carbon, nitrogen and the compounds they form are released because of their activities.
- The nutrients are absorbed in solution by plants and pass to animals as the result of feeding.

G–E

The carbon cycle

- Carbon is recycled as carbon dioxide (CO_2) through **respiration** and **photosynthesis** in the carbon cycle.
- Plants absorb CO_2 from the environment, which enables their cells to produce sugars by photosynthesis. The sugars can be used to form the other carbohydrates, lipids (fats and oils) and proteins that build plant bodies.
- When plants are eaten by animals, the carbon in the plant tissues becomes part of the animal bodies. This transfer of carbon continues as animals are eaten by other animals.
- During respiration, organisms release carbon dioxide into the atmosphere. It is also released by the burning of fossil fuels.
- Chalk is formed from the fossilised remains of sea creatures. Exposed to rain (which is slightly acid), the chalk dissolves and more carbon dioxide is released.

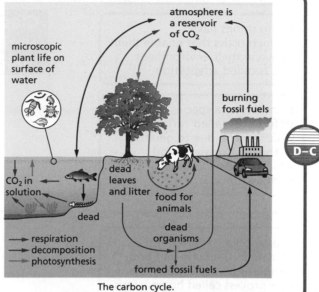

The carbon cycle.

D–C

The nitrogen cycle

- Nitrogen is an essential part of the proteins which build bodies, but most living things cannot use nitrogen in the air.
- **Nitrogen-fixing bacteria** in the root nodules of leguminous plants or in the soil can use this gaseous nitrogen, though. They 'fix' (to make part of) gaseous nitrogen as ammonia (NH_3), which forms ammonium compounds.
- Plants absorb the compounds in solution from the soil through their roots.
- Animals obtain nitrogen from food they eat.
- Proteins are a major part of the remains of dead animals and plants and animal wastes. Decomposers convert these proteins and urea into ammonia.
- **Nitrifying bacteria** convert the ammonia from decaying and waste matter to nitrates, which are also absorbed by plants.
- Lightning breaks apart nitrogen molecules in the atmosphere, and the nitrogen reacts with atmospheric oxygen to form nitrogen oxides. These dissolve in raindrops.
- Nitric acid (HNO_3) and nitrous acid (HNO_2) are produced. These acids react with compounds in the soil and form nitrates.
- Nitrates not absorbed by plants are converted by **denitrifying bacteria** in the soil to nitrogen gas, which is released to the atmosphere.

EXAM TIP

Different processes recycle nitrogen between the environment, the dead and the living: nitrogen fixation → decomposition → nitrification → denitrification. Remembering these terms and the role of the bacteria at each stage will help you to recall the nitrogen cycle.

B–A*

Improve your grade

Carbon dioxide levels

Foundation: Scientists around the world are monitoring the amount of carbon dioxide in the atmosphere and collecting data such as that shown in the graph opposite.

Describe the trend in the graph and suggest a reason for it.

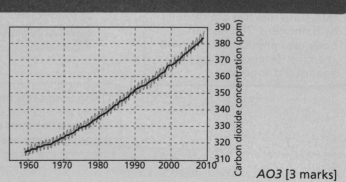

Atmospheric carbon dioxide measured at Mauna Loa, Hawaii.

AO3 [3 marks]

B1 Summary

Classification, variation and inheritance

Living things are put into groups according to how closely they are related to one another. The kingdom is the largest group and species the smallest. Each species is given a two-part (binomial) name.

Living things have characteristics that enable them to live in their environment. This is called adaptation.

Members of a species show differences called variation, which can be due to genetics or the environment, and which can be continuous or discontinuous.

Darwin's theory of natural selection explains how living things evolved due to variation and survival of the best-adapted.

Alternative forms of genes called alleles are passed from parents to offspring. A dominant allele masks the presence of a recessive allele.

Disorders such as sickle cell disease and cystic fibrosis are inherited. Genetic diagrams help us to analyse who in a family is at risk of having the disorder or being a carrier of the affected allele.

Responses to a changing environment

The amount of water and glucose in the body and body temperature are kept constant by a process called homeostasis.

Sensory neurones carry electrical impulses to the central nervous system from receptors in sense organs. Motor neurones carry messages from the CNS to effectors (muscles and glands).

Hormones are produced in endocrine glands and are transported in the blood to target organs, where they have an effect.

Insulin is a hormone that regulates the amount of glucose in the blood. People with Type 1 diabetes do not make insulin so have to inject it. Type 2 diabetes is caused by resistance to insulin, and can be controlled via diet.

Plant hormones cause their shoots to grow towards the light (positive phototropism) and roots to grow towards gravity (positive geotropism).

Plant hormones can be used commercially as selective weedkillers, to grow seedless fruit, to ripen fruit and as rooting powder.

Problems of, and solutions to a changing environment

Cigarette smoke contains chemicals that are damaging to health including tar (causes cancer), nicotine (an addictive stimulant) and carbon monoxide (reduces the amount of oxygen in the blood).

Infectious diseases are caused by pathogens. The human body uses physical barriers and chemical defences in the fight against pathogens.

Living things rely on each other – they are interdependent. Some relationships are parasitic, others are mutualistic.

Energy is wasted at each stage in a food chain, which results in less biomass at each trophic level. This is represented as a pyramid of biomass.

An increasing human population contributes to air and water pollution and a growing demand for resources.

Both nitrogen and carbon can exist in different compounds in the air and bodies of living things. These elements are continually recycled.

Seeing cells and cell components

- Most cells are too small to be seen with the naked eye but can be seen using a light microscope. In a light microscope, a beam of light is passed through the cells.

- Electron microscopes pass a beam of **electrons** through the cells. They enable us to see much more detail.

G–E

- You can work out the magnification of a light microscope by carrying out this calculation:

  total magnification = magnifying power of eyepiece lens × magnifying power of objective lens

- The greater the **resolving power** of a microscope, the clearer the image it forms. Electron microscopes have better resolving powers than light microscopes.

D–C

- Resolving power depends on the wavelength of the electromagnetic radiation (e.g. light or electrons) that is used:

  $$\text{Resolving power} = \frac{\text{wavelength}}{2}$$

B–A*

Cells – components and their functions

- **Multicellular** organisms are made up of many different types of cell.

- Each cell contains components. Some of these are found in all cells, others are only found in a few types.

EXAM TIP

Make sure you remember which components are found in both plant and animal cells and which are found only in plant cells.

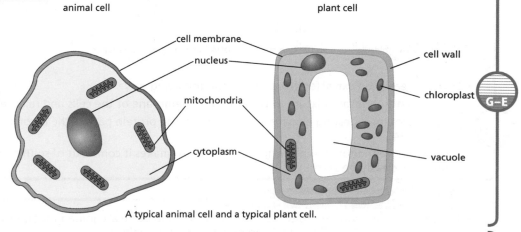

animal cell   plant cell

cell membrane
nucleus
cell wall
chloroplast
mitochondria
cytoplasm
vacuole

A typical animal cell and a typical plant cell.

G–E

- The table below shows the components of animal cells and their functions.

Cytoplasm	Where most of the cell's chemical reactions take place
Cell membrane	Controls the movement of chemicals in and out of the cell
Nucleus	Contains chromosomes
Mitochondria	Where sugars are broken down, releasing energy

- Plant cells have the same components as animal cells, plus those in the table below.

Cell wall	Made of cellulose to strengthen the cell
Vacuole	Keeps the cell turgid
Chloroplasts	Contain chlorophyll and used in photosynthesis

D–C

- Bacterial cells are different to plant and animal cells:

  – They do not have mitochondria, chloroplasts or a nucleus (chromosomal DNA and loops of DNA called **plasmids** lie loose in the cytoplasm).

  – The cell wall is not made of cellulose.

  – Some bacteria have a **flagellum**, which propels the cell through liquid.

⦿ Improve your grade

Looking at cells

Foundation: The image opposite shows some cells as seen down a microscope.

Are these animal or plant cells? Explain how you decided.

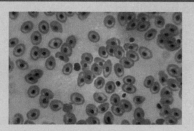

AO2 [2 marks]

DNA

DNA and its structure

- A cell's nucleus contains **chromosomes**, which are made up of **deoxyribonucleic acid** (DNA) wrapped around a core of protein molecules.
- DNA is a double-stranded molecule twisted into a spiral called a double helix.
- The strands of DNA may consist of thousands of building units called **nucleotides**.
- Each nucleotide includes one of four **bases**: guanine (G), thymine (T), adenine (A) and cytosine (C).
- The bases can be in any order, but A always bonds with T and G always bonds with C.
- A **gene** is a section of a strand of DNA which carries the information in the sequence of its bases, enabling a cell to make a protein or part of a protein.

first twist   second twist

Part of a DNA molecule showing the two strands of DNA joined through their bases.

- DNA can easily be extracted from cells in the following way:
  - Use a detergent/salt mixture to break up the membrane of the cells and release the chromosomes.
  - Use a protein-digesting enzyme to break down the protein part of the chromosomes, releasing their DNA.
  - Add cold methanol, which precipitates the DNA (makes it come out of solution). The strands can now clearly be seen.

Complementary base pairing

- The arrangement where the two strands of a DNA molecule are joined together through their bases (A to T and G to C) is called **complementary base pairing**.
- Weak hydrogen bonds join a base with its complementary partner. Because the bonds are weak, they are easily broken, enabling a DNA molecule to separate into two strands.
- This is important when cells make protein and when they divide to form new cells.

Discovering DNA

- The structure of DNA was discovered by scientists building upon each other's work.
- In 1952, Rosalind Franklin and Maurice Wilkins used X-ray crystallography to discover the arrangement of the atoms of DNA molecules.
- In 1953, James Watson and Francis Crick used this information to propose that the structure of a molecule of DNA is a double helix.

The genetic code

- The **sequences** of three bases is called a **codon**. Each codon specifies a particular amino acid.
- This **genetic code** is universal. It works in the same way in the cells of all living things.

How science works

You should be able to:
- describe how we came to our current understanding about the structure of DNA through scientists collaborating and developing each other's ideas.

Improve your grade

Cell division
Higher: Why must the two strands in the DNA molecule separate during cell division?   *AO2* [2 marks]

Genetic engineering and GM organisms

Genetic engineering using bacteria

- **Genetic engineering** involves transferring genes from one type of organism to another.

- A gene controlling the production of a useful protein such as human insulin can be inserted into bacterial cells.

- The cells grow in a solution called a **culture** in huge containers called **fermenters**. They produce large amounts of the protein very quickly.

- The genes are transferred from cell to cell using bacterial enzymes. The diagram opposite shows the process.

- Genetically engineered insulin has several advantages:
  - It is cheap to produce and is available in large quantities.
  - It is human insulin, so users experience no allergic reactions or intolerances.
  - It does not use animal products and so is not a problem for certain religious groups or vegetarians.

- However, some people are concerned that GM organisms could have unknown and unforeseen effects on other organisms, including humans.

strand of DNA carrying a gene which enables cells to produce useful proteins

bacterial cell
bacteria have pieces of circular DNA called plasmid DNA

the DNA strand is cut out using a **restriction enzyme** to isolate the 'useful' gene

plasmid is cut open with the same enzyme used to cut out the 'useful' gene

the 'useful' gene is inserted into the plasmid using **ligase** enzyme

plasmid is put back into the bacterial cell

Remember!
An organism that has had genes from another type inserted into its cells is called a **genetically modified** (GM) organism.

bacteria multiply and produce millions of identical clones, all with the DNA coding for the required protein

bacteria grow in special tanks called fermenters. The end product is removed from the fermenter

A summary of producing genetically engineered bacteria.

 G–E

D–C

 B–A*

Feeding the world

- Developments in GM crops include those that can:
  - grow in places with low rainfall
  - produce their own chemicals to kill insects that damage them
  - resist diseases
  - resist the effects of **herbicides** (weedkillers); farmers can then destroy competing weeds without destroying the crop
  - produce their own fertiliser.

- Some people are concerned about GM crops. Their worries are that:
  - it's not natural
  - eating GM food may affect our health
  - GM crops may harm wildlife
  - pollen from crops modified to resist herbicides may transfer to weeds.

G–E

- **Vitamin A** deficiency is common among millions of poorer people and can cause blindness.

- Golden rice is a variety that has been genetically modified to produce more **beta-carotene** in the rice grain. This is converted to vitamin A in our cells.

- There are arguments for and against using golden rice.

For	Against
Human cells convert beta-carotene into vitamin A very efficiently.	Aiming for a balanced diet is a better solution to vitamin A deficiency.
Golden rice is not meant to be the only solution to the vitamin A problem.	Trying to deal with the vitamin A problem with a single GM solution is too limiting.
We are only trying to deal with a part of the bigger problem of making a healthy balanced diet available to poor people.	Golden rice has never undergone trials to check that it is safe for people to eat.

D–C

How science works

You should be able to:

- discuss the benefits and drawbacks of genetic modification in the contexts of using bacteria to make human insulin, the use of golden rice and herbicide-resistant crops.

Improve your grade

GM protest
Foundation: Fifty people turned up to protest on a piece of land that was growing GM wheat. Outline the concerns they have.

AO1 [3 marks]

Mitosis and meiosis

Dividing cells

- When parent cells divide, new cells called **daughter cells** are formed.

- The nucleus of the parent cells can divide in one of two ways: **mitosis** or **meiosis**.

- Mitosis results in two daughter cells with identical chromosomes to the parent cells. If the parent cells have two sets of chromosomes (**diploid**) then the daughter cells will also be diploid.

- Mitosis is used in order for organisms to:
  - *repair damage*: damaged or old skin cells are replaced by mitosis with identical new skin cells.
  - *grow*: the mass of a plant root increases because the existing root cells produce more by mitosis.

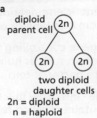

a
diploid parent cell (2n)

two diploid daughter cells
2n = diploid
n = haploid

b diploid parent cell (2n)

four haploid daughter cells

Reproduction

- Asexual reproduction also happens via mitosis. It involves only one parent, which produces new cells to form offspring. The offspring are therefore genetically identical to each other and the parent. They are **clones**.

- Meiosis is the process that is used to form **gametes**.

- In meiosis, the four daughter cells each have half the number of chromosomes of the parent cell, resulting in genetically different **haploid** gametes.

Remember!
The symbol **2n** represents diploid cells, **n** represents haploid cells.

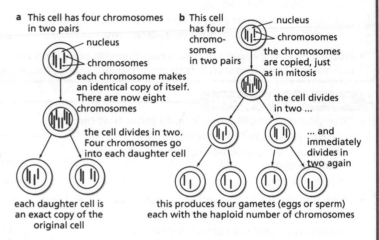

a This cell has four chromosomes in two pairs

nucleus

chromosomes

each chromosome makes an identical copy of itself. There are now eight chromosomes

the cell divides in two. Four chromosomes go into each daughter cell

each daughter cell is an exact copy of the original cell

b This cell has four chromosomes in two pairs

nucleus

chromosomes

the chromosomes are copied, just as in mitosis

the cell divides in two ...

... and immediately divides in two again

this produces four gametes (eggs or sperm) each with the haploid number of chromosomes

a Mitosis and **b** meiosis: the pattern of chromosomes compared.

- When a parent cell is about to divide, its chromosomes **replicate** (copy) themselves.

- This results in chromosomes with two identical strands called **chromatids**.

- During meiosis, each chromosome pairs up with its corresponding partner along the centre of the cell.

- The pairs of chromosome copies exchange pieces of DNA with one another before the cell divides into two.

- Another division then takes place, where the chromatids are split in half. Each daughter cell receives a different chromosome. This results in gametes that are haploid and genetically different from one another.

- During fertilisation, a haploid male gamete (sperm) fuses with a haploid female gamete (egg). The chromosomes of each cell combine.

- The result is a diploid **zygote** (fertilised egg). This has inherited a new combination of chromosomes – and therefore genes – contributed 50:50 from the parents.

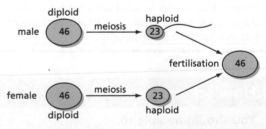

diploid
male 46 →meiosis→ haploid 23

fertilisation 46

female 46 →meiosis→ 23
diploid haploid

Fertilisation restores the diploid number.

Improve your grade

Zygote to foetus

Foundation: Explain why mitosis is an essential process in the formation of a baby from a fertilised egg.

AO1 [3 marks]

Cloning plants and animals

Vegetative reproduction

- In flowering plants, parts of the root, leaf or stem can grow into new plants. This type of asexual reproduction is sometimes called **vegetative reproduction**.

- It produces new plants which are genetically identical (clones) to the parent plant. This is useful to gardeners and farmers who want stocks of plants with preferred characteristics such as disease resistance, fruit colour, flower shape, and so on.

- A simple type of vegetative reproduction is to take **cuttings**.

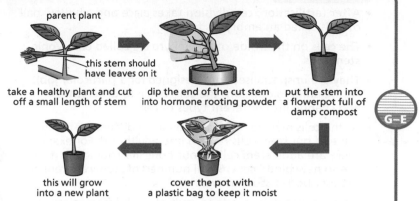

parent plant

this stem should have leaves on it

take a healthy plant and cut off a small length of stem

dip the end of the cut stem into hormone rooting powder

put the stem into a flowerpot full of damp compost

this will grow into a new plant

cover the pot with a plastic bag to keep it moist

Making a cutting.

G–E

Tissue culture

- **Tissue culture** is a process that involves cutting small pieces of tissue from the parent that is to be cloned.

- The pieces are grown in a sterile liquid or gel, which provides all the substances needed for their development.

D–C

Embryo transplants

- **Embryo** transplants begin in the laboratory and end in normal births:
  - **Donor eggs** are taken from female animals and fertilised in the laboratory.
  - Each embryo that forms is split up into its separate cells.
  - Some of the separated cells are transplanted into the womb of a **host mother**, where they develop into identical embryos.
  - The host mother later gives birth to several genetically identical youngsters. They are clones.

EXAM TIP

You may be asked to discuss the advantages, disadvantages and risks of cloning mammals, so make sure you can apply some of the points here to different situations.

G–E

Cloning

- The benefits and dangers of cloning animals have been hotly debated.
  - Cloning allows scientists to produce animals with desirable characteristics quickly and reliably.
  - It helps to build up populations of rare animals which might otherwise be threatened with extinction. Host mothers can carry transplanted embryos of different species even after the original parents have died, as fertilised eggs can be kept frozen for many years.
  - There is evidence that cloned animals have medical issues. Should we produce clones that may have short and painful lives?

- Although it is currently illegal, it is possible to clone human cells. Healthy cells from a sick person can be cloned and used to repair that person's damaged tissues.

- A person needing a transplant could use a brain-dead clone of themselves as a source of tissues and organs for transplant. The transplanted material would not be rejected because it is genetically identical.

- Dolly the sheep was the first cloned mammal. She suffered medical issues that eventually caused her to be put down.

- Such points raise issues about the safety and ethics of cloning.

D–C

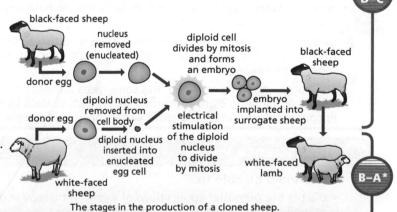

black-faced sheep

nucleus removed (enucleated)

donor egg

diploid cell divides by mitosis and forms an embryo

black-faced sheep

diploid nucleus removed from cell body

donor egg

electrical stimulation of the diploid nucleus to divide by mitosis

embryo implanted into surrogate sheep

diploid nucleus inserted into enucleated egg cell

white-faced sheep

white-faced lamb

The stages in the production of a cloned sheep.

B–A*

Improve your grade

Cloning Daisy

Foundation: Daisy the cow produces the most milk in her herd. Her farmer is considering using her eggs for embryo transplants.

Explain to him why cloning her might be an even better idea.

AO2 [3 marks]

Stem cells and the human genome

Stem cells

- After fertilisation, cell division takes place and a hollow ball of cells called an embryo is formed.
- The cells on the inside of an embyro are called **embryonic stem cells**.
- They are unspecialised (not designed for a particular job). As the embryo develops the cells begin to **differentiate** and change into different types of cell.
- As the cells mature, they can no longer differentiate, but some of our stem cells remain into adulthood. For example, there are adult stem cells in our bone marrow which give rise to new blood cells. Small numbers of stem cells remain in other body tissues as well.
- If stem cells can be made to multiply and differentiate, we would have an unlimited supply of different types of cells, which could be transplanted into people whose tissues are damaged. This is called stem cell therapy.

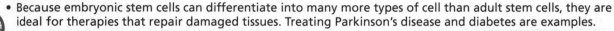

Remember!
Once stem cells mature, they lose their ability to differentiate and their role stays with them for life.

hard bone   bone marrow stem cells

red blood cells (transport oxygen)

different types of white blood cells (defend the body against disease-causing microbes)

platelets (help to clot blood, stopping bleeding)

Section of a long bone cut lengthways. The bone marrow is a source of stem cells, which differentiate into red blood cells, white blood cells and platelets.

Stem cell therapy

- Because embryonic stem cells can differentiate into many more types of cell than adult stem cells, they are ideal for therapies that repair damaged tissues. Treating Parkinson's disease and diabetes are examples.
- Sourcing embryonic stem cells destroys the embryo they come from. Some people think that it is unethical to destroy embryos. Using adult stem cells is less controversial.

- Risks of stem cell therapy include:
  - rejection of the embryonic stem cells
  - the possibility that adult stem cells may carry genetic mutations for disease or may become defective
  - the appearance of side-effects and complications in the recipient; stem cell therapy may trigger adverse immune responses or the development of cancers
  - claims of the effectiveness of the therapies and treatments offered are sometimes from unregulated sources, whose treatment may be dangerous.
- Scientists are researching new ways of producing stem cells that offer the benefits of embryonic stem cells but do not destroy embryos.

The human genome

- **Genome** refers to all of the DNA in each cell of an organism.
- The **Human Genome Project** (HGP) began in 1989. A group of scientists from all over the world collaborated to work out the human genome. The result was announced in April 2003.
- The scientists broke up chromosomes of cells to get at their DNA. Thousands of copies of the pieces of DNA were placed inside machines called **sequencers**, which display the most likely order of the bases.
- Powerful computers were used to help match the base sequences of genes with the proteins for which they are the code.
- Understanding the human genome enables scientists to look at how genes control our vulnerability to particular diseases and to personalise drug treatments to work with an individual's genome (**pharmacogenomics**).
- The human genome revealed that some races are more or less vulnerable to certain diseases than others. Some people are concerned that if genetic data identifying ethnicity were available it might encourage discrimination against certain groups of people.

How science works

You should be able to:
- discuss the potential applications of stem cell therapy and the results of the Human Genome Project, as well as their potential benefits and drawbacks.

Improve your grade

Future applications of the HGP

Higher: In the future we may be able to use an individual's genome to calculate the likelihood of them developing diseases such as cancer.

Evaluate this potential application of the Human Genome Project.   *AO2* [4 marks]

Protein synthesis

Making proteins

- Cells make, or synthesise, **proteins** by joining together amino acid units in the correct order.

- Proteins have a complicated shape that helps them to carry out their jobs. Protein molecules that are the wrong shape cannot perform their functions correctly.

- DNA makes sure that the correct number of amino acid units joins together in the right order.

- The more amino acid units joined together, the larger the molecule: **peptides** are chains of 2–20 amino acids; **polypeptides** contain 21–50; proteins contain more than 50 amino acid units.

Ribonucleic acid

- **Ribonucleic acid (RNA)** is a chemical like DNA. However, RNA is a single strand and has U base instead of T.

- RNA has two roles in protein synthesis:
  - **Messenger RNA (mRNA)** carries the protein-making information from the DNA inside the nucleus of the cell to the **ribosomes** in the cytoplasm, where the protein is made.
  - **Transfer RNA (tRNA)** carries the amino acids needed to form the protein to the ribosomes.

Protein synthesis

- There are two stages in protein synthesis: transcription and translation. Each step numbered below is shown in the diagram.

- **Transcription**
  1. The strands of DNA separate.
  2. Strands of mRNA form as the bases of RNA nucleotides combine with their complementary bases of the single-stranded DNA.
  3. The strands of mRNA separate from their respective complementary strands of DNA. They pass from the nucleus through gaps.

- **Translation**
  4. Each strand of mRNA binds to a ribosome, forming an mRNA–ribosome complex.
  5. Each type of tRNA molecule binds to its particular type of amino acid dissolved in the cytoplasm, depending on the triplet of bases (codon) it carries.
  6. tRNA/amino acid combinations pass to the mRNA–ribosome complex. The exposed bases of each tRNA bind to their complementary bases on the mRNA. Chemical bonds form between the amino acids next to each other
  7. Once the bonds form, each tRNA separates from its amino acid and the mRNA strand.
  8. The linked amino acids form a polypeptide.

> **Remember!**
> RNA has the base uracil (U) instead of thymine (T). Therefore during transcription, any A bases on the DNA strand will be paired up with a U to form the mRNA.

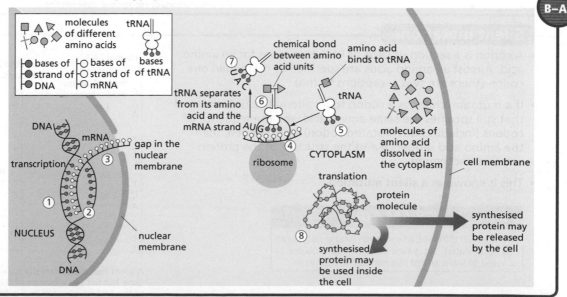

Protein synthesis simplified.

Improve your grade

Sickle cell mutation

Higher: Haemoglobin is a protein that is found in red blood cells. Sickle cell disease is a genetic illness where the red blood cells have a distorted shape. It is caused by a mutation in the gene for haemoglobin that converts a GAG codon into a GTG.

Explain how this causes a change in the shape of the hameoglobin molecule.   *AO2* [4 marks]

Mutations

Variation and mutation

- If a sperm or an egg carries a mutated gene, the **mutation** will be inherited by offspring after fertilisation has taken place.
- Mutations can be harmful, as altering the proteins produced can disturb the activity of cells. Affected organisms are therefore less likely to survive.
- Some genes that are now 'normal' were once mutants. The mutations added genetic variation that happened to be beneficial.
- This meant that the organisms carrying the mutated genes survived. Their descendants inherited the genes and now they are the normal versions.
- Some mutations are neutral – they do not affect an organism's chances of survival one way or another.

How mutations change DNA

- Mutations occur because of copying errors in the sequence of bases during DNA **replication**.
- A base may be deleted or inserted. This changes the sequence of bases along the gene from where the mutation occurs.
- The diagram opposite shows a normal DNA base sequence mutated by a deletion (removal of a base) and an insertion (addition of a base).
- The order of amino acid units from where the mutation occurs changes in each mutated gene. This affects the structure of the protein.
- The structure (and therefore shape) of a protein affects its function. Changes in structure can therefore affect how well the protein works or may even prevent it from working at all.

normal DNA base sequence:
- C, A, A — Val (normal amino acid sequence)
- T, T, C — Lys
- T, C, A — Ser
- A, T, A — Tyr

example 1: mutated DNA base sequence caused by deletion:
- C, A, A — Val
- *deletion*
- T, C, T — Arg (changed amino acid sequence)
- C, A, A — Val
- T, A —

example 2: mutated DNA base sequence caused by insertion:
- C, A, A — Val
- *insertion*
- G, T, T — Glu (changed amino acid sequence)
- C, T, C — Glu
- A, A, T — Leu
- A (first base of next codon)

The deletion or insertion of a base causes a mutation. The amino acid units and their sequence controlled by the particular short sequence of DNA bases shown here are named as their internationally recognised abbreviations.

Silent mutations

- A codon is a section of DNA or RNA that codes for an amino acid. Almost all amino acids are specified by more than one codon (there are two exceptions to this).
- If a mutation changes a codon to an alternative (substitution) that still specifies the same amino acid, and the sequence of codons (including the mutated codon) is unchanged, then the amino acid sequence and the structure of the protein remains unchanged.
- This is known as a **silent mutation**.

normal DNA base sequence:
- C, A, A — Val (normal amino acid sequence)
- T, T, C — Lys
- T, C, A — Ser
- A, T, A — Tyr

mutated DNA base sequence:
- C, A, A — Val
- T, T, C — Lys (unchanged amino acid sequence)
- T, C, G — Ser (*substitution*)
- A, T, A — Tyr

A silent mutation alters the base sequence of a gene but not the sequence of the amino acid units of the protein encoded by the gene.

EXAM TIP

There are 20 different amino acids whose names can be abbreviated, e.g. valine becomes val. You don't need to learn any of the names for your exam.

Improve your grade

Sex cell mutation

Foundation: A mutation occurs in the DNA of an organism's sperm cells.

Explain how this mutation will be passed to its offspring. *AO1* [2 marks]

Enzymes

Rates of reaction

- **Enzymes** are biological **catalysts**. They increase the rate of chemical reactions inside and outside cells.

- Enzymes are specific in their action – each enzyme only catalyses a particular chemical reaction or type of chemical reaction.

- During digestion, enzymes break down large insoluble food molecules into smaller, soluble ones that can dissolve into the blood.

How enzymes work

- The diagram opposite shows an enzyme working. The **substrate** binds to a part of the enzyme called the **active site**.

- Notice how they fit together like a lock and key. An enzyme will only catalyse a particular reaction when the shape of its active site matches the shape of the substrate molecule.

- The enzyme catalyses the breakdown of the substrate into the **products**, which then leave the enzyme. The enzyme is then free to join to another substrate molecule.

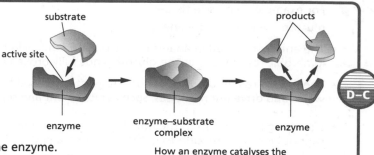

How an enzyme catalyses the breakdown of a molecule.

- Two examples of enzymes at work in the body include:
  - **DNA polymerase**, which breaks up the double helix before DNA replication. It is also involved in checking the copying of the DNA strand.
  - Speeding up the rate of joining together the individual amino acids during protein synthesis.

Factors affecting enzymes

- Because most enzymes are proteins, they are sensitive to changes in temperature and pH.

- The activity of enzymes increases as the temperature goes up. When the activity of the enzyme is at a maximum, we say this is the optimum temperature. This is around 37 °C for enzymes that exist in the human body. After this point, as the temperature increases the activity will decrease.

- Different enzymes have different optimum pHs (the pH at which their activity is highest). For example, the enzyme **amylase** has an optimum pH of around 8.

- Enzyme activity is also affected by substrate concentration, as shown in the graph.

  1 When there is more than enough enzyme, the **rate of reaction** is proportional to the concentration of substrate.

  2 When all of the enzyme's active sites are filled with substrate molecules, the rate of reaction levels off.

  3 Adding more enzyme increases the rate of reaction because more active sites are available to substrate molecules, which fill them.

EXAM TIP

The activity of an enzyme increases with temperature because a rise in heat energy increases the movement of the enzyme and substrate molecules, so there is more chance of them colliding and forming an enzyme-substrate complex.

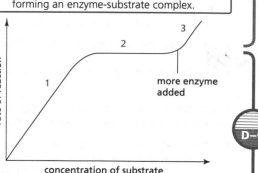

Enzyme activity varies with the concentration of substrate.

Protein synthesis

- At extremes of temperature or pH an enzyme will become **denatured**.

- Denaturing is a permanent change in shape of the protein molecule. It is caused by the breaking up of the hydrogen bonds that hold the structure together.

- A change in a protein's shape will affect its activity because the active site will change, so the substrate will no longer fit.

Improve your grade

Understanding enzymes

Higher: Biological washing tablets contain enzymes that help break down stains found on clothes. Mark's clothes were very dirty so he decided to wash them at a much higher temperature than recommended on the instructions. The stains did not come off his clothes.

Use your understanding of enzymes to explain why.

AO2 [3 marks]

Respiring cells and diffusion

Respiration

- We exhale less oxygen and more carbon dioxide than we inhale.

- This is the result of cells carrying out the chemical reaction **aerobic respiration**. This uses oxygen to break down glucose molecules to release energy, carbon dioxide and water:

  glucose + oxygen ⟶ carbon dioxide + water + energy

- All of the available energy is released from a glucose molecule during aerobic respiration.

- **Anaerobic respiration** does not use oxygen.
  For example, in muscle tissue:

  glucose ⟶ lactic acid + energy

- The energy released in anaerobic respiration is much less than in aerobic respiration, as the glucose is not fully broken down.

- The energy released during aerobic respiration is used to keep us warm and drive life processes, such as movement and reproduction.

Remember!
All living cells carry out respiration. A common misconception is that plants carry out photosynthesis instead, but in fact they carry out both processes when it is light, and only respiration when it is dark.

- Exercising hard results in the muscles carrying out anaerobic respiration, as the heart and lungs cannot work fast enough to get the required amount of oxygen for aerobic respiration.

- The lactic acid produced in anaerobic respiration accumulates in the muscles and makes them tired and sore.

- After exercise, rapid breathing draws more air into the lungs and the fast-beating heart sends the oxygen to the muscle cells, where it helps break down lactic acid into carbon dioxide and water.

- This is called **excess post-exercise oxygen consumption (EPOC)**.

- The time taken for the lactic acid to be removed and breathing and heart rates to return to normal is the **recovery period**.

What is diffusion?

- Molecules in liquids and gases are in constant random motion. Some molecules will spread from areas where they are highly concentrated to areas of lower concentration. As a result there is a net movement of molecules in this direction. This is called **diffusion**.

- The greater the difference between the regions of high and low concentration, the faster the substance's rate of diffusion.

- Substances move into and out of cells of living things by diffusion.

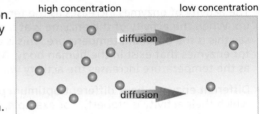

Diffusion: molecules of a substance move from where the substance is in high concentration to where it is in low concentration.

- **Capillary vessels** supply body tissue with blood via the circulatory system.

- Substances such as glucose, oxygen and hormones pass between the blood and tissues via diffusion.

- Gaseous exchange is the process by which oxygen enters the blood and carbon dioxide leaves. It takes place across the walls of the alveoli (air sacs) in the lungs.

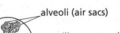

Gas exchange to and from blood in the alveoli

Concentration gradients

- The **concentration gradient** is the difference in concentration of a substance between high and low concentration regions.

- The greater the difference between the regions, the greater the concentration gradient. As a result the rate of diffusion is maximised.

- The diffusion of glucose and oxygen to respiring cells relies on a high concentration gradient between the cells and the blood capillaries.

- This is maintained because the cells are constantly respiring, breaking down the glucose and oxygen and lowering their concentration inside the cells.

Improve your grade

The race

Foundation: You take part in a race. At first you sprint off feeling full of energy. However, half way through your legs start to ache and you have to stop. Explain why this happened.   *AO1* [3 marks]

Effects of exercise

Exercise and rates

- Exercise increases the **heart rate** and breathing rate.

- During exercise aerobic respiration in the muscle cells releases energy, enabling muscles to contract quickly and strongly.

- Respiration produces carbon dioxide, which passes from the muscle cells to the blood, raising its acidity and lowering its pH below 7.0.

- The lowered pH stimulates an increase in the heart rate and breathing rate.

- Heart and breathing rates remain high for some minutes after exercising. As a result, more blood – with its load of carbon dioxide – passes to the lungs, where it is exhaled.

- The concentration of carbon dioxide in the blood decreases, restoring the pH of the blood to its normal value of 7.4. Heart rate and breathing rate return to normal.

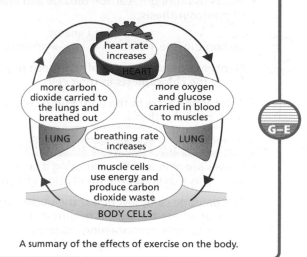

G–E

A summary of the effects of exercise on the body.

Investigating the effects of exercise

- Breathing rate can be investigated by using limewater to test for carbon dioxide.

- An easy way to measure your carbon dioxide output is to place a straw in limewater and note the time taken for the limewater to turn cloudy as you breathe out through the straw.

- Counting the number of times your back rises and falls in a minute gives the breathing rate.

- Taking your **pulse** is an easy way of measuring heart rate (the number of pulses per minute).

D–C

Breathing and heart rate

- The more air we breathe in, the more oxygen reaches the muscles and the more energy is released through aerobic respiration.

- The muscles contract more vigorously, enabling us to exercise more.

- The increased rate of aerobic respiration in the muscle cells produces more carbon dioxide. The increased breathing rate rapidly removes the carbon dioxide from the lungs.

- The larger the volume of air moving in and out of our lungs with each breath, the higher the volume of oxygen that can reach the muscles.

- The following equation can be used to calculate this:

  $$\text{number of breaths in a minute (breathing rate)} \times \text{volume of air per breath} = \frac{\text{volume of air}}{\text{exchanged per minute}}$$

- One complete contraction and relaxation of the heart produces one **heartbeat**.

- The volume of blood pumped from the heart each minute (called the cardiac output) depends on the heart rate and volume of blood pumped out with each beat (the stroke volume).

- Heart rate, stroke volume and cardiac output measure the heart's effectiveness and fitness. Each can be calculated using the equation:

  cardiac output = stroke volume × heart rate

B–A*

Remember!
Both the volume of air per breath, the number of breaths in a minute and heart rate will increase with exercise.

Improve your grade

Stroke volume

Higher: Ben is a professional runner. Through training he has been able to increase the stroke volume of his heart.

Explain how this enables Ben to run faster.

AO2 [5 marks]

Photosynthesis

Light and photosynthesis

- Plants use sunlight, carbon dioxide and water to produce glucose, through the process of **photosynthesis**.

$$\text{carbon dioxide + water} \xrightarrow{\text{light energy}} \text{glucose + oxygen}$$

- Plants are green because of the green pigment **chlorophyll** inside the **chloroplasts** in their cells.

- Chlorophyll absorbs the light energy required to drive photosynthesis.

> **EXAM TIP**
>
> Notice that the chemical equation for photosynthesis is the same as the one for respiration, but in reverse. This means that if you only learn one for the exam, you will also know the other.

Leaves and photosynthesis

- Leaves are thin and flat, exposing a large surface area. This maximises the absorption of light.

- The palisade cells, just under the upper surface of the leaf where the light is brightest, are packed with chloroplasts containing chlorophyll for a maximum rate of photosynthesis.

- Air spaces enable gases including water vapour to circulate within the leaf, so the reactants for photosynthesis can reach the cells that need them.

- Oxygen, carbon dioxide and water vapour diffuse between the leaf's air spaces and the atmosphere through the gaps called **stomata** that perforate the underside of the leaf.

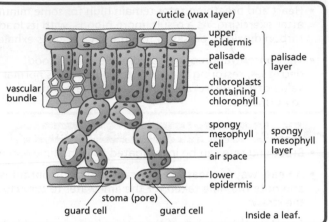

Inside a leaf.

The rate of photosynthesis

- The rate at which plants make glucose is affected by conditions of:
  - temperature
  - light intensity
  - carbon dioxide
  - water.

- If levels of any of these factors drop too low, photosynthesis slows even if the others are in abundant supply. This is called a **limiting factor**.

Limiting factors

- You can measure the rate of photosynthesis using the technique shown in the diagram opposite.

- The rate is measured by counting the number of bubbles of oxygen produced by the water-weed in a given time.

- This set-up works for investigating the effects of changing light intensity, carbon dioxide concentration and temperature on the rate of photosynthesis.

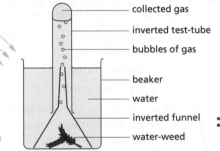

An experimental set-up to investigate the rate of photosynthesis. What is the gas collected?

- Most plants grow best in warm, light conditions and when there is a high concentration of carbon dioxide and plenty of water. Growing plants in greenhouses can help maximise these conditions.

- The higher the temperature, the faster the rate of photosynthesis and the faster the production of materials that enable plants to grow.

- If the temperature continues to increase beyond an optimum, photosynthesis slows because the enzymes controlling the different reactions of photosynthesis are **denatured**.

- Plants grow more vigorously in bright sunlight because high light intensity maximises the rate of photosynthesis. The rate of increase is up to a maximum value (called the optimum light intensity). Even though light intensity increases further, the rate of photosynthesis does not.

Improve your grade

Weed removal

Foundation: Suzanne has weeds growing in her flowerbeds and notices that her plants are not growing. Explain why this is.

AO2 [3 marks]

Transport in plants, osmosis and fieldwork

Transport systems

- **Xylem** (zy-lem) tissue consists of columns of hollow, dead cells. It carries water and dissolved mineral salts from the roots, through the stem and out into every leaf and flower. **G–E**

- **Phloem** (flow-em) tissue runs by the side of the xylem. Its tube-like cells carry dissolved glucose and other substances to all parts of the plant.

- Water evaporates from stomata in a process called **transpiration**. **D–C**

- As water is lost from the leaves, more is drawn up through the xylem tissue from the roots, which absorb more water from the soil. This continuous movement of water is called the transpiration stream.

Osmosis

- **Osmosis** is the movement of water from a high water concentration to a lower one across a partially permeable membrane (one that will only let water molecules across). **G–E**

- You can investigate osmosis by studying cells under a microscope. Cells left in a concentrated salt or sugar solution will lose water by osmosis. They will become **flaccid** (limp).

- Visking tubing is a partially permeable membrane that can be used to investigate the movement of substances into and out of cells.

- In experiment **a** opposite there is a higher water concentration outside the tubing than inside. Over time the water will enter the tubing by osmosis, causing the pressure inside to increase and the tubing to become **turgid**. **D–C**

- Experiment **b** shows the opposite happening, as water flows out of the tubing, leaving it flaccid.

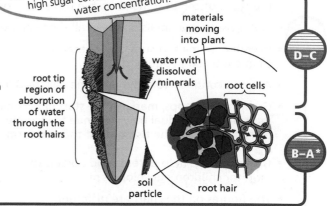

The Visking tubing experiment.
a Filled with a sugar solution and standing in water.
b Filled with water and standing in a sugar solution.

Remember!

A solution that is highly concentrated, e.g. has a high sugar concentration, will have a low water concentration.

Root hairs

- Root hair cells are fine, hair-like extensions of a root.

- Water flows into root hair cells by osmosis. Their large surface area is an adaptation that enables plants to maximise their absorption of water from the soil. **D–C**

Water absorption through a plant's root system.

- Root hairs also take up mineral salts in solution. The solutions are much more concentrated in the cells of root tissue than in the soil. Therefore, mineral salts cannot pass into the roots by diffusion. **Active transport** is used, which requires energy from aerobic respiration. **B–A***

Organisms and their environment

- Fieldwork investigations are designed to find out more about where organisms live (their distribution) and why they live where they do.

- Techniques used include:
  - pooters, nets and traps to collect animals in order to estimate their distribution **D–C**
  - quadrats to count the number of animals or plants in a known area
  - probes to measure temperature, pH and light intensity.

- **Ecosystems** and their **populations** are usually too large for us to study everything about them. Instead we study small parts called samples.

- Errors in sampling techniques can be reduced by taking a number of random samples and standardising the samples taken (e.g. at the same time of day or in the same season/weather conditions). **B–A***

Improve your grade

Preserving fish

Foundation: Covering fish with salt and leaving it causes the fish to dry out and become hard. Use osmosis to explain why this happens.

AO2 [3 marks]

Fossil record and growth

The fossil record

- **Fossils** are the remains or impressions made by dead organisms. They are usually found in **sedimentary rocks**. They are preserved over millions of years as rock particles from ancient seas fell on dead organisms on the seabed.

- Each layer of fossils records life on Earth at the time the layer formed. This helps us trace the history and evolution of life. Fossils provide evidence for Darwin's theory of **natural selection**.

- Fossil records do not show a continuous series of changes between ancestors and their descendants. There are gaps. This is because:
  - most organisms **decompose** quickly when they die; not all of them find their way into an environment where they will be preserved and so only a small number of fossils form
  - many fossils are yet to be found
  - even if a fossil forms, it may not survive geological cycles.

- Most vertebrates today have a **pentadactyl limb** – a forelimb with five 'fingers' or 'toes'.

- The discovery of pentadactyl limbs in fossils has led scientists to believe that all vertebrates directly descend from a **common ancestor**. They use this as evidence of **evolution**.

Growth

- Growth is measured as an increase in an organism's size, length and mass. The increase is the result of:
  - cell division: the number of cells increases
  - cell elongation: the length of cells increases
  - synthesis of organic materials (**carbohydrates**, proteins, fats and oils): the mass of cells increases.

- Plants grow throughout their life from cell division in tissues called **meristems**.

- Behind the meristems are regions where cells elongate and increase in size by water and other organic materials flowing into them.

- These cells are undifferentiated. As growth continues, differentiation of cells begins producing the types of cell that make up the tissues and organs of the plant.

- Cell division in animals occurs in all the tissues of the body.

- In young animals, tissues grow because cell division produces more cells than die through age or damage.

- Animals continue to grow until the gain of cells balances the loss of cells.

- Growth then stops, marking the start of becoming an adult. However, the mass of an individual may continue to increase as protein synthesis adds more mass to cells and the tissues that the cells form.

> **Remember!**
> Plant growth involves cell division, cell elongation and differentiation. Cell division and differentiation contribute to growth in animals.

- Growth charts help parents and doctors to monitor children's development.

- The figures on the chart are different **percentiles**, each representing the spread of values for the characteristic (such as height or mass) selected. The 50th percentile represents the average value.

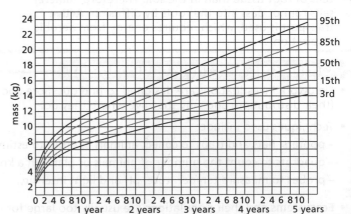

A growth chart for mass of boys aged up to 5 years
(Source: WHO Child Growth Standards.)

Improve your grade

Growth chart

Higher: Adam is four years old and has a weight of 13 kg.

Use the growth chart above to comment on his weight.

AO3 [3 marks]

Cells, tissues, organs and blood

Cells join together

- Cells are the building blocks from which humans and all other living things are made

- **Tissues** are a group of similar cells with a particular function.

- An **organ** is a group of different tissues that work together. An organ has a particular function.

- An organ system is a group of different organs that work together. Organ systems also have specific functions.

  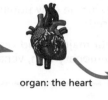 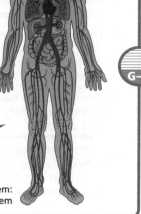

cells: cardiac cells   tissue: cardiac muscle   organ: the heart

organ system: the circulatory system

Cells form tissues which form organs which form organ systems.

G–E

Differentiation

- All the different types of human cell are the result of cell division (mitosis) and **differentiation** during the development of the egg to the **embryo** to the **foetus**.

- Each type of human cell is specialised to enable it to carry out a particular function. For example, neurones (nerve cells) transmit nerve impulses and muscle cells contract (shorten).

D–C

- The process of mitosis produces daughter cells. These are genetically identical to one another and to their parent cell. However, most adult cells are differentiated.

- Genetically the cells may be the same, but the pattern of genes switching on and off (gene activity) is different.

- This process occurs in the development of all living things.

B–A*

Blood

- Blood contains the following components:
  - *Red blood cells* contain a red pigment called **haemoglobin**. They do not have a nucleus.
  - *White blood cells* have a nucleus. They contain cytoplasm, which allows them to access tissues so they can protect the body by attacking and destroying bacteria and viruses.
  - *Platelets* are fragments of cells with no nucleus. Platelets contain proteins.
  - *Plasma* is a straw-coloured liquid that transports carbon dioxide, soluble food products and urea (and waste products from the liver) in solution. Plasma also circulates the heat released by the chemical reactions in body cells, and this helps to maintain body temperature.

G–E

- The function of red blood cells is to transport oxygen from the lungs to respiring tissues.

- In the lungs, where oxygen concentration is high, haemoglobin combines with oxygen to form **oxyhaemoglobin**.

- Oxyhaemoglobin breaks down to release oxygen to respiring tissues where the concentration of oxygen is low.

D–C

> **Remember!**
> Blood containing a lot of oxyhaemoglobin is called **oxygenated blood** and is bright red. Blood with little oxyhaemoglobin is called **deoxygenated blood** and is a deep red-purple colour (not blue!).

- When platelets are damaged by a cut or torn tissue, they release a substance that starts a chain of chemical reactions in the blood. These reactions end with the soluble plasma protein called fibrinogen changing into insoluble **fibrin**.

- Fibrin forms a mesh of fibres across the wound and traps red blood cells, forming a clot.

- The clot plugs the wound and stops bleeding. It also prevents bacteria and viruses from entering the body.

B–A*

Improve your grade

The heart

Foundation: Why is the heart classed as an organ?

AO1 [2 marks]

The heart and circulatory system

The heart pumps blood

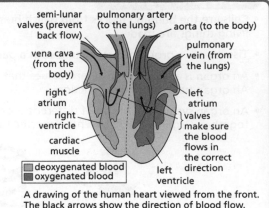

- The heart lies inside the chest cavity, protected by the rib cage. Much of the wall of the heart is made of **cardiac muscle**. This muscle contracts and relaxes to pump blood through the circulatory system.

- The heart has four chambers: two **atria** (singular atrium) and two **ventricles**.

- The wall of the left ventricle is thicker than that of the right ventricle because it has to pump blood to all parts of the body. The right ventricle only pumps blood to the lungs, so less effort is required.

- The heart also has four major blood vessels: pulmonary artery, pulmonary vein, vena cava and aorta.

semi-lunar valves (prevent back flow) — pulmonary artery (to the lungs) — aorta (to the body)
vena cava (from the body) — pulmonary vein (from the lungs)
right atrium — left atrium
right ventricle — valves make sure the blood flows in the correct direction
cardiac muscle — left ventricle

☐ deoxygenated blood
■ oxygenated blood

A drawing of the human heart viewed from the front. The black arrows show the direction of blood flow.

Blood circulates

- The heart is a double pump: each side pumps blood along a different route.

- The diagram opposite shows a simplified arrangement of the circulation of blood:
  - The left atrium and ventricle pump **oxygenated blood** from the lungs around the rest of the body.
  - The right atrium and ventricle pump **deoxygenated blood** to the lungs, where it can be oxygenated.

- When the muscular walls of the heart relax, blood fills the chambers. When the muscles contract, blood is forced from the chambers.

- The valves control the flow of blood through the heart and into the arteries leading from the heart, preventing backflow (flow in the opposite direction).

Remember!

Arteries carry blood away from the heart, veins carry blood towards it. Arteries (apart from the pulmonary artery) carry oxygenated blood; veins (apart from the pulmonary vein) carry deoxygenated blood.

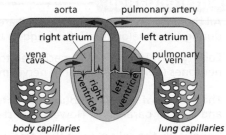

☐ deoxygenated blood
■ oxygenated blood

aorta — pulmonary artery
right atrium — left atrium
vena cava — pulmonary vein
right ventricle — left ventricle
body capillaries — lung capillaries

The heart and circulatory system, showing the lung circuit and the head and body circuit.

The circulatory system

- The circulatory system is a network of tube-like vessels called **arteries** and **veins**.

- The heart pumps blood through arteries to body tissues.

- Blood drains from the tissue through the veins, back to the heart.

- Smaller vessels branch from arteries and veins. The smallest are called **capillaries**. They link arteries and veins.

- Blood in veins flows more slowly than blood in arteries because it is at lower pressure. The large diameter of a vein enables the blood to flow easily.

- Blood flow through veins is helped by the contractions of the muscles in the arms and legs through which the veins pass.

- The heart pumps blood into arteries at high pressure, as the blood needs to reach the extremities of the body.

- Elastic fibres in the artery wall help maintain the flow of blood away from the heart and prevent backflow, so no valves are needed.

- Capillaries form dense networks, called **capillary beds**, in the tissues of the body.

- They provide a large surface area for the efficient exchange of materials between the blood and tissues.

- The blood is at a higher pressure at the artery end of the capillary bed. The higher pressure forces plasma through the thin capillary walls. The liquid, called **tissue fluid**, carries nutrients and oxygen to the surrounding cells.

⊙ Improve your grade

Capillaries

Higher: Capillaries are blood vessels that are very thin (0.005 mm in diameter) and whose walls are only one cell thick. Explain how these features enable them to carry out their function.   *AO2* [4 marks]

The digestive system

Make-up of the digestive system

- The digestive system is made up of the alimentary canal (a muscular tube through which food passes from mouth to anus), liver and **pancreas**.

- The different parts of the digestive system are:
  - *Mouth*: food is taken in (ingestion), chewed into smaller pieces and mixed with saliva. This begins the breakdown of food.
  - *Oesophagus*: this muscular tube pushes food into the stomach.
  - *Stomach*: muscles in the stomach wall contract and relax to mix food with digestive juices.
  - *Pancreas*: produces pancreatic juice containing digestive enzymes that pass to the small intestine.
  - *Small intestine*: where digestion (breaking food down into soluble products) and absorption (diffusion of soluble products into the blood) takes place.
  - *Liver*: processes the nutrients from the small intestine and produces **bile**, which helps digest fat.
  - *Large intestine*: absorbs water from the remaining indigestible food matter.
  - *Anus*: undigested food is removed as faeces (egestion).

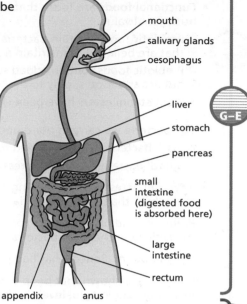

The different parts of the human digestive system.

- Contraction and relaxation of the muscle layers in the wall of the alimentary canal moves food through the digestive system. This muscular action is called **peristalsis**.

- The **gall bladder** is a small sac-like structure connected to the small intestine by the bile duct. It stores the greenish alkaline liquid called bile produced by the liver.

circular muscles contract, squeezing food into the next section where the circular muscles are relaxed

when longitudinal muscles contract the canal shortens, pushing the food along

food passing through the alimentary canal

Peristalis – the alimentary canal narrows when the circular muscles contract and shortens when the longitudinal muscles contract.

G–E

D–C

B–A*

Digestion and absorption

- **Enzymes** break down large, insoluble molecules of carbohydrates, fats and protein into smaller molecules which the body can absorb, as summarised in the table below.

Remember!
Digestion is both chemical (the action of enzymes) and mechanical (teeth grinding food, muscle contraction).

Enzyme group	Examples	Location	Food component acted on	Substance produced
Carbohydrases digest carbohydrates	**Amylase**	Mouth and small intestine	Starch	Glucose (simple sugars)
Proteases digest proteins	**Pepsin**	Stomach	Proteins	Amino acids
Lipases digest fats and oils	Lipase	Small intestine	Fats and oils	Fatty acids and glycerol

G–E

- Visking tubing (see page 35) can be used as a model of the small intestine.

- Add a mix of enzyme(s) and large food molecule(s) into the tubing and suspend it in warm water (which acts like the blood supply). You can then detect the presence of the soluble products of digestion in the water.

D–C

Bile and villi

- Bile breaks down fats into small droplets (**emulsification**), which increases the surface area, speeding up the action of lipase.

- Bile also neutralises the stomach acid present in the food, which enters the small intestine to allow enzymes to work at their optimum pH of around 8.

- Tiny projections called **villi** (singular villus) line the small intestine. They increase its surface area, allowing more efficient absorption of the soluble products of digestion.

B–A*

Improve your grade

Enzyme experiment

Foundation: Dipesh mixed some starch with amylase in a beaker and left the mixture in a water bath at 37 °C. After 30 minutes he tested the mixture to see if there was any starch present. Predict what Dipesh will find and give a reason for your answer.

AO3 [3 marks]

Functional foods

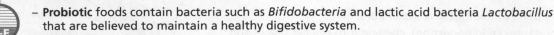

What are functional foods?

G–E

- **Functional foods** are foods that have health-promoting benefits over and above their basic nutritional value.
  - **Probiotic** foods contain bacteria such as *Bifidobacteria* and lactic acid bacteria *Lactobacillus* that are believed to maintain a healthy digestive system.
  - **Prebiotic** foods contain added sugars called oligosaccharides. These cannot be digested, but act as a food supply to the 'good' bacteria in the alimentary canal.
  - **Plant stanol esters** have been clinically proven to reduce the absorption of harmful **cholesterol**.

D–C

- The bacteria we carry in our digestive system can be divided into:
  - 'bad' bacteria, which can lead to diseases of the alimentary canal
  - 'good' bacteria, which suppress the activities of the 'bad' bacteria.
- Poor diet, stress, food poisoning and the use of antibiotics can disturb the balance so there are more 'bad' than 'good' bacteria.
- Prebiotics and probiotics are foods which aim to boost the numbers of 'good' bacteria that suppress the activity of the 'bad' bacteria.
- The margarine Benecol has had plant stanol ester added.
- Studies have shown that people who include plant stanols in their diet over a year might expect the cholesterol levels in their blood to fall by up to 10%.
- Lowering blood cholesterol reduces a person's risk of heart disease.
- Manufacturers of functional foods say that they prevent, treat or cure disease, but many scientists think that the foods should be tested in the same way that new drugs are, to ensure that these claims are valid.

Evaluating functional foods

B–A*

- In the past, the public's interest has led to food producers making health claims for their products that were unsupported by clear scientific evidence.
- However, there are concerns about the effectiveness of functional foods. The use of *Lactobacillus* and *Bifidobacterium* bacteria in some dairy products is an example. There are concerns involving:
  - how well the bacteria survive the manufacture and storage of probiotics before sale
  - their passage through the digestive system
  - competition with the trillions of other microorganisms already in the gut.
- The health claims made for most functional foods remains in doubt. Much more research is still needed.

> **Remember!**
> It is the balance of 'good' bacteria and 'bad' bacteria that influences the health of the digestive system. Both probiotics and prebiotics aim to increase the number of 'good' bacteria which suppress the activity of the 'bad'.

How science works

You should be able to:

- understand why there are doubts about the effectiveness of functional foods
- evaluate claims made by functional food producers that they promote good health.

Improve your grade

Probiotics vs prebiotics

Higher: Explain why some people believe that prebiotics are more likely to affect the health of the gut than probiotics.

AO1 [5 marks]

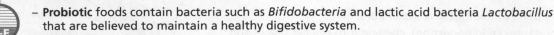

B2 Summary

Light and electron microscopes are used to magnify cells so we can study them and their components.

Cells can be animal, plant or bacterial. They all share the components of a cell membrane, cytoplasm and DNA, but have their own unique components that enable the cells to carry out processes, e.g. chloroplasts in plant cells carry out photosynthesis.

Advances in the understanding of DNA and cells have led to the Human Genome Project, cloning mammals, stem cell therapy and the production of GM organisms. Technology such as this has advantages and disadvantages.

The building blocks of cells

Mitosis results in two daughter cells identical to the parent cell. It is used for growth, repair and asexual reproduction. Gametes (sex cells) are produced during meiosis.

A DNA molecule consists of two strands linked by bases (a double helix). The bases always pair up in the same way (A with T and G with C), and their order determines the order of amino acids in the protein the DNA codes for.

Enzymes are biological catalysts whose activity is affected by temperature and pH. They are specific in their action.

Respiration takes place in all cells and is the release of energy from glucose.

The word equation for aerobic respiration is:
glucose + oxygen ⟶ carbon dioxide + water + energy

During exercise, heart and breathing rate increases to maximise the rate of respiration.

Oxygen diffuses from the alveoli into the blood. Carbon dioxide diffuses in the opposite direction.

A lack of oxygen in the muscles will result in anaerobic respiration:
glucose ⟶ lactic acid + energy

Sampling techniques such as quadrats are used to investigate the distribution of organisms in an ecosystem.

Organisms and energy

Plants produce glucose by photosynthesis (water + carbon dioxide ⟶ oxygen + glucose). The rate is controlled by limiting factors: temperature, light intensity and carbon dioxide concentration.

Osmosis is the diffusion of water and is how water moves into plant roots. The uptake of minerals by the roots requires energy and is known as active transport.

Plant tissues called xylem and phloem transport the reactants and products of photosynthesis. Leaves are adapted to maximise the rate.

Fossils provide evidence for Darwin's theory of evolution by natural selection.

All organisms grow. There are differences between how plants and animals grow, but both involve mitosis and differentiation.

Percentile charts help us to monitor growth in children.

Functional foods are those that have health benefits over their basic nutritional value. Examples include prebiotics, probiotics and foods that contain plant stanol esters such as Benecol margarine.

Common systems

Blood contains red blood cells (for the transport of oxygen), white blood cells (for defence against pathogens), plasma (which transports substances in solution) and platelets (for clotting).

The function of the digestive system is to break down food into simple soluble products. This is done by mechanical processes, such as the chewing of food, and the chemical breakdown of food molecules catalysed by enzymes.

The heart and blood vessels (arteries, veins and capillaries) make up the circulatory system. Its function is to transport blood to every cell in the body.

The early atmosphere and oceans

The early atmosphere

- The early **atmosphere** of the Earth was formed from gases that escaped from volcanoes.
- We can tell that the main gases in the early atmosphere were water vapour (H_2O) and carbon dioxide (CO_2) because these gases are still produced by modern volcanoes.

Remember!
There was very little oxygen in the atmosphere when the Earth first formed.

- Rocks contain a record of the gases that were in the atmosphere at the time they formed.
- Scientists analyse rocks that formed at different times to provide evidence for the way the atmosphere has changed.
- The atmosphere of volcanic planets such as Mars provides evidence for the early atmosphere of Earth.

- Oxygen is one of the most reactive gases in the atmosphere.
- **Oxidation reactions** can be seen in rusting metals and the maintenance of fire.
- The type of iron compounds found in rocks leaves a permanent record of the oxygen content of the atmosphere at the time the rocks were formed.
- More recent rocks contain iron oxide, but in older rocks the iron is not oxidised. This provides evidence that oxygen levels in the atmosphere have increased.
- Nitrogen has also increased and carbon dioxide has dramatically decreased.

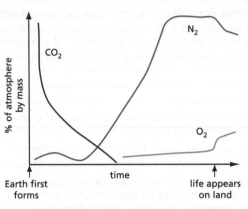

Earth's changing atmosphere.

The early oceans

- As the Earth cooled, water vapour in the atmosphere **condensed** and fell as rain to make the oceans.
- Carbon dioxide from the atmosphere **dissolved** into the oceans, causing the percentage in the atmosphere to drop.
- Early sea plants used the dissolved carbon dioxide for **photosynthesis**, and released oxygen into the atmosphere.
- Some sea animals took in carbon dioxide to make their shells. Shells from dead animals collected on the sea floor and over many years turned into **carbonate** rocks.

> ### EXAM TIP
> Use scientific terminology in your answers – condensed, dissolved, oxidised.

- The chemical symbol for water is H_2O.
- H_2O can exist in different states – solid, liquid or gas.
- When water vapour condenses, intermolecular forces hold the molecules together.

Life on Earth

- The evolution of life caused major changes to the atmosphere.
- Photosynthesis decreased the level of carbon dioxide and increased the level of oxygen in the atmosphere:

$$6CO_2 \text{ (g)} + 6H_2O \text{ (l)} \longrightarrow C_6H_{12}O_6 \text{ (aq)} + 6O_2 \text{ (g)}$$

carbon dioxide + water $\longrightarrow$ glucose + oxygen

- State symbols tell us whether a chemical is a solid (s), a liquid (l), a gas (g) or dissolved in water (aq).

⊙ Improve your grade

The changing atmosphere

Foundation: Explain how the early atmosphere changed to become the atmosphere we have today.

AO1 [4 marks]

Today's atmosphere

Investigating change

- The modern atmosphere contains:
  - 78% nitrogen
  - 21% oxygen
  - 1% argon
  - 0.03% carbon dioxide
  - tiny amounts of other gases.
- Iron oxide is formed when iron reacts with oxygen in the air.
- We can measure how much iron oxide is formed from a known volume of air. This lets us calculate the percentage of oxygen in the air.
- The gases in air can be separated by fractional distillation.
- Scientific instruments can measure accurately the amount of each gas in the atmosphere. This allows any small changes to be detected.
- The gases in the atmosphere change when volcanoes erupt, when humans burn **fossil fuels** and when forests are cut down.

G–E

The changing atmosphere

- Burning fossil fuels (which are mainly **hydrocarbons**) for energy releases lots of carbon dioxide into the atmosphere.
- Additional carbon dioxide is released into the atmosphere when volcanoes erupt.
- Forests carry out more photosynthesis than fields. This means that the amount of carbon dioxide taken in by photosynthesis decreases when humans cut down large areas of trees to plant crops (**deforestation**).
- Atmospheric carbon dioxide levels have increased over the last 300 years due to human activity.
- Farming with fertilisers increases the amount of nitrogen oxides in the atmosphere, while cattle farming increases the amount of methane.

D–C

Chemical equations

- In a chemical reaction, no atoms are created or destroyed.
- Chemical equations can be shown as formulae equations or word equations.
- An example of a word equation is:

  iron + oxygen ⟶ iron oxide

- Formulae equations must be balanced – there must be the same number of atoms on the **reactant** side as on the **product** side.

> **Remember!**
>
> When you balance an equation, the formulae of the reactants and products doesn't change. Never change the subscripts!

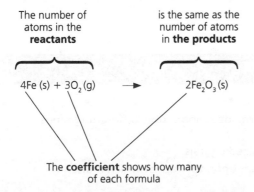

The number of atoms in the **reactants** is the same as the number of atoms in **the products**

$$4Fe\ (s) + 3O_2\ (g) \longrightarrow 2Fe_2O_3\ (s)$$

The **coefficient** shows how many of each formula

B–A*

- Change the **coefficients** to balance the equation.

Improve your grade

Measuring gases

Higher: The percentage of gases in the atmosphere has remained the same for thousands of years, but recently there have been small changes.

Explain how we can tell when the atmosphere changes and suggest some reasons for the change.

AO1 [4 marks]

Types of rock

Different rocks

- Different types of rock have different properties. These properties reveal how the rocks formed.

- **Igneous rock** forms when molten **magma** or lava cools and becomes solid.
  - Igneous rock that has solidified on the surface of the Earth has small crystals. Basalt is an example.
  - Igneous rock that has solidified underground has large crystals. Granite is an example.

- **Sedimentary rock** forms when tiny particles (sediment) settle on to the ocean floor. Over a long time, more and more layers of sediment settle on top. The pressure causes the layer to harden.
  - Chalk and limestone are examples of sedimentary rock. They are made from the shells of tiny sea creatures that fell to the bottom of the ocean.

- **Metamorphic rock** is formed from sedimentary rock that has come under heat and pressure underground. This makes the arrangement of the crystals in the rock change.
  - Limestone and chalk change to marble when they come under heat and pressure.

Remember!
Magma is molten rock under the ground and *lava* is magma that has reached the Earth's surface.

G–E

Cooling down

- Scientists learn about how igneous rock has formed by looking at the crystals within it.

- **Intrusive** igneous rock cools slowly under the ground because it is insulated by the surrounding rock. It forms rock with large crystals.

- **Extrusive** igneous rock cools quickly above the ground. It forms rock with small crystals.

Remember!
When metamorphic rock forms, the minerals line up and crystals form, but it never becomes liquid.

D–C

Crystal formation

- The rate of cooling affects the crystal size in igneous rocks.

- Crystals are formed by atoms or molecules fitting together in rigid structures with regular lines and particle layers.

- The amount of **kinetic** energy (energy of movement) in atoms and molecules is related to their temperature.

- As molten rock cools, the particles within it have less and less kinetic energy. Eventually they bond together to form solid crystals.

- If cooling is slow, just a few crystals start to form, and gradually more particles join on to each crystal. A few large crystals are made.

- If cooling is rapid, many crystals start to form at once. Each crystal collects only a small share of the cooling particles, so many small crystals are made.

B–A*

How science works

You should be able to:

- understand how scientists use their observations to form ideas about how different igneous rocks were made.
  - They *observe* that different rocks contain different sized crystals.
  - This leads to a *hypothesis* that the crystal size depends upon the rate at which molten rock cools.
  - The hypothesis can be used to make a *prediction* – cooling molten salol slowly and quickly will produce different sized crystals.
  - An *experiment* then provides some data which will confirm or refute the prediction.

Improve your grade

Igneous rock formation

Foundation: Scientists can gain a great deal of information by looking at the properties of rocks.

What is igneous rock, and what does the appearance of igneous rock tell scientists about the way it has formed?

AO1 [5 marks]

Sedimentary rock and quarrying

Sedimentary rock formation and fossils

- Weathering breaks rocks into fragments, which are transported to the ocean by rivers. The fragments fall to the bottom of the ocean as sediments.

- The sediment particles become squashed (**compaction**) and are cemented together by minerals. After a very long time, sedimentary rock is formed.

- Sedimentary rock changes into metamorphic rock when it comes under heat and pressure. This happens because it gets buried deep underground and comes close to hot molten rock.

- The particles in **chalk** and **limestone** are rearranged to form **marble**.

- Fossils form in sedimentary rock when:
  - organisms fall into sediment
  - there is not enough oxygen to cause the organism to decay
  - the organism has hard enough parts to leave a caste in the forming rock
  - water filtering through the rock deposits minerals in the caste.

- Chalk is full of microscopic fossils because the sediment from which it is made comes from the shells of microscopic sea organisms.

Remember!
Fossils are only found in sedimentary rock, not in igneous or metamorphic rock.

Erosion

- Rocks may be **eroded** by wind and water, producing small fragments.

- The eroded rock fragments build up in layers at the bottom of oceans and eventually form layers of rock.

- After long periods of time, the layers may be uplifted and exposed to wind and water.

- The different exposed layers erode at different rates because of their varying properties, such as solubility or hardness.

Quarrying limestone

- The chemical in limestone, chalk and marble is calcium carbonate ($CaCO_3$).

- Limestone is dug up (quarried) for use in making glass, cement and concrete.

- Problems associated with quarrying limestone include:
  - the fact that it creates noise and dust, and damages animal habitats
  - transporting limestone contributes to traffic congestion and road damage.

- Advantages of quarrying limestone include:
  - it provides jobs for those near the quarry
  - the industries using limestone are important for both the local and national economy.

> ### EXAM TIP
> It is always good to give the composition of substances where relevant, for example the chemical composition of limestone – $CaCO_3$.

- Limestone is an important raw material. It is processed to make the following materials:
  - Glass is made by heating limestone, sodium carbonate and sand together.
  - Cement is made by heating limestone and clay.
  - Concrete is made by mixing together sand, gravel, cement and water.

- The negative impacts of essential quarrying can be minimised by:
  - restricting the size of quarries
  - only blasting in the quarries at specific times
  - using water sprays to reduce the amount of dust created
  - creating earth barricades to reduce the noise impact of blasting
  - factoring in the reclamation of the land once the quarry closes.

Improve your grade

Uses of limestone

Foundation: The limestone industry plays an important part in the British economy but it also causes environmental problems.

Explain what limestone is and how it is used. Discuss whether you think that the damage to the environment means we should no longer use limestone.

AO2 [4 marks]

Atoms and reactions

Fundamentals of chemistry

- **Atoms** are the smallest particles of an **element** that take part in a chemical reaction.
- All the atoms of an element are the same. Atoms from different elements have different properties.
- Atoms cannot be created or destroyed in chemical reactions.
- In a chemical reaction, the atoms of the reactants are rearranged to make the products. Reactants and products have different properties.
- The mass of all the reactants added together at the start of a reaction is the same as the mass of all the products at the end of a reaction. This is called the **conservation of matter**.

Products and reactants

- **Chemical properties** describe how something behaves in a chemical reaction. Examples are how easily something burns or the way something reacts with acid.
- The chemical properties of the reactants in a chemical reaction are always different to the chemical properties of the products.
- Examples of **physical properties** are solubility, melting point, density, colour and magnetism.
- Two dissolved chemicals reacting to form a solid is called **precipitation**.
- If a reaction takes place in a sealed container, there will be no change in mass before and after the reaction.

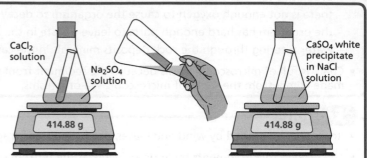

CaCl$_2$ solution

Na$_2$SO$_4$ solution

CaSO$_4$ white precipitate in NaCl solution

414.88 g   414.88 g

The conservation of matter. Note that the mass remains the same.

Remember!
In a *physical* change the chemical properties of the substance do not change.

Conservation of matter

- The number of each type of atom in the reactants is the same as the number of each type of atom in the products.
- Formula equations must be balanced to show this, for example:

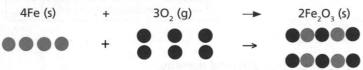

$$4Fe\ (s) \quad + \quad 3O_2\ (g) \quad \rightarrow \quad 2Fe_2O_3\ (s)$$

Forming iron oxide.

- The equation above shows the formation of iron oxide.
  - Four iron atoms react with six oxygen atoms.
  - They produce two iron oxide molecules, each made up of two iron atoms and three oxygen atoms.
- The law of conservation of mass can be used to calculate the amount of oxygen in iron oxide:
  - Weigh a sample of iron.
  - Heat with a known volume of air.
  - The mass of iron oxide produced minus the mass of iron reactant = mass of oxygen used as reactant.

Improve your grade

Rearranging atoms

Higher: Prakash and John are discussing where all the nitrogen gas in the atmosphere came from. Prakash thinks that the carbon dioxide in the early atmosphere was converted into nitrogen.

$$CO_2\ (g) \quad \rightarrow \quad N_2\ (g)$$

carbon dioxide   nitrogen

John disagrees. Who do you agree with and why?   *AO2* [3 marks]

Thermal decomposition and calcium

Calcium carbonate

- In a **thermal decomposition** reaction, a chemical breaks down into more than one substance when heated.
- Calcium carbonate decomposes to make calcium oxide and carbon dioxide.
- Other metal carbonate compounds follow the same pattern:

  zinc carbonate ⟶ zinc oxide + carbon dioxide
  copper carbonate ⟶ copper oxide + carbon dioxide
- Calcium oxide is useful for making glass, cement and **limewater**.

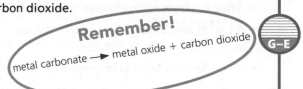

Remember!
metal carbonate ⟶ metal oxide + carbon dioxide

G–E

Calcium oxide

- The reaction between calcium oxide and water produces calcium hydroxide (slaked lime). It gives off a lot of heat, causing the water to boil and spit.

  CaO (s) $\quad + \quad$ H_2O (l) $\quad \longrightarrow \quad$ $Ca(OH)_2$ (s)
  calcium oxide $\quad + \quad$ water $\quad \longrightarrow \quad$ calcium hydroxide

- Limewater is calcium hydroxide solution. When limewater reacts with carbon dioxide, a cloudy precipitate of calcium carbonate forms, because calcium carbonate is **insoluble**.

  $Ca(OH)_2$ (aq) $\quad + \quad$ CO_2 (g) $\quad \longrightarrow \quad$ $CaCO$ (s)
  calcium hydroxide $\quad + \quad$ carbon dioxide $\quad \longrightarrow \quad$ calcium carbonate

D–C

Metal carbonates

- Many metal carbonates undergo thermal decomposition to metal oxides and carbon dioxide, but some decompose more easily (at a lower temperature) than others.
- Calcium carbonate decomposes at 825 °C, zinc carbonate at 300 °C and copper carbonate at 200 °C.
- Uses of metal oxides include:
  - zinc oxide (ZnO) – rubber, concrete, medicines, cosmetics
  - copper oxide (CuO) – pigments, semiconductors.

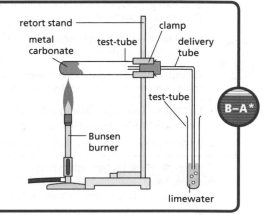

retort stand — clamp
metal carbonate — test-tube — delivery tube
test-tube
Bunsen burner
limewater

B–A*

Apparatus to carry out thermal decomposition.

Calcium and neutralisation

- Neutral soil (**pH 7**) is best for growing crops.
- Acid soil can be made less acidic by mixing it with calcium carbonate, oxide or hydroxide.
- Burning coal produces sulfur dioxide gas. If this gets into the atmosphere it makes **acid rain**.
- Calcium compounds are used to prevent sulfur dioxide getting into the atmosphere.

G–E

- Both calcium oxide (lime) and calcium hydroxide (slaked lime) are irritants and cause severe burns.
- Three times more limestone than lime is needed to neutralise the same amount of soil.
- Lime and slaked lime neutralise soil faster, but limestone acts for longer.

D–C

- Calcium oxide neutralises acidic sulfur dioxide to produce calcium sulfate:

  CaO (aq) + SO_2 (g) ⟶ $CaSO_3$ (s)

- When coal is burned in power stations, the sulfur dioxide gas is removed from the waste gases by passing them thorough a scrubber.
- Calcium sulfate is a useful product of calcium oxide and can be sold.
- Nitrogen oxides are also acidic gases made in power stations that can cause acid rain. They can be removed in a similar way.

B–A*

⦿ Improve your grade

Limestone reactions

Foundation: Explain the reaction that occurs when limestone is heated and why this makes it more useful.

AO2 [4 marks]

Acids, neutralisation and their salts

Neutralising acids

- A **base** is a compound that can neutralise an acid. The reaction is called **neutralisation**.
- After neutralisation there is no acid left. A neutral solution has a pH of 7.
- All acids and bases react to form the same type of products.
  acid + base → salt + water
- If you neutralise an acid using a carbonate, carbon dioxide is produced.
  acid + carbonate → salt + water + carbon dioxide
- Bases are compounds of metals – either metal oxides (e.g. calcium oxide) or metal hydroxides (e.g. potassium hydroxide) or metal carbonates (e.g. sodium carbonate).
- Not all bases will dissolve to make solutions. Bases that *are* soluble are called **alkalis** (e.g. sodium hydroxide).
- Hydrochloric acid (HCl) is made naturally in our stomachs. It is important for killing bacteria and for activating some enzymes for digestion.
- Too much stomach acid causes indigestion. **Antacids** are medicines taken to relieve indigestion. They work by neutralising excess stomach acid.

Antacids

- Antacids are indigestion remedies that contain bases.
- You can work out how effective indigestion remedies are by measuring how much acid they can neutralise.
- You can find out how acidic something is by measuring its pH. Acids have a pH less than 7 while alkalis have a pH more than 7.
- To tell when an acid is neutralised you can measure the pH. As the acid is neutralised, the pH rises. At pH 7 all the acid has been neutralised.
- **Titration** is slowly adding alkali to acid (or acid to alkali) until exactly the right amount has been added to neutralise it.
- When all the acid is neutralised, only salt and water remain.

Acids and their salts

- A salt is a compound formed when an acid is neutralised.
- The name of the salt formed depends upon the acid and base used.
- The first word of the name comes from the name of the metal in the base, e.g. *calcium* carbonate, *sodium* hydroxide, *magnesium* oxide.

Remember!
A salt is neutral, neither an acid nor a base.

- The second word in the name comes from the acid. Hydrochloric acid forms salts whose name ends in chloride, sulfuric acid forms sulfates and nitric acid forms nitrates.

Acid	Base	Salt
Hydrochloric acid	Sodium hydroxide	Sodium chloride
Hydrochloric acid	Calcium carbonate	Calcium chloride
Nitric acid	Calcium oxide	Calcium nitrate
Sulfuric acid	Calcium carbonate	Calcium sulfate

Examples of acids, bases and the salts produced.

Improve your grade

Neutralising acid spills

Higher: Commercial spill kits are available to help deal with acid spills in the workplace. They often contain sodium carbonate.

Explain how such kits would help to prevent damage from a spill of sulfuric acid. Include any reactions that occur in your answer.

AO2 [5 marks]

Electrolysis and chemical tests

Electrical energy and charged solutions

- **Electrolysis** is when a compound is split up using electrical energy. An electric current is made to pass through a solution of the compound.

- Electrolysis of water produces oxygen and hydrogen.

- Electrolysis of hydrochloric acid produces hydrogen gas at the negative **electrode** (the **cathode**) and chlorine gas at the positive electrode (the **anode**).

- Chlorine can also be produced by the electrolysis of seawater.

- Chlorine is important for the manufacture of poly(chloroethene) and of bleach.

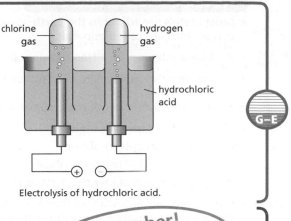

Electrolysis of hydrochloric acid.

- The two electrodes used in electrolysis have opposite charges.

- Negative **ions** are attracted to the positive electrode and positive ions are attracted to the negative electrode. In this way the ions in the solution separate and move to opposite electrodes.

- Gases produced at the electrode are less dense than liquid and so bubble up. Gases can be captured when they displace water from an upturned container.

Remember!
Electrolysis requires a direct current (d.c.) so that one electrode remains negative and the other positive.

- Electrolysis of hydrochloric acid produces an equal volume of gas at each electrode. There are equal amounts of chlorine and hydrogen in hydrochloric acid (HCl).

- Twice as much hydrogen as oxygen is produced when water is electrolysed. This tells us that there is twice as much hydrogen as oxygen in water (H_2O).

Chemical testing

- Chemical tests help us to identify some gases:
  - A lighted splint causes a popping sound when plunged into a tube of hydrogen, but goes out in a tube of chlorine.
  - A glowing splint relights in oxygen.
  - Damp indicator paper is bleached to white when it comes into contact with chlorine.

- In chemical tests, scientists must consider all the evidence before coming to a conclusion.

- Physical and chemical properties are useful in identifying gases. For example:
  - Chlorine is pale green with a distinctive pungent smell.
  - Oxygen and hydrogen are colourless and odourless.

Gases and their uses

- Oxygen is essential for breathing. It is supplied commercially for medical use and for underwater diving tanks.

- Oxygen is essential for combustion. Many fire-fighting techniques depend on removing oxygen to put out fires.

- Hydrogen is required for ammonia manufacture, **cryogenics** and hydrogen balloons. It must be used with caution because it is very flammable.

- Chlorine is toxic, which makes it hazardous to handle but useful in destroying microorganisms. It is used in bleach and **disinfectant** manufacture.

How science works

You should be able to:
- assess the risks involved when conducting experiments with toxic substances such as chlorine
- adopt appropriate safety precautions.

Improve your grade

Creating chlorine
Foundation: Describe how you could produce a test tube full of chlorine from hydrochloric acid.

AO2 [4 marks]

Metals – sources, oxidation and reduction

Rocks, minerals and ores

- Most metals are found in the Earth as part of a chemical compound. Naturally occurring compounds are called **minerals**.
- Rocks are mixtures of different minerals. Some rocks are metal **ores**.
- Very unreactive metals like gold can be found in the Earth as elements.

- Metal ores contain metal compounds and unwanted rock. The metal must be extracted from the ore.
- The cost of the metal is higher if:
  - the ore is difficult to obtain
  - the method of extraction is expensive
  - the ore contains a low percentage of the metal.
- Iron is cheap because iron ore (haematite) is plentiful and inexpensive to extract. Gold has a high price because the ore is rare. Aluminium ore (bauxite) is plentiful but aluminium is expensive because it costs a lot to extract.
- Some metals can be extracted by heating with carbon, e.g. extracting iron.

  iron oxide + carbon ⟶ carbon dioxide + iron

- Other metals must be extracted by **electrolysis**, e.g. extracting aluminium.

  electricity
  aluminium oxide ⟶ aluminium + oxygen

- Electrolysis is expensive because of the high cost of the electricity.

Methods of extraction

- Metal ores often contain metal oxides. To extract the metal, oxygen must be removed from the oxide.
- The method of extraction chosen depends on how reactive the metal is.
- The **reactivity series** shows which metals are more reactive than carbon and must be extracted by electrolysis, and which are less reactive and can be extracted with carbon.

Remember!
The method of extraction of a metal depends on the properties of the metal.

Oxidation and reduction

- Metals can gain oxygen atoms and become oxides. This is called **oxidation**, e.g. iron + oxygen ⟶ iron oxide. This is what happens when iron rusts.
- Oxidation of metals is called corrosion.
- Metal oxides can lose oxygen and become metals. This is called **reduction**. Reduction takes place when iron is extracted from its ore.

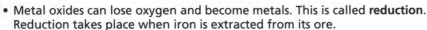

- The equation for the reduction of iron ore with carbon is:

  $2Fe_2O_3$ (s) + $3C$ (s) ⟶ $4Fe$ (s) + $3CO_2$ (g)
  iron oxide + carbon ⟶ iron + carbon dioxide

- Carbon is oxidised and iron oxide is reduced.
- If iron is not protected it will corrode by reacting with oxygen from the air.

  $4Fe$ (s) + $3O_2$ (g) ⟶ $2Fe_2O_3$ (s)
  iron + oxygen ⟶ iron oxide

Reactivity and corrosion

- The more reactive a metal is, the more rapidly it becomes oxidised.
- Many metals form an impermeable layer of metal oxide, which creates a barrier between oxygen and the metal underneath. They do not continue to corrode.
- Iron corrodes when it comes into contact with water and oxygen. You can prevent this by coating the metal, to protect it.
- **Galvanising** is when iron is coated with zinc, a more reactive metal. The zinc corrodes first, protecting the iron. This process is called **sacrificial protection**.

Improve your grade

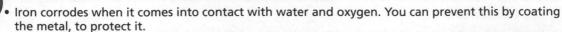

Pricing metals

Higher: Aluminium is the most common metal in the Earth's crust, yet 100 g of aluminium costs about ten times more than 100 g of iron. Explain the reasons for this difference in value. *AO1* [5 marks]

Metals – uses and recycling

Properties dictate uses

- The different properties of metals makes them useful for different purposes.

Metal	Properties	Uses
Aluminium	Low density, strong for its weight, corrosion resistant, highly reflective	Aircraft, drinks cans, overhead electricity cables, bicycle frames, window frames, mirrors and reflectors
Copper	Excellent conductor of heat and electricity, does not corrode in water, **malleable** and **ductile**	Plumbing, electrical wiring
Gold	Excellent electrical conductor, does not corrode, very reflective	Jewellery, dentistry, electronic connectors, reflectors and radiation shields

Alloys

- An alloy is a mixture of different metal atoms. Alloying changes the properties of the metal, e.g. steel is an alloy of iron.

Element added	Change in properties
Carbon	Harder, stronger, more brittle
Chromium	Very corrosion resistant (stainless steel)
Titanium	Stronger, lighter, more corrosion resistant

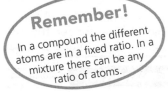

> **Remember!**
> In a compound the different atoms are in a fixed ratio. In a mixture there can be any ratio of atoms.

In pure metals the atoms are aligned in a very regular pattern (top left), but in alloys the atoms of different substances disrupt the alignment of the metal atoms (top right).

- Pure metals have atoms of identical size. The layers of atoms easily slide over each other, making the metal soft and malleable.

- Mixing different sized atoms interferes with this and makes the metal stronger.

- Pure gold is too soft to be useful, so it is alloyed with silver and copper. The carat and **fineness system** indicates the percentage of gold.

% gold	99.9%	75%	50%	37.5%
Carat rating	24	18	12	9
Fineness rating	999	759	500	370

- Shape memory alloys return to their original shape after being deformed. For example, nitinol is an alloy of nickel and titanium. Uses include:
  - medical stents to keep blood vessels open
  - dental braces to pull teeth into position
  - spectacle frames.

- Scientists are learning to design metals like shape memory alloys to meet specific requirements.

Recycling

- Metal ores are a **non-renewable** resource.

- Recycling of metal brings both economic and environmental advantages:
  - reduces energy use in mining and transport
  - causes less environmental damage from mining
  - reduces the amount of land used up by dumping waste rock and waste metal
  - preserves the supply of metal ore.

Improve your grade

The right metal

Higher: A town council is trying to decide on the best metal to choose as a building material for a bridge over the local river.

Explain what factors they should take into account and suggest a suitable metal.
Give reasons for your choice.

AO2 [5 marks]

Hydrocarbons and combustion

Hydrocarbons and fractional distillation

- **Crude oil** is a mixture of lots of different **hydrocarbon** molecules. Hydrocarbons are made from hydrogen and carbon atoms only.
- Different molecules have different lengths, and this gives them different boiling temperatures. We can use the different boiling points to separate the molecules.

- In fractional distillation, crude oil is vaporised and the vapour is slowly cooled. As the temperature drops, first the larger and then the smaller molecules turn to liquid and can be collected. This separates the molecules into groups (fractions) with similar sizes.
- The longer the average carbon chain in the fraction, the higher the boiling point and viscosity and the lower the flammability.
- The different fractions are much more useful than crude oil.
- Products from fractional distillation include gases for heating and cooking, petrol for cars, kerosene for aircraft fuel, diesel oil for some cars and trains, fuel oil for ships and power stations, and bitumen for tarmac and roofing.

> **Remember!**
> The fractions from fractional distillation are still *mixtures* of different molecules.

- A **homologous series** is a family of similar **compounds**. Each successive member of the family has one more CH_2 group than the previous member.
- The molecules in liquid fractions of crude oil are held together by intermolecular forces. As the molecules get longer, the intermolecular forces get stronger. The stronger the force the more kinetic energy (higher temperature) needed to boil the fraction.

Combustion

- Combustion is burning, a type of **oxidation reaction** that produces heat and light.
- Burning hydrocarbon fuels produces carbon dioxide and water.

- Carbon dioxide turns limewater cloudy because it forms solid calcium carbonate.

$$Ca(OH)_2 \text{ (aq)} \quad + \quad CO_2 \text{ (g)} \quad \rightarrow \quad CaCO_3 \text{ (s)} \quad + H_2O \text{ (l)}$$
calcium hydroxide (limewater) + carbon dioxide → calcium carbonate + water

- Living organisms use oxidation reactions inside cells to transfer energy from food. This is why we need a constant supply of oxygen and we breathe out carbon dioxide.

Incomplete combustion

- When hydrocarbons burn in plenty of oxygen they produce carbon dioxide (CO_2).
- When there is not enough oxygen they produce **carbon monoxide** (CO). This is called **incomplete combustion**.
- Carbon monoxide is very poisonous. It is difficult to detect because it has no odour or colour.
- It is very important to use all heaters that burn hydrocarbons in a well-ventilated space. Then there is plenty of oxygen and no carbon monoxide is made.

- A yellow, sooty flame means incomplete combustion. The soot is particles of carbon (C).

- Carbon monoxide combines permanently with **haemoglobin** in **red blood cells**. This prevents them from carrying oxygen to the body cells.
- Carbon particles released from incomplete combustion can cause breathing problems, especially for asthmatics.
- Particles in the atmosphere also result in global dimming and encourage cloud formation.

🔘 **Improve your grade**

Making crude oil
Foundation: Explain how and why kerosene is made from crude oil.   AO2 [4 marks]

Acid rain and climate change

Destruction from the sky

- Hydrocarbon fuels naturally contain some atoms of sulfur. When the fuels burn, the sulfur atoms are oxidised to sulfur dioxide.

- Sulfur dioxide reacts with oxygen and water vapour and becomes sulfuric acid. This forms **acid rain**.

- The sulfur dioxide gas is usually carried away in the air and falls as acid rain far away from where it was made.

- Acid rain damages buildings, trees, life in streams and ponds, washes valuable minerals from the soil and releases toxic metals.

- When acid rain falls on trees it damages the leaves. The leaves do not photosynthesise efficiently and eventually the tree dies.

- Acid rain causes damage to the surface of limestone structures because it reacts with the calcium carbonate in limestone.

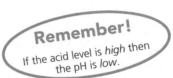
Remember!
If the acid level is *high* then the pH is *low*.

- The UK government has pledged to reduce sulfur dioxide emissions. Methods include:
  - removing sulfur from fuels before they are burned
  - trapping sulfur dioxide released after burning fuels
  - swapping to fuels with lower sulfur content, such as low sulfur coal and methane.

Climate change and the greenhouse effect

- Some gases in the atmosphere act like a blanket around the Earth and prevent heat from escaping into space. This warming effect is essential for life on the planet.

- Gases that can do this are called **greenhouse gases**. Important examples are carbon dioxide, water vapour and methane.

- The temperature of the planet may be affected by changes in solar activity and changes in the amounts of greenhouse gases in the atmosphere.

- Humans have increased the amount of carbon dioxide in the atmosphere by burning hydrocarbons and by cutting down and burning forests.

- Scientists are experimenting with ways to reduce the amount of carbon dioxide and other greenhouse gases.

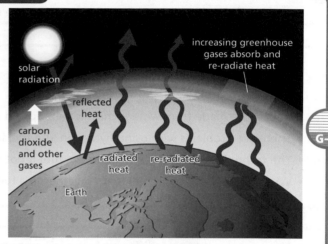
The greenhouse effect.

- The warm Earth radiates **infrared radiation** out towards space. Greenhouse gases absorb this radiation and so trap the energy on the planet. This increases the global temperature (the **greenhouse effect**).

- There is a correlation between the amount of carbon dioxide in the atmosphere and the temperature of the Earth.

- Most scientists consider that the rise in the Earth's temperature has been caused by human activities, which have increased the level of carbon dioxide in the atmosphere.

- Some scientists believe that the global temperature rise has a different cause.

Removing carbon dioxide from the atmosphere

- **Algae** remove carbon dioxide from the atmosphere by **photosynthesis** and by forming carbonate shells. Growth of algae is usually limited by a lack of iron in ocean water. Scientists are experimenting with seeding the oceans with iron to increase algal growth.

- Carbon dioxide can be reduced back into hydrocarbons using experimental **nanotechnology** and energy from the Sun.

- A sure way of preventing further increases in atmospheric carbon dioxide is to reduce energy consumption.

Improve your grade

Causes of climate change

Higher: Discuss the evidence that the activities of humans have resulted in climate change.   *AO2* [5 marks]

Biofuels and fuel cells

Fuels from plants

G–E

- Petrol, diesel and kerosene are **fossil fuels** made from crude oil. Fossil fuels are a **non-renewable resource** that will eventually run out.
- **Biofuels** are made from plants. Plants are a **renewable resource** because more can be grown.
- Ethanol is a biofuel that can be made from sugar beet or sugar cane. It can be mixed with petrol so that less petrol is needed.
- If fuel crops are grown instead of food, this may put up food prices.
- When fuel crops grow they take in the carbon dioxide that will be released when the fuel is burned.

> **Remember!**
> Biofuels are not carbon neutral, because energy is used to make fertiliser and to process the fuel.

D–C

- A good fuel will:
  - be easy to ignite and keep alight
  - not produce much ash, smoke or polluting gases
  - release a lot of energy per kg
  - be easy to store and transport
  - be cheap to produce and use.

B–A*

- **Fermentation** is the name given to the process that uses the **enzymes** in yeast cells to convert plant carbohydrate into ethanol. The overall reaction is:

$$C_6H_{12}O_6 \text{ (aq)} \rightarrow 2C_2H_5OH \text{ (aq)} + 2CO_2 \text{ (g)}$$

- Enzymes are biological **catalysts** that speed up the rate of reactions.

Cells and electricity

G–E

- Hydrogen makes a good fuel because:
  - it releases a large amount of energy
  - it produces only water as a waste product
  - it can be made from water, which is a cheap and plentiful.
- There are problems with using hydrogen as a fuel for cars because it is flammable and explosive.
- The technology for using hydrogen as a fuel is not completely ready yet.
- When oxygen and hydrogen react together in a **fuel cell**, the energy released is captured as electricity. This can be used to power a car.

Comparing fuels

D–C

- The energy released by different fuels can be compared using a burner and a **calorimeter**.
- The temperature of the water is measured before and after burning the fuel.
- It is important to keep the control variables the same in all experiments. These are the volume of water and the mass of fuel burned.
- The independent variable is the type of fuel and the dependent variable is the temperature rise.

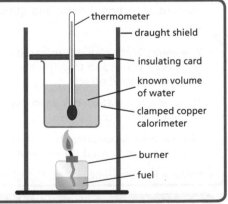

- thermometer
- draught shield
- insulating card
- known volume of water
- clamped copper calorimeter
- burner
- fuel

A basic calorimeter.

Where does the hydrogen come from?

B–A*

- Generating hydrogen by electrolysis of water requires electricity, which is mainly produced by burning fossil fuels. This means that using hydrogen-fuelled cars will result in the burning of fossil fuels and carbon dioxide emissions, unless renewable methods of generating electricity are used.
- Supplying the hydrogen to vehicles using fuel cells is a problem. There is no network of refuelling stations for hydrogen. One solution is to produce hydrogen in the vehicle by reacting petrol with steam.

Improve your grade

Energy from ethanol

Foundation: Describe an experiment that you could do to decide whether ethanol or hexane releases the most energy. Ethanol and hexane are both liquids.

AO2 [4 marks]

Alkanes and alkenes

Natural gas

- Most of the **hydrocarbons** in crude oil are **alkanes**.

- All alkanes:
  - have the formula $C_nH_{(2n+2)}$
  - have a name which ends in -ane
  - are **saturated** (all carbons atoms have four single bonds to four different atoms).

- The first three alkanes in the family are methane, ethane and propane. They are found in natural gas.

Hydrocarbon	Formula and structure	No. of carbon atoms	State
Methane	CH_4	1	Gas
Ethane	C_2H_6	2	Gas
Propane	C_3H_8	3	Gas

The structure and formulae of methane, ethane and propane.

- Refinery gas is a mixture of methane (70–90%), ethane and propane.

- The methane is separated and used in the home for heating and cooking. It releases more energy and less carbon dioxide than other fuels.

- Ethane is mostly processed further then used to make either **polymers** or other chemicals.

- Propane turns to a liquid easily when put under pressure in a cylinder. It is used in LPG fuel for cars, and for cooking and heating where piped gas is not available.

Covalent bonding

- Carbon atoms form four **covalent bonds** because this makes them stable.

- Atoms are made from a central **nucleus** containing **protons** and neutrons and surrounded by **electrons**. Carbon has six protons in its nucleus and six electrons arranged in two **shells** around the nucleus.

- A covalent bond is formed when a pair of electrons is shared between two different atoms. The nuclei of both atoms are attracted to the shared electrons, so the atoms are held together.

- When carbon shares four electrons to make four covalent bonds, it has eight electrons in its outer shell. This makes it chemically stable.

Remember!
Bromine water changes from brown to *colourless*, not from brown to *clear*. Clear means transparent – the opposite of cloudy.

Unsaturated hydrocarbons

- **Alkenes** are hydrocarbons that have a carbon-carbon double bond. This is two covalent bonds between the same two carbons.

- Molecules with a double bond are called **unsaturated**.

- Bromine water changes from brown to colourless when it reacts with an alkene. Alkanes do not decolourise bromine water.

The structure of ethene and propene.

Organic compounds

- **Organic compounds** are molecules that contain carbon.

- The hydrocarbons in crude oil make very good starting points from which chemists can manufacture other useful organic molecules, such as medicines.

- Organic chemicals can be named in a systematic way so that the name tells us the structure of the compound.

- The prefix to the name gives the number of carbon atoms.

- The suffix gives the family or **homologous group** to which the molecule belongs.

Prefixes used in naming organic compounds.

No. of carbon atoms	Prefix
1	Meth-
2	Eth-
3	Prop-
4	But-
5	Pent-

G–E

D–C

B–A*

G–E

D–C

B–A*

Improve your grade

Ethane and ethene

Higher: Most of the ethane obtained from natural gas is converted into ethene and made into polymers.

How could you distinguish between ethane and ethane? Illustrate your answer with drawings of the molecules concerned.

AO2 [5 marks]

Cracking and polymers

Cracking

- Crude oil from different sources contains different amounts of each fraction. The shorter-chain fractions are the most valuable.

- **Cracking** is a process that splits long-chain alkanes into shorter alkanes and alkenes. To crack an alkane you pass the alkane vapour over a heated **catalyst**.

- Cracking is a **thermal decomposition** reaction.

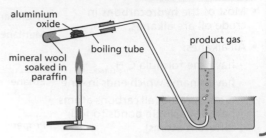

Laboratory equipment to crack hydrocarbons.

- Cracking uses a catalyst of aluminium oxide or zeolites.

- There is a mismatch between supply and demand for the fractions of crude oil. Cracking increases the supply of shorter-chain alkanes and provides alkenes for the polymer industry.

Making polymers

- Alkenes are used to make **polymers**. Many alkene molecules (**monomers**) link together to make a very long polymer molecule. This is **polymerisation**.

- The name of the polymer tells you the name of the monomer. Poly(*ethene*) is made from ethene. Poly(*propene*) is made from propene.

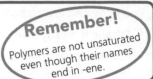

Remember!
Polymers are not unsaturated even though their names end in -ene.

- To make ethene into poly(ethene) you need a high temperature and pressure, and a catalyst.

- One bond from the double bond breaks and is used to join one alkene monomer to another in a long chain. This process of adding the monomers together is called **addition polymerisation**.

- The longs chains of the polymer can be spun into threads or moulded into any shape.

- Different alkenes can be used as monomers for polymerisation. This gives polymers with different properties.

- The equation below shows polymerisation for PTFE.

fluoroethene   a strand of poly(tetrafluoroethene)

Disposing of polymers

- Disposing of unwanted polymers is a problem. Polymers put into landfill remain there for a very long time because they are not **biodegradable**. Burning polymers often releases toxic fumes.

- **Recycling** solves the problem of what to do with unwanted polymers. It also saves oil, energy used in making new polymers, and space in landfill sites.

- Scientists have produced new polymers that are biodegradable.

- Polymers must be sorted into different types before they can be recycled. Melting a mixture of polymers gives a poor-quality product with few uses.

- International recycling codes make it easier to identify and sort polymers.

- Not all polymers are easily recycled, and the quality of the recycled product is not as good as the original.

- Adding biodegradable components into the polymer mixture (like starch grains) helps the polymer break down more easily.

- Polymers made from **starch** and **cellulose** are renewable and biodegradable. There is potential to develop these so they have the required properties to replace oil-based polymers.

Improve your grade

Polymers and energy

Higher: Explain why making bin-liners from new polymers uses so much more energy than using recycled polymers. Suggest why only a small proportion of polymers are recycled.   *AO2* [4 marks]

C1 Summary

The early atmosphere was formed from volcanic activity and comprised mainly carbon dioxide and water vapour.

Evidence for the early atmosphere is found in gases trapped in rocks and in the oxidation of iron in rocks.

The Earth's sea and atmosphere

Photosynthesis eventually caused decreased CO_2 and increased O_2 in the atmosphere.

The modern atmosphere is mostly nitrogen and oxygen with tiny amounts of carbon dioxide.

Early oceans formed as water vapour condensed.

They absorbed carbon dioxide to make carbonate rock.

Igneous rock solidifies from molten rock. Intrusive igneous rock cools below the surface, forming large crystals; extrusive rock cools on the surface, forming small crystals.

Metamorphic rock forms when other rocks come under heat and pressure.

Limestone becomes marble.

Sedimentary rock forms from eroded rock fragments cemented together.

It is relatively soft and easily eroded, and may contain fossils.

In thermal decomposition, a metal carbonate decomposes to form a metal oxide and carbon dioxide.

Materials from the Earth

Limestone is quarried for building, making cement and glass, and neutralising soil.

Quarrying destroys habitats and causes pollution, but provides jobs.

Acids are neutralised by bases, which are metal oxides, hydroxides or carbonates.

Neutralising acids produces salts.

Indigestion remedies are used to neutralise stomach acid.

Acids

Electrolysis is the splitting up of compounds using electricity (d.c.).

HCl splits to H_2 (g) and Cl_2 (g); H_2O splits to H_2 (g) and O_2 (g).

Metals are extracted from ores in the Earth.

Less reactive metals are reduced from their compounds with carbon; more reactive metals are reduced by electrolysis.

Recycling preserves supplies of ore and reduces environmental damage from mining and extraction.

Alloying changes the properties of a metal, e.g. it becomes harder or stronger.

Steel is iron + carbon + other metals.

Obtaining and using metals

Alternative fuels include:
- Biofuels made from plants (renewable).
- Ethanol made by fermenting sugar cane and beet.
- Hydrogen, which is non-polluting.

Carbon dioxide is a greenhouse gas. Greenhouse gases trap heat on Earth.

Human activity has increased the level of greenhouse gases, and this may be causing climate change.

Polymers are made by polymerising alkenes (unsaturated molecules formed by cracking alkanes).

Artificial polymers are not biodegradable.

Recycling and producing biodegradable polymers reduces landfill usage.

Hydrocarbon fuels are distilled from crude oil. They are mostly alkanes (C_nH_{2n+2}).

Complete combustion gives CO_2 + H_2O; incomplete combustion gives toxic CO + C (soot).

Fuels

Atomic structure and the periodic table

Matter and mass

- **Atoms** are the smallest building blocks of **elements**. They are made from three types of subatomic particle: **protons**, **neutrons** and **electrons**.

- Protons and neutrons are held tightly together in a *tiny* space in the centre of the atom called the **nucleus**. Electrons move around in a *large* area surrounding the nucleus.

- Protons have a positive charge. Electrons have a negative charge. Neutrons are neutral (they have no charge).

- There are the same number of negative electrons as positive protons. This means that an atom has no overall charge.

- Elements are made from only one type of atom. Different elements have different numbers of protons in the nucleus of their atoms.

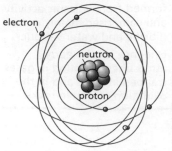

The structure of an atom. This shows the position of the particles, but really the nucleus is thousands of times smaller than the space that the electrons move in.

- The mass of a proton is 1 **atomic mass unit**. The mass of other particles are compared to this.

Name of particle	Relative mass	Relative charge
Proton	1	+1
Neutron	1	0
Electron	1/1836	−1

Remember! Atoms are too tiny to be seen or weighed. The mass numbers are *relative*.

The periodic table

- In 1869, Dmitri Mendeleev arranged the elements known at that time in a table in order of increasing atomic mass. He put elements with similar properties underneath each other.

- Mendeleev left gaps in the table to make the elements fit the pattern. He correctly predicted that new elements would be discovered to fill the gaps. He was able to predict the properties of these elements by looking at the properties of the elements around them.

The modern periodic table. Rows are referred to as periods and columns are called groups.

- The modern **periodic table** shows each element like this:

  Mass number. The number of protons + neutrons

  Atomic number. The number of protons

  $$^{12}_{6}C$$

  A capital letter or capital followed by a small letter is the symbol for the element

- Elements appear in order of their **atomic number**. Elements in the same group have similar properties.

- Non-metal elements are at the top right. All the other elements are metals.

Isotopes

- All atoms of the same element have the same number of protons in their nucleus.

- **Isotopes** are atoms of the same element with different numbers of neutrons in their nucleus. They have the same number of protons and electrons, and the same chemical properties, but a different mass.

- Hydrogen has three isotopes: ^3H, ^2H, ^1H

- **Relative atomic mass** is the average mass of an atom of the element, taking into account the number and abundance of all the different isotopes.

- Because of the existence of isotopes, not all relative atomic masses are whole numbers.

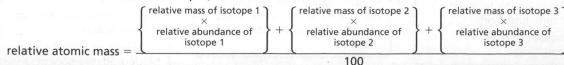

$$\text{relative atomic mass} = \frac{\left\{\begin{array}{c}\text{relative mass of isotope 1}\\ \times\\ \text{relative abundance of isotope 1}\end{array}\right\} + \left\{\begin{array}{c}\text{relative mass of isotope 2}\\ \times\\ \text{relative abundance of isotope 2}\end{array}\right\} + \left\{\begin{array}{c}\text{relative mass of isotope 3}\\ \times\\ \text{relative abundance of isotope 3}\end{array}\right\}}{100}$$

Improve your grade

Atomic structure

Foundation: Describe the structure of an oxygen atom.

AO1 [4 marks]

Topic 1: 1.1, 1.2, 1.3, 1.4, 1.5, 1.6, 1.7, 1.8, 1.9, 1.10, 1.11

Electrons

Completing the picture

- The number of electrons in an atom is the same as the number of protons in the nucleus.

- The electrons are arranged in **shells** around the nucleus. Electrons in the same shell have the same energy.

- The nearer the nucleus, the lower the energy of the electrons in the shell. Electrons always go to the lowest energy shell they can.

- The first shell can hold up to two electrons, the second can hold eight electrons, the third holds eight electrons before the fourth shell begins to fill up.

- The position of electrons in an atom is shown using the 2.8.8 convention. This is called the **electron configuration**, e.g. sodium has 11 electrons: two in the first shell, eight in the second shell and one in the third shell. The electron configuration of sodium is 2.8.1.

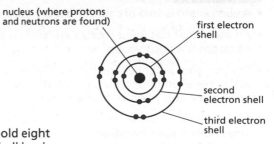

nucleus (where protons and neutrons are found)

first electron shell

second electron shell

third electron shell

The arrangement of electrons in an atom of argon.

- The table below shows the electron configurations of the first three periods of the periodic table.

G–E

Element	Atomic number	Electron configuration	Element	Atomic number	Electron configuration
Hydrogen	1	1	Neon	10	2.8
Helium	2	2	Sodium	11	2.8.1
Lithium	3	2.1	Magnesium	12	2.8.2
Beryllium	4	2.2	Aluminium	13	2.8.3
Boron	5	2.3	Silicon	14	2.8.4
Carbon	6	2.4	Phosphorus	15	2.8.5
Nitrogen	7	2.5	Sulfur	16	2.8.6
Oxygen	8	2.6	Chlorine	17	2.8.7
Fluorine	9	2.7	Argon	18	2.8.8

Remember!

Electrons always fill the lowest energy shell available.

Location, location, location

- Elements in the same group of the periodic table have the same number of electrons in their outer shell.

- The number of shells shows which period an element is in. The number of electrons in the outer shell shows which group it is in.

D–C

Periodicity

- Elements with the same number of electrons in their outer shell react in a similar way in chemical reactions. For example, all group 2 elements react with oxygen to form a compound with the formula XO (put any group 2 element in place of X).

- Mendeleev arranged the elements in order of increasing atomic mass. He noticed that at regular intervals (periodically), elements showed the same properties. We now know that this is because they have a similar electronic structure. This repeating nature is called **periodicity**.

B–A*

How science works

You should be able to:

- understand the process by which scientific hypotheses are developed, proven and accepted. For example, Mendeleev used his arrangement to make predictions about the properties of new elements that had yet to be discovered. When these elements were discovered and his predictions found to be correct, Mendeleev's periodic table was accepted.

Improve your grade

Similar reactions

Higher: Lithium and potassium both react in a very similar way when they are added to water.

Use your knowledge of atomic structure to explain why lithium and potassium react in a similar way, even though they are different elements.

AO2 [4 marks]

Ionic bonds and naming ionic compounds

Bonding

- A compound is two or more different atoms that are chemically bonded together.
- An atom that has gained extra electrons is a negative **ion**. An atom that has lost electrons is a positive ion.
- Positive and negative ions form when one atom transfers electrons to a different atom. For example, sodium (Na) transfers an electron to chlorine (Cl) to make a sodium ion (Na$^+$) and a **chloride** ion (Cl$^-$).
- Positive and negative ions are held together by **electrostatic** attraction. This is called **ionic bonding**.
- Compounds held together by ionic bonding are called **ionic compounds**.

Making ions

- An atom that has a full outer shell of electrons is very unreactive (stable).
- Elements in group 0 of the periodic table (the noble gases) have full outer shells of electrons and are very stable.
- When atoms form ions, they give or receive electrons to gain a full outer shell of electrons.
- The electronic configuration of an atom tells us what type of ion it will form:
  - Oxygen 2.6 will gain two electrons to fill up the second shell, so there are eight electrons. It forms an O^{2-} ion.
  - Magnesium 2.8.2 will lose two negative electrons so that the full second shell becomes the outer shell. It forms an Mg^{2+} ion.
- Negative ions change the ending of their element name to -ide, e.g. a fluorine atom becomes a fluoride ion.

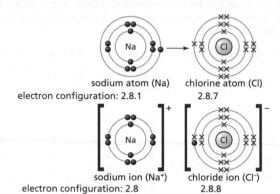

sodium atom (Na)
electron configuration: 2.8.1

chlorine atom (Cl)
2.8.7

sodium ion (Na$^+$)
electron configuration: 2.8

chloride ion (Cl$^-$)
2.8.8

Sodium and chloride ions being formed.

Remember!
Forming ions only involves electrons. The number of protons and neutrons in the nucleus does not change.

Ionic bonds

- All metal atoms form positive ions by giving away electrons. Positive ions are called **cations**.
- Non-metal atoms form negative ions by receiving electrons. Negative ions are called **anions**.
- Ions can also be compounds. Examples are OH$^-$, NO$_3^-$, HCO$_3^-$, CO$_3^{2-}$, SO$_4^{2-}$, NH$_4^+$.

Naming ionic compounds

- The name of an ionic compound tells us which ions it contains.
- When the negative ion is an element, the name ends in -ide

potassium bromide

The first word gives the name of the positive ion K$^+$

The second word gives the name of the negative ion Br$^-$

- When the negative ion is a compound containing oxygen, the ending (usually) changes to -ate.

magnesium carbonate

The first word gives the name of the positive ion Mg^{2+}

Shows the negative ion contains carbon

-ate shows the negative ion is a compound with oxygen CO$_3^{2+}$

Improve your grade

Ionic compounds

Higher: Explain the meaning of 'ionic compound'. Illustrate your answer with a diagram that shows an example of how ionic compounds form.

AO1 [4 marks]

Topic 2: 2.1, 2.2, 2.3, 2.4, 2.5

Writing chemical formulae

Symbols for ions

- The formula for an ion shows the symbol and the charge.
- Monoatomic ions are formed from atoms of only one element.
  - Monoatomic cations have the same name as the element from which they were made. A sodium atom becomes sodium ion.
  - Monoatomic anions have the same name as their atoms but with the ending changed to -ide. A chlorine atom becomes chloride ion.
- **Polyatomic** cations are formed from atoms of more than one element. The different symbols show the different elements from which they are formed.
- When ions join to form ionic compounds, the total number of negative charges always equals the total number of positive charges.
- This means that the number of positive and negative ions is not necessarily the same. Some examples are shown in the table below.

	cations			anions	
1+ ions	2+ ions	3+ ions		2- ions	1- ions
H^+ hydrogen					
Li^+ lithium	Be^{2+} beryllium			O^{2-} oxide	F^- fluoride
Na^+ sodium	Mg^{2+} magnesium	Al^{3+} aluminium		S^{2-} sulfide	Cl^- chloride
K^+ potassium	Ca^{2+} calcium				Br^- bromide

Some monoatomic ions (made from one atom).

cations	anions	
1+ ions	1- ions	2- ions
NH^+ ammonium	OH^- hydroxide	CO_3^{2-} carbonate
	NO_3^- nitrate	SO_4^{2-} sulfate

Cation	Anion	Compound
Na^+	Cl^-	NaCl
Na^+	S^{2-}	Na_2S
Mg^{2+}	Cl^-	$MgCl_2$
Mg^{2+}	S^{2-}	MgS

Some polyatomic ions (made from more than one atom).

Writing formulae for ionic compounds

- The **chemical formula** of an ionic compound shows how many and which type of ions it contains.
- When the compound contains more than one polyatomic ion, the whole ion has brackets around it. For example:

$Ca(NO_3)_2$ means $2 \times NO_3$ so $Ca(NO_3)_2$ contains $1 \times Ca$, $2 \times N$ and $6 \times O$ atoms

- To work out the formula of an ionic compound:
  1 Write the name of the compound.   Calcium nitrate
  2 Write the ions it contains. Ca^{2+} NO_3^-
  3 Multiply the number of ions so that the number of positive charges equals the number of negative charges. $1 \times Ca^{2+}$ $2 \times NO_3^-$
  $= 2 \times +$ $= 2 \times -$
  4 Write the formula. Show the number of each type of ion as a subscript. Remember to put brackets around the compound ions before writing the subscript. $Ca(NO_3)_2$

> **Remember!**
> The formula of an ionic compound never has a charge.

Balancing equations

- To convert a word equation into a formula equation:
  1 Write the **word equation**.
  2 Work out the formula of all reactants and products.
  3 Count the number of each type of atom in the reactants. Add coefficients to balance the numbers of each atom in the reactants with the numbers in the products.

EXAM TIP

Many students try to balance ionic equations that contain compound ions like SO_4^{2-} by breaking up the sulfate ion to form oxygen. In fact, in most reactions studied at GCSE, sulfate ions in the reactants appear as sulfate ions in the product. Always try to balance an equation first by keeping compound ions intact.

Improve your grade

Formulae equations

Higher: Potassium iodide and lead nitrate react together to form lead iodide. Lead forms a Pb^{2+} ion.

Decide the name of the other product and write a balanced formula equation for this reaction.

AO2 [4 marks]

Ionic properties and solubility

Substances made of ions

G–E
- All ionic compounds have similar properties because their particles are joined together in the same way. Positive and negative ions are arranged one after the other in three dimensions.
- This means they often form crystals.

D–C
- Ionic compounds:
  - have high melting points and boiling points
  - are brittle and hard
  - conduct electricity only when dissolved or molten
  - may be soluble in water.

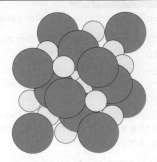

A model of a small part of a crystal of sodium chloride. The green balls are chloride ions and the yellow balls are sodium ions.

Structure of ionic compounds

B–A*
- Ionic substances form giant **lattice** structures with strong **electrostatic forces** between positive and negative ions.
- A large amount of energy is needed to break strong ionic bonds, so the melting point is high.
- The regular arrangement of millions of ions gives a crystalline structure.
- Ionic substances conduct electricity when molten or dissolved because the charged ions are able to move towards oppositely charged electrodes. They do not conduct electricity when solid because the ions are not able to move.
- The strength of the ionic bonding is higher if the ions have a higher charge. For example, magnesium oxide, MgO (Mg^{2+} and O^{2-}), has stronger bonding than sodium chloride, NaCl (Na^+ and Cl^-).

Remember!
There are no individual molecules in an ionic compound, just a giant ionic lattice containing millions of ions. The formula shows the ratio of ions.

Solubility of ionic compounds

G–E
- There are patterns in the solubility of ionic compounds, as shown in the table below.

General rule	Compounds that break the rule
Compounds are *soluble* if the positive ion is: sodium; potassium; ammonium	None
Compounds are *soluble* if the negative ion is: nitrate; chloride; sulfate	Silver chloride, lead chloride, lead sulfate, barium sulfate, calcium sulfate
Compounds are *insoluble* if the negative ion is: carbonate; hydroxide	Sodium carbonate, potassium carbonate, ammonium carbonate
	Sodium hydroxide, potassium hydroxide, ammonium hydroxide

Precipitates

D–C
- Sometimes two soluble salts form an insoluble salt when mixed together. This is called a **precipitation reaction**.
- **State symbols** in an equation show if a reactant or product is dissolved (aq), a solid (s), a liquid (l) or a gas (g). For example:

magnesium sulfate (aq) + sodium carbonate (aq) → sodium sulfate (aq) + magnesium carbonate (s)

B–A*
- To decide if a precipitation reaction will occur:
  - Look at the ions in each of the soluble reactants.
  - If any combination of these ions would make an insoluble compound (see table above), then a **precipitate** will form.

Improve your grade

Ionic compounds
Foundation: Jon has a test tube containing white crystals. He wonders if the white crystals are ionic.

What properties will the crystals have if they are ionic? Suggest a simple experiment Jon could carry out to test for one of these properties.   *AO2* [5 marks]

Preparation of ionic compounds

Making an insoluble salt

- An insoluble **salt** can be made by a precipitation reaction in the following way:

Step	Example	
Name the insoluble salt you want to make and decide which ions are in it.	silver chloride	
	silver ions	chloride ions
Use the solubility rules to choose a soluble compound for each ion.	*silver* nitrate	sodium *chloride*
Mix solutions of the two soluble compounds to make a precipitate of the insoluble salt.	silver chloride	

- To make a pure sample of an insoluble salt, follow this method:
  - Mix solutions of the reactants.
  - Filter to separate the insoluble salt.
  - Wash the residue in the filter paper to remove impurities.
  - Allow to dry.

Remember!
The solid trapped in the filter paper is the **residue**, the liquid that passes through the filter paper is the **filtrate**.

Using salts in X-rays

- Barium sulfate is insoluble and X-rays cannot pass through it. Doctors can use these properties to diagnose gut problems.
- The patient drinks a suspension of barium sulfate (**barium meal**). Once it has passed into the gut the patient is X-rayed. The X-ray shows the silhouette of the gut and any abnormalities can be seen.
- Barium sulfate is toxic, but because it is very insoluble it is not absorbed into the bloodstream and so is safe to drink.

Testing for ions

- Metal ions make colours when they are put into a flame. Different metals make different colours, helping to identify which metals are in a compound. Some examples are shown in the table opposite.
- To conduct a **flame test**:
  - Dip a nichrome wire in acid.
  - Dip it into the compound to be tested.
  - Put the wire into a Bunsen flame.

Metal ion	Colour
Na^+	Yellow/orange
K^+	Lilac
Ca^{2+}	Brick red
Cu^{2+}	Blue/green

Testing for anions

- Flame tests do not work for negatively charged anions, but other tests can be carried out.
- To test for a carbonate, add a dilute acid. If a gas bubbles off, test it with limewater. If it turns the limewater cloudy then the gas is carbon dioxide and the carbonate is confirmed.

  acid + carbonate ⟶ salt + water + carbon dioxide

- To test for sulfate ions, add hydrochloric acid and barium chloride solution. If a white precipitate forms then the solution contains sulfate ions.

  sulfate ions (aq) + barium chloride (aq) ⟶ barium sulfate (s) + chloride ions (aq)

- To test for chloride ions, add silver nitrate. If a white precipitate forms it contains chloride.

  chloride ions (aq) + silver nitrate (aq) ⟶ silver chloride (s) + nitrate ions

Spectroscopy

- Emission **spectroscopy** can be used to detect the presence of individual elements.
- Each element produces a unique and characteristic spectrum when heated, like a chemical fingerprint.
- The **emission spectrum** of an unknown substance can be compared with standard spectra of elements to see which one it matches. This led to the discovery of rubidium and caesium.
- The method can detect very small amounts and can also measure the quantity of each element.

Improve your grade

Preparing copper carbonate

Higher: Describe how a pure sample of copper carbonate could be made.

AO2 [5 marks]

Covalent bonds

Covalent molecules

G–E
- Covalent molecules are groups of atoms joined by **covalent bonds.** A covalent bond is a shared pair of electrons.
- Covalent substances can be elements like oxygen (O_2) or compounds like water (H_2O).
- Covalent compounds are made from non-metals.

D–C
- Atoms are held together in a covalent bond because the nuclei of two different atoms are both holding onto the same pair of electrons.
- No electrons are transferred from one atom to another.
- Atoms share electrons because they are more stable when they share. The outer shell of each atom in the molecule fills up with shared electrons. A full outer shell of electrons makes atoms stable.

Stable configurations

B–A*
- The number of covalent bonds that an atom can form depends on how many more electrons are needed to fill the outer shell. For example:
  - Carbon needs four electrons, so it can form four bonds.
  - Oxygen needs two electrons, so it can form two bonds.
- We can predict the formula of a covalent molecule by looking at how many covalent bonds each atom can form.
- Sometimes two atoms share two or three pairs of electrons to form **double bonds** and **triple bonds**.
- Some elements form **diatomic** molecules (molecules containing two atoms), e.g. H_2, O_2, N_2, F_2, Cl_2, Br_2, I_2.

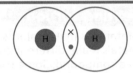

Remember!
No electrons are transferred when covalent bonds form, and there are no charged particles.

Drawing covalent bonds

G–E
- We can show covalent bonds in molecules using **dot and cross diagrams**.
- Dot and cross diagrams represent the electron configuration in the outer shells of the combining atoms.
- Dots represents the electrons in the outer shell of the left-hand atom. Crosses represents the electrons in the outer shell of the other atom.
- Overlapping areas show the shared electrons.

Two hydrogen atoms combining to form a hydrogen molecule (H_2). Hydrogen has just one electron so there is one dot and one cross.

D–C
- Carbon has four electrons in its outer shell. It can share them to form four covalent bonds. Hydrogen has one electron to share in a covalent bond.
- Four hydrogen atoms share with one carbon atom to form methane. Carbon now has eight electrons in its outer shell and each hydrogen atom has two electrons.
- A hydrogen chloride molecule contains one hydrogen atom and one chlorine atom. How would this molecule be drawn as a dot and cross diagram?

Bonding between carbon and hydrogen atoms to form methane (CH_4).

Double bonds

B–A*
- Oxygen atoms share two pairs of electrons to form an oxygen molecule with a double bond.
- Carbon forms double bonds with two oxygen atoms to form carbon dioxide.
- Carbon atoms can bond together to make chains and rings. This allows the formation of the large molecules needed for life to exist.

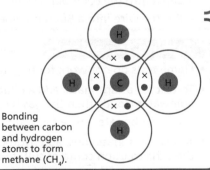

Covalent bonding in an oxygen molecule and a carbon dioxide molecule.

Improve your grade

Bonding in water

Foundation: Describe the type of bonding that holds the atoms together in a molecule of water. Illustrate your answer with a diagram.

AO2 [4 marks]

Properties of elements and compounds

Classifying substances

- Covalent and ionic substances have different properties. The properties of an unknown substance can help you to decide if it is ionic or covalent.

- Properties that can identify a substance as covalent or ionic include:
  - solubility in water
  - **melting point** and **boiling point**
  - electrical conductivity.

- The methods for investigating some of these properties are shown in the diagrams opposite.

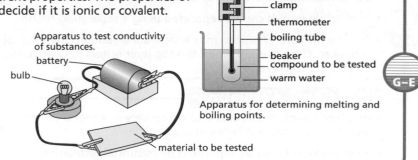

Apparatus to test conductivity of substances.

Apparatus for determining melting and boiling points.

G–E

Properties of compounds

- Most covalent substances form small molecules. Their structure is called simple molecular. Examples are water, methane and hexane.

- The covalent bonds between the atoms of the molecules are strong, but the forces holding the molecules together are weak.

- Simple molecular substances have low melting and boiling points because the weak forces between molecules are easily broken. They are often gases or liquids at room temperature, or they melt easily.

- Simple molecular substances do not conduct electricity because there are no charged particles.

- Most simple molecular substance do not dissolve in water. There are some exceptions, such as sucrose (sugar).

- Some covalent substances form giant networks of atoms. Their structure is known as giant molecular. Diamond, graphite and sand are examples.

- Giant molecular substances have high melting points and do not dissolve in water or conduct electricity.

- The table below shows the properties of substances with different structures.

D–C

Ionic	Covalent	
Giant lattice	Simple molecular	Giant molecular
High mpt	Low mpt	High mpt
Often soluble	Usually insoluble	Never soluble
Conducts electricity when molten or dissolved	Non-conductor	Non-conductor (except graphite)

EXAM TIP

Many students confuse the covalent bonds *within* simple covalent molecules and the weak forces *between* molecules. When simple covalent molecules boil, the covalent bonds do not break. H_2O (g) contains exactly the same two covalent bonds as H_2O (l).

Diamond and graphite

- Diamond and graphite are both covalent substances with giant molecular structures. They have different properties because their bonds are arranged differently.

- The electrons in the weak bonds between the layers of graphite can move when a voltage is put across them. So, graphite conducts electricity.

Diamond	Graphite
Four strong covalent bonds	Three strong covalent bonds + one weak bond
Three-dimensional network structure, strong in all directions	Layered structure
Very hard	Layers slide and rub off easily
Poor electrical conductor	Good electrical conductor
Used in cutting tools	Used to make electrodes and to lubricate moving parts

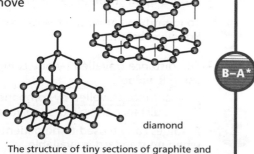

graphite

diamond

The structure of tiny sections of graphite and diamond lattices.

Remember!
Diamond and graphite are both the element carbon.

B–A*

Improve your grade

Using graphite

Higher: Sodium is extracted from sodium chloride by electrolysis. A graphite electrode is inserted into the molten compound and then electricity is passed through it.

Describe the properties of graphite that make it particularly useful as an electrode for this purpose, and explain why graphite has these properties.   *AO2* [5 marks]

Separating solutions

Separating immiscible liquids

G–E

- **Immiscible** liquids are liquids that do not mix but separate into layers. Oil and water are examples.
- Immiscible liquids can be separated using a separating funnel.
- Put the mixture of liquids into a separating funnel. Open the tap and collect the bottom layer in a beaker. Close the tap when the top layer enters it. Change the beaker then open the tap to collect the top layer.

Fractional distillation

- Miscible liquids are liquids that completely mix together. Ethanol and water are miscible, as are the different alkanes in crude oil.
- Miscible liquids can be separated by **fractional distillation**, as shown in the diagram opposite.

D–C

- Each liquid in the mixture has a different boiling point.
  - The temperature in the flask rises until it reaches the boiling point of liquid 1.
  - Liquid 1 turns to gas and moves into the condenser, where it cools and turns to liquid on the walls of the condenser.
  - When no more liquid is collected, the container is changed.
  - The temperature in the flask rises until it reaches the boiling point of the next liquid.

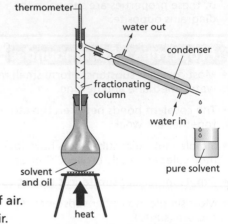

Laboratory fractional distillation apparatus.

B–A*

- Oxygen and nitrogen gas are produced by fractional distillation of air.
  - Dust, water vapour and carbon dioxide are removed from the air.
  - Air is cooled to –200 °C and all gases become liquids.
  - The temperature is gradually allowed to rise, and at about –196 °C nitrogen vaporises and is collected as gas.
  - The remaining liquid is mostly oxygen.

Chromatography

G–E

- **Chromatography** can be used to separate and identify components in substances such as colours in foods:
  - Draw a pencil line about 2 cm up from the bottom of a sheet of chromatography paper.
  - Put a spot of the mixture of colours on the line.
  - Put the paper in a container with solvent and a lid; make sure the solvent does not cover the spot.
  - Wait for the solvent to rise up the paper, then remove the paper and let it dry.
- Different colours move different distances up the paper, so you can see which colours were in the mixture.

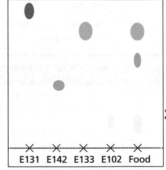

A chromatogram of food dyes. Which colouring in this food needs further investigation?

D–C

- A developed **chromatogram** shows a spot for each substance in a mixture.
- To identify what was in the mixture, compare the spots with known substances.

B–A*

- The distance travelled by one component of the mixture can be described using **R_f values**.

$$R_f = \frac{\text{distance travelled by component}}{\text{distance travelled by solvent}}$$

- R_f values can be used to help identify a component when no standard is available.
- Forensic science and the food industry make use of chromatography.

Remember!
Scientists often use chromatography to identify colourless substances like amino acids and sugars. The invisible spots can be seen by spraying the paper with a locating reagent. This reacts with the spots to produce a colour.

Improve your grade

Identifying colours

Foundation: A sweet manufacturer wants to know what colours have been used in a chewy sweet.

Describe a method that they could use to find this out.

AO2 [5 marks]

Classifying elements

Properties of substances

- Scientists can use physical properties to find out about the structure of a substance.
- **Physical properties** include:
  - **Melting point** (the temperature when it changes from a solid to a liquid).
  - **Boiling point** (the temperature when it changes from a liquid to a gas).
  - **Solubility** (how easily it dissolves in water).
  - **Electrical conductivity** (how well it conducts electricity).
- **Chemical properties** describe how a substance reacts in a chemical reaction.
- Elements and compounds are classified as:
  - **Ionic compounds**, which contain positively charged metal ions and negatively charged non-metal ions.
  - **Simple molecular covalent compounds**, which are small molecules consisting of only non-metallic atoms.
  - **Giant molecular covalent compounds**, which are large structures with many non-metallic atoms held together with covalent bonds.
  - **Metallic compounds**, which are individual metal ions surrounded by electrons.

- A high melting point substance has strong forces holding its particles together. A low melting point substance has weak forces holding its particles together.
- Each type of element and compound has relative properties, as shown in the table below.

Type	Relative melting point	Relative boiling point	Relative solubility in water	Electrical conductivity
Ionic	High	High	Soluble	Good conductors in aqueous solutions or when molten
Simple molecular covalent	Low	Low	Insoluble	Non-conductors as solids, liquids and in solutions
Giant molecular covalent	Very high	Very high	Insoluble	Non-conductors (except graphite)
Metallic	High	High	Insoluble	Good conductors as solids or liquids

Classifying elements into groups

- Elements are grouped by similar properties in the **periodic table**.
- Some groups are given special names, including **alkali metals**, **transition metals**, **halogens** and **noble gases**.
- Metals are **malleable** and can conduct electricity.
- Most of the metal elements are transition metals. They have high melting points.
- When transition metals react with other elements they often make coloured compounds.

- All metals have the same structure – a regular arrangement of positive metal ions surrounded by a sea of delocalised electrons.
- Metals are malleable because the layers of positive ions can slide over each other in the sea of electrons.
- Metals conduct electricity because the delocalised electrons can move when a voltage is applied.

Remember!
When molten or dissolved, ionic substances conduct electricity. However, it is not electrons that carry the charge but the ions themselves.

- The sea of delocalised electrons in metals is formed from the outer-shell electrons of the metal atoms. They are described as delocalised because they are no longer associated with just one atom, but can move between the metal ions.
- The metal atoms are held together in a strong metallic bond because all the ions are attracted to the same sea of delocalised electrons.
- Lots of energy is required to break these bonds so the melting point is high.

Improve your grade

Electrical conductivity

Higher: Use ideas about structure and bonding to explain why solid copper oxide cannot conduct electricity while solid copper can.

AO2 [5 marks]

Alkali metals

- **Alkali metals** are in group 1 of the periodic table. They have one electron in their outer shell. This gives them similar properties.

- Compared to transition metals, alkali metals are soft and have low melting points.

- They are very reactive and must be stored in oil to prevent them reacting with the air.

- As you go down the group from lithium to caesium the metals:
  - get more reactive
  - become softer
  - have a lower melting point
  - have one more shell of electrons.

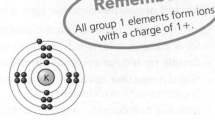

Remember!
All group 1 elements form ions with a charge of 1+.

The electron shells of the first three alkali metals.

Reactivity

- When alkali metals are added to water they float and skid across the surface. Hydrogen fizzes off and the alkali metal gradually disappears. A metal hydroxide is formed, which dissolves into the water. The water becomes alkaline and turns universal indicator purple.

  alkali metal + water $\longrightarrow$ metal hydroxide + hydrogen

- The reactions of the first three alkali metals – lithium, sodium and potassium – with water demonstrate that they get more reactive going down the group.

- Lithium moves slowly across the surface of the water and gradually gets smaller as it turns into lithium hydroxide and hydrogen.

- Sodium moves quickly across the surface. It melts and forms a ball, then rapidly gets smaller as it turns to sodium hydroxide and hydrogen.

- Potassium melts and moves very fast across the water. The hydrogen produced often catches fire. The flame is lilac because of the potassium present.

- The reaction of the alkali metals with water all have the same equation. (Just put the symbol for one of the alkali metals in the place of X.)

  $2X\ (s) + 2H_2O\ (l) \longrightarrow 2XOH\ (aq) + H_2\ (g)$

- The reaction becomes more vigorous as you go from lithium to caesium because:
  - the outer-shell electron gets further from the positive nucleus
  - there are more inner shells of electrons to shield it from the positive pull
  - it is therefore held to the atom less tightly
  - and is thus more easily lost, to allow the atom to become a stable ion.

- The trend in reactivity going from lithium to potassium allows us to make predictions about other members of the group.

- Caesium, at the bottom of the group, reacts explosively with water. We can predict that francium would be even more reactive, but it is very rare.

How science works

You should be able to:

- predict how reactive certain metals would be when added to water

- describe the safety precautions you would take when testing your predictions, and explain how you would minimise the risks

- understand and describe the trends you see in the results from testing your predictions.

Improve your grade

Observing a reaction

Foundation: Sarah's teacher demonstrates the reaction of potassium with water. She carefully cuts a small piece of potassium and drops it into a large trough of water.

Describe what Sarah would see.

AO2 [5 marks]

Halogens and noble gases

Group 7 elements

- **Halogens** are in group 7 of the periodic table. Each only needs one more electron to have a full outer shell.

- The colours and states of the first four halogens at room temperature are described in the table below.

Halogen	State	Colour
Fluorine	Gas	Pale yellow
Chlorine	Gas	Pale green
Bromine	Liquid	Red/brown
Iodine	Solid	Grey

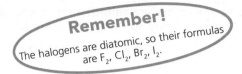

Remember!
The halogens are diatomic, so their formulas are F_2, Cl_2, Br_2, I_2.

- Halogens get less reactive going down the periodic table from fluorine to iodine.

- Halogens react with hydrogen to form hydrogen halides.

- Hydrogen halides form acids when they dissolve in water.

- Halogens react with metals to form ionic compounds as the halogens take electrons from the metal.

- In the reaction, the halogen becomes negatively charged because of the electrons it gains and the metal becomes positively charged because it has lost electrons to the halogen.

- The ions then form ionic bonds, making metal halides. For example:

  iron + bromine ⟶ iron bromide

  $2Fe\ (s) + 3Br_2\ (l) \longrightarrow 2FeBr_3\ (s)$

Halogen displacement

- A more reactive halogen will displace a less reactive halogen from its compound in solution. This is called a **displacement reaction**.

- Halogen compounds contain halide ions. These are colourless, but halogen atoms are coloured.

- When a displacement reaction occurs, there is a colour change because a different halogen is produced.

  chlorine water + potassium iodide ⟶ iodine + potassium chloride

  green   colourless   brown   colourless

  $Cl_2\ (aq) + 2KI\ (aq) \longrightarrow I_2\ (aq) + 2KCl\ (aq)$

Noble gases

- The **noble gases** (helium, neon, argon, krypton and xenon) form group 0 of the periodic table. They are monoatomic.

- Helium is less dense than air, so it can be used to make balloons and airships float. It is non-flammable, which makes it safer to use than hydrogen.

- Argon is used in filament bulbs and in welding because it is very **inert**. This prevents the unwanted reactions that can occur if air is used.

- The noble gases are inert because they have a full outer shell of electrons. They do not become any more stable by losing, gaining or sharing electrons.

- The noble gases were discovered in 1894 as a result of observation, **hypothesis** and experimentation.
  - Observation: nitrogen produced from air had a different **density** from nitrogen produced in a chemical reaction.
  - Hypothesis: the nitrogen produced from air also contains other unknown gases.
  - Experiment: fractional distillation of air separated the noble gases from nitrogen.

- The boiling point and density of elements increase going down group 0 of the periodic table.

- The boiling point and density of any noble gas will be midway between those of the gases above and below it in the group.

Improve your grade

Extracting bromine

Higher: Bromine is extracted from the sea on a commercial basis. The first step is to bubble chlorine through sea water. Sea water contains sodium bromide.

Write a symbol equation for the reaction that occurs. Explain what you would see and why. *AO2* [4 marks]

Endothermic and exothermic reactions

Energy changes

G–E

- In a chemical reaction, the atoms of the reactants are rearranged to form the products. There is always an energy change.

- Most reactions are **exothermic**. In exothermic reactions, **chemical energy** is transferred to heat energy, so the temperature rises.

- Some reactions are **endothermic**. In endothermic reactions, heat energy is taken in and transferred to chemical energy, so the temperature falls.

- All reactions start by breaking the bonds between the reactant atoms. Heat energy is taken in when bonds are broken.

- Next, the new bonds between atoms of the product are made. Heat energy is given out when bonds are made.

- In exothermic reactions, less energy is taken in to break the reactant bonds than is given out when the product bonds are made.

- In endothermic reactions, more energy is taken in than is given out.

Measuring change

D–C

- To measure the energy change in a reaction:
  - Place a solution of reactant 1 in an insulated cup.
  - Record the temperature of the solution.
  - Add a solution of reactant 2 and stir (or add powdered reactant 2).
  - Wait for the temperature to finish changing, then record the new temperature.
  - Calculate the difference in temperature before and after the reaction.

- An increase in temperature means the reaction is exothermic. A decrease means it is endothermic.

- The insulated cup reduces energy transfer in and out of the cup. Stirring makes an even temperature throughout the mixture.

- Examples of exothermic reactions include: combustion, explosions, metal displacement reactions, precipitation reactions, many neutralisation reactions.

- Examples of endothermic reactions include: photosynthesis, thermal decomposition reactions, many salts dissolving in water (e.g. ammonium nitrate).

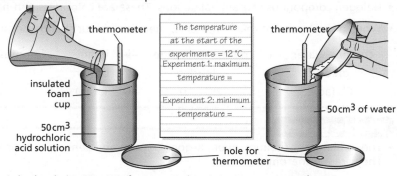

thermometer

The temperature at the start of the experiments = 12 °C
Experiment 1: maximum temperature =
Experiment 2: minimum temperature =

thermometer

insulated foam cup

50 cm³ hydrochloric acid solution

50 cm³ of water

hole for thermometer

A simple calorimeter set-up for two experiments to measure energy change.

Energy profile diagrams

B–A*

- The breaking and making of bonds in a chemical reaction can be described using an **energy profile diagram**.

- For exothermic reactions, the energy of the products is less than the energy of the reactants. The temperature rise in the experiment comes from the energy released.

- For endothermic reactions, the energy of the products is more than the energy of the reactants. The temperature fall in the experiment comes from energy being removed from the surroundings.

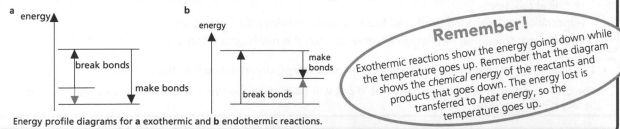

a energy — break bonds, make bonds

b energy — make bonds, break bonds

Energy profile diagrams for **a** exothermic and **b** endothermic reactions.

Remember!
Exothermic reactions show the energy going down while the temperature goes up. Remember that the diagram shows the *chemical energy* of the reactants and products that goes down. The energy lost is transferred to *heat energy*, so the temperature goes up.

Improve your grade

Proving exothermic reactions

Foundation: Andy says that adding iron filings to copper sulfate solution is an exothermic reaction. Describe an experiment he could do to show this.

AO2 [4 marks]

Reaction rates and catalysts

Understanding reaction rates

- Chemical reactions happen when particles collide.

- Reactions happen at different speeds, from very slow to extremely fast. The rate of a reaction is how fast the products are made or how fast the reactants disappear.

- Four factors can increase the rate of a reaction:
  - Increasing the concentration of reactants.
  - Increasing the temperature of the reaction.
  - Breaking a large solid reactant into smaller pieces to increase the surface area.
  - Using a catalyst.

- You can measure the rate of a reaction by recording how much product is made per second. The diagram below shows methods of measuring temperature, concentration and surface area on the **rate of reaction** of marble chips and hydrochloric acid.

- A faster rate of reaction occurs if the concentration of hydrochloric acid is increased, the temperature is higher or the same mass of marble is broken into smaller pieces.

- A slower rate of reaction occurs when a lower acid concentration, lower temperature or smaller surface area is used.

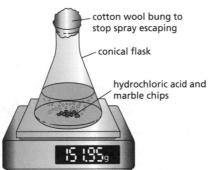

cotton wool bung to stop spray escaping

conical flask

hydrochloric acid and marble chips

An experimental set-up to study the rate of reaction of hydrochloric acid and calcium carbonate.

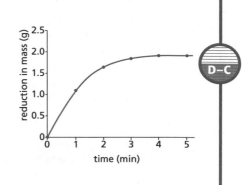

Collision theory

- Not all collisions lead to a reaction. To react, they must have sufficient energy and have the correct orientation.

- The rate of a reaction depends on the frequency of effective collisions.

- When the concentration of reactants increases, there are more particles per cm^3. This means there are more collisions per second.

- When the temperature increases, the reactants have more energy so a higher percentage of the collisions are effective. The reactants also collide more frequently.

- Increasing the surface area of a solid reactant increases the frequency of collisions.

Remember!
Increasing the *volume* of reactants does not increase the rate of reaction.

Catalysts

- A **catalyst** speeds up the rate of a reaction, but is not used up by the reaction.

- Catalytic converters in vehicles reduce the amount of carbon monoxide and unburned fuel that come from the exhaust pipe.

- The platinum in a catalytic converter is spread in a thin layer over a honeycomb ceramic structure to give a large surface area. This increases the number of reactions catalysed per second.

- Catalysts work better at higher temperatures. Catalytic converters do not work well until the car warms up.

- Two important reactions that are catalysed in a catalytic converter are:

  carbon monoxide + oxygen ⟶ carbon dioxide

  hydrocarbons + oxygen ⟶ carbon dioxide and water

⊙ Improve your grade

Collision theory

Higher: In glow sticks, two chemicals react to form a product that glows for a short time. Use collision theory to explain why manufacturers recommend keeping glow sticks in the fridge to make them last longer.

AO2 [4 marks]

Mass and formulae

Calculating mass

- To find the **relative formula mass** (M_r) of a compound, such as calcium carbonate, $CaCO_3$:

1 Work out the number of each type of atom	$Ca \times 1$	$C \times 1$	$O \times 3$
2 Look up the relative atomic mass of each	40	12	16
3 Multiply the mass by the number	$40 \times 1 = 40$	$12 \times 1 = 12$	$16 \times 3 = 48$
4 Add them all together	$40 + 12 + 48 = 100$		

- To find the **percentage composition** of a compound, such as the percentage oxygen in water, divide the mass of oxygen (16) by the total mass of water (18) and multiply by 100 = 88.89%.

- **Empirical formula** is the simplest ratio of atoms, e.g. formula of ethane = C_2H_6 and empirical formula of ethane = CH_3.

- To calculate the empirical formula from experimental results:

1 Write down the symbol for each element	Fe	Cl
2 Write down the experimental mass of each element	22.4	42.6
3 Divide the mass of each element by its A_r	$\frac{22.4}{56}$	$\frac{42.6}{35.5}$
The result is the ratio of atoms in the compound	0.4 : 1.2	
4 Make it into a whole number ratio by dividing the biggest number by the smallest	$\frac{Cl}{Fe} \frac{1.2}{0.4} = 1:3$	
5 Write the formula	$FeCl_3$	

Calculating yields

- You never get 100% of the possible **yield** because some reactants may not have reacted, the product may stick to the apparatus or otherwise get lost, or because unwanted products may be created by side reactions.

- The **theoretical yield** is the maximum amount of product that could be made if all the reactants were converted into product.
- **Percentage yield** is: $\frac{\text{actual yield}}{\text{theoretical yield}} \times 100\%$

Reacting masses

- You can calculate the masses of all reactants and products if you know the equation and the mass of one substance.

- What mass of Na_2SO_4 will be made from 10 g of NaOH? The mass of NaOH is known and the mass of Na_2SO_4 is unknown.

1 Write the equation.	$H_2SO_4 + 2NaOH \longrightarrow Na_2SO_4 + 2H_2O$
2 Calculate the M_r of known and the unknown	40 142
3 Multiply the M_r by the balancing number	$2 \times 40 = 80$ $1 \times 142 = 142$
4 Divide the unknown by the known and multiply by the mass of the known.	$\frac{142}{80} \times 10 = \textbf{17.8 g}$ of Na_2SO_4 will be made

Commercial chemistry

- In industry, the unwanted products of a reaction must be disposed of. This can be costly and must avoid damaging the environment or causing problems to people living nearby.

- Economic considerations drive industry to search for reactions that have a high percentage yield, make no unwanted products and occur at a suitable rate.

 Improve your grade

Calculating mass

Foundation: Which compound contains the highest percentage by mass of nitrogen, KNO_3 or NH_4Cl? Show your working.

AO1 [4 marks]

C2 Summary

Atomic structure and the periodic table

Atomic number = number of protons. All atoms of an element contain the same number of protons.

Mass number = number of protons + neutrons.

Isotopes have the same number of protons but a different number of neutrons.

The periodic table lists all the elements in order of increasing atomic number.

Rows are called periods. Columns are called groups and contain elements with similar properties.

Ionic compounds and analysis

Ionic compounds form when the outer-shell electrons from metal atoms are transferred to the outer shell of non-metals.

Ionic compounds have high melting points and only conduct electricity when dissolved or molten.

Insoluble salts are made as precipitates by mixing two suitable solutions.

Many cations give coloured flames. Sodium is yellow, potassium is violet, calcium is brick red, copper is green.

Anions can be identified by chemical tests.

Covalent molecules form when atoms share electrons to make covalent bonds.

Simple molecular covalent substances have low melting points. Giant molecular covalent structures have high melting points. Neither conduct electricity.

Covalent compounds and separation techniques

Immiscible liquids can be separated using a separating funnel.

Miscible liquids can be separated by distillation.

Mixtures in solution can be separated by paper chromatography.

Most elements are metals. Metals consist of regular arrangements of metal ions in a sea of delocalised electrons.

Groups in the periodic table

Transition metals have a high melting point and form coloured compounds.

The noble gases are very unreactive because they have full outer shells of electrons

Halogens react with hydrogen to form acids and with metals to form metal halides. A more reactive halogen will displace a less reactive halogen.

Alkali metals are soft and react with water to form hydroxides and hydrogen.

Chemical reactions

Exothermic reactions give out energy; endothermic reactions take in energy.

A reaction is exothermic if the energy needed to break bonds is less than the energy released when bonds are made. The reverse is true for endothermic reactions.

The rate of a reaction increases if the concentration of reactants, the surface area or the temperature increases, or if a catalyst is used.

Catalytic converters in cars catalyse reactions to oxidise carbon monoxide and unburned hydrocarbons.

Relative formula mass is the sum of the atomic masses of atoms in a compound.

Percentage composition is the atomic mass divided by the relative formula mass × 100%.

The formula equation can be used to calculate the masses of reactants and products.

Yield is the amount of product. Percentage yield is the actual yield divided by theoretical yield × 100%.

Quantitative chemistry

Disposing of waste products from reactions may cause economic, environmental and social problems.

Industrial chemists must consider the percentage yield, the rate of reactions and the use of all products to make a process economically viable.

The Solar System

Models of the Solar System

G–E

- The ancient Greeks used a **geocentric model** of the **Solar System**, proposed by Ptolemy.
- In the geocentric model, the Earth was in the centre and everything else (planets, the Moon, the Sun and the stars) moved around the Earth in circular orbits.
- Ptolemy explained the occasional **retrograde motion** of Mars using epicycles.
- Ptolemy's model of the Solar System was very complicated and did not accurately predict the position of the planets.

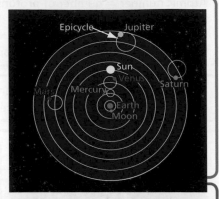

The geocentric model of the Solar System.

D–C

- In the 14th century, Nikolaus Copernicus proposed a **heliocentric model**. This stated that:
  - The Sun was at the centre of the Solar System.
  - The planets moved around the Sun in circular orbits and the Moon orbited the Earth.
  - The planets further away from the Sun travelled more slowly than those close to it.
  - The stars were fixed in a dome beyond Saturn.
- The heliocentric model explained Mars's retrograde motion. However, it still did not accurately predict the position of the planets because Copernicus used circular orbits.
- In the 1600s, Johannes Kepler realised that the planets had elliptical orbits. This model forms the basis for the one we use today.

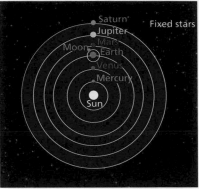

The heliocentric model of the Solar System.

B–A*

- In the 17th century, Galileo Galilei first observed the night sky using a **telescope**.
- He saw mountains on the Moon, and four moons orbiting Jupiter.
- These observations provided proof that the heliocentric model of the Solar System was correct.

Remember!
Ideas about the Universe are still changing as astronomers and scientists make new discoveries.

Observing the Universe

G–E

- Early astronomers observed the Universe with the naked eye.
- Stars are visible because they emit light, but planets and moons reflect light from the Sun.
- The invention of the telescope made it possible to see more distant objects, because telescopes gather more light, making objects brighter and more magnified.
- In the 1850s, **photography** enabled astronomers to make permanent records of their observations.
- Today, waves other than light are used to look deeper into the Universe.

D–C

- The Sun is at the centre of our Solar System, and is orbited by the planets Mercury, Venus, Earth, Mars, Jupiter, Saturn, Uranus and Neptune.
- An asteroid belt lies between the orbits of Mars and Jupiter.
- The Kuiper belt consists of several frozen bodies orbiting beyond Neptune. It is believed to be the source of comets.
- The outermost region of the Solar System is known as the Oort cloud. This consists of billions of lumps of rock and ice.

B–A*

- The Hubble Space Telescope takes photographs using visible light as well as **infrared** and **ultraviolet waves**.
- Some space probes take images of distant galaxies and the Milky Way using **X-rays**.
- Some space observatories use infrared to produce images. Cooler objects emit more infrared than visible light. Infrared astronomy was used to discover the Kuiper belt.

Improve your grade

Jupiter's moons

Higher: Galileo observed four moons orbiting Jupiter.
Explain how this observation did not fit with the geocentric model of the Solar System.   *AO2* [3 marks]

Reflection, refraction and lenses

Reflection

- All waves can be **reflected**:
  - Sound waves are reflected off walls and buildings (echoes).
  - Dolphins and bats use ultrasound (**echolocation**) to monitor their surroundings.
  - Radio waves are reflected from surfaces within the atmosphere such as the ionosphere.
- Light waves will reflect from shiny surfaces following the laws of reflection:
  - The angle of incidence (i) equals the angle of reflection (r).
  - The incident ray, the reflected ray and the normal are all in the same plane.

Remember!
Always measure the angle from the normal to the ray.

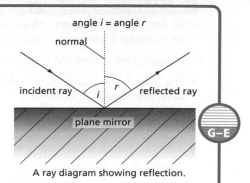

A ray diagram showing reflection.

G–E

Refraction

- **Refraction** is the bending of light rays at a surface, for example when entering or leaving a glass block.
- When light travels from air to glass, the direction of the ray bends towards the normal. The angle of incidence is larger than the angle of refraction.
- All waves are refracted at the boundary between different materials.

A ray of light is refracted towards the normal when it travels from air into glass.

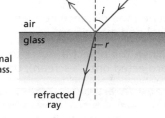

D–C

- Refraction occurs because the **speed** of the wave changes as it passes through different materials.
- The speed of light depends on the **density** of the material. The speed of light in air is almost 3.0×10^8 m/s, but in water it is closer to 2.0×10^8 m/s.
- When light travels from a less dense material to a more dense material it gets slower. This causes it to bend towards the normal.
- When light travels from a more dense material to a less dense material it speeds up. This causes it to bend away from the normal.

B–A*

Understanding lenses

- There are two types of lens:
  - A **converging** (or convex) **lens** is fatter in the middle.
  - A **diverging** (or concave) **lens** is thinner in the middle.
- Converging lenses are used to form images in telescopes, cameras, projectors, binoculars and our eyes.
- Rays of light that are parallel to the principal axis of a converging lens are refracted inwards. They converge on the **principal focus** (focal point, F).
- The distance between the centre of the lens and the focal point is called the **focal length** (f) of the lens.
- The fatter the lens, the shorter the focal length.

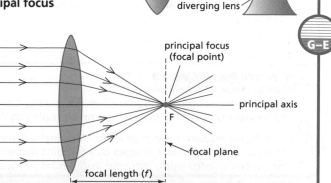

G–E

Improve your grade

Refraction

Higher: Light travels more slowly in water than it does in air.

Explain why this makes it difficult to pick up an object from the bottom of a swimming pool. *AO2* [4 marks]

Lenses in telescopes

Determining focal length

- To find the focal length of a converging lens you obtain an image of a distant object (tree, building, etc.) on a screen and measure the distance between the lens and the focused image.

- This works because the rays of light from a distant object are parallel and will converge at the focal point.

- The image you obtain is inverted (upside down), **diminished** (smaller than the object) and **real** (can be projected onto a screen).

- On ray diagrams:
  - Draw a ray parallel to the principal axis. It will refract through the focal point.
  - Draw a ray straight through the centre of the lens (it is not deviated).
  - The image is formed where the two rays cross.

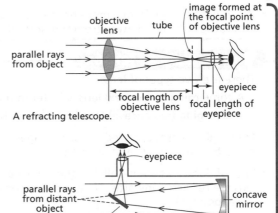

A ray diagram showing how to find the properties of an image.

B–A*

- A converging lens can produce both **magnified** and diminished images, depending on the position of the object.

- When the distance between the object and the lens is greater than 2f the image will be inverted, real and diminished.

- When the distance between the object and the lens is between f and 2f, the image will be inverted, real and magnified.

- When the object is closer to the lens than the focal length, the image is **virtual** (it cannot be projected onto a screen), upright and magnified. This is a magnifying glass.

> ## EXAM TIP
> Always use a ruler to draw rays of light – they travel in straight lines.

Types of telescopes

G–E

- Early telescopes had a converging lens at the front (**objective lens**) and a diverging lens (**eyepiece**) to look through.

- Lenses have different focal points for different colours of light. This makes the image a bit blurred.

- A clearer image can be formed by using a concave (parabolic) mirror in place of the objective lens.

D–C

- In a modern **refracting telescope**, both the objective lens and the eyepiece are converging lenses.

- The objective lens produces an image of a distant object at its focal point, and the eyepiece magnifies the image.

- **Reflecting telescopes** can be much larger and are easier to manoeuvre than refracting telescopes. This is because a mirror has much smaller mass than a bulky lens.

- A simple reflecting telescope uses a large concave mirror, a plane mirror and a converging lens.

- The concave mirror forms an image of a distant object, which is then reflected towards the eyepiece using the plane mirror.

A refracting telescope.

A reflecting telescope.

Telescopes in space

B–A*

- Modern telescopes are huge. They are usually housed in observatories on high mountains, where there is little light pollution from cities. The air is also cooler, so there is less interference from the atmosphere.

- The Hubble space telescope is orbiting Earth. There is no atmospheric interference in space, and the telescope can produce images of very faint distant objects up to magnifications of about 5000. It also produces images using infrared and ultraviolet waves.

Improve your grade

Advantages of modern telescopes

Foundation: Early telescopes used two lenses. Modern astronomical telescopes are reflecting telescopes.

Describe **two** advantages of reflecting telescopes over refracting telescopes.

AO1 [4 marks]

Waves

Understanding waves

- Waves transfer energy and information from one place to another using vibrations, but they do not transfer matter in the direction they are travelling.

- The **wavelength** (in metres) of a wave is the distance the wave travels in one complete cycle – the distance between two adjacent **peaks** or two adjacent **troughs**.

- The **amplitude** (in metres) of a wave is the maximum **displacement** of a wave. It is measured from the top of a peak to the centre line, or from the bottom of a crest to the centre line.

- The **frequency** (in hertz) of a wave is the number of complete waves passing a point in one second.

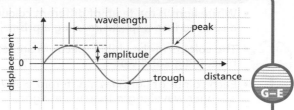

Important things you need to know about a wave.

G–E

Wave equations

- There are two ways to calculate the **speed** of a wave:

speed of a wave (m/s) = distance travelled (m) / time taken (s) or $v = x/t$

speed of a wave (m/s) = frequency (Hz) × wavelength (m) or $v = f \times \lambda$

D–C

- An example of how to calculate a wavelength is given below.
  - Sound waves travel through air at a speed of 340 m/s. A particular note of sound has a frequency of 260 Hz. To calculate the wavelength:

  speed = frequency × wavelength

  wavelength = speed / frequency

  wavelength = 340 / 260 = 1.3 m

B–A*

Types of waves

- All waves belong to one of two groups:
  - In **transverse waves**, the vibration is at right angles to the direction of wave travel.
  - In **longitudinal waves**, the vibration is back and forth along the direction of wave travel.

- Electromagnetic waves and water waves are transverse waves.

- Sound waves are longitudinal waves.

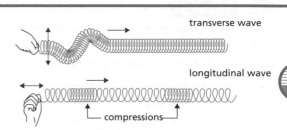

You can use a Slinky to create both transverse and longitudinal waves.

G–E

Seismic waves

- Longitudinal waves such as sound waves travel as a series of **compressions** (areas of higher pressure) and **rarefactions** (areas of lower pressure).

- **Seismic waves** are produced by earthquakes and explosions. There are two main types:
  - Primary (P) waves are slower-moving longitudinal waves.
  - Secondary (S) waves are faster-moving transverse waves.

Remember!
The wavelength of a longitudinal wave is the distance from one compression to the next compression.

D–C

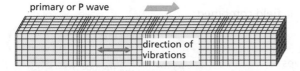

Two types of seismic waves.

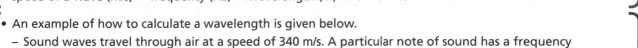

- Seismic waves are detected using a seismograph or **seismometer**.

- The vibration of the Earth's surface is recorded by its motion relative to a heavy pendulum.

B–A*

⊙ Improve your grade

Thunder and lightning

Higher: During a thunderstorm you always see the lightning before you hear the thunder. Light travels so fast that the lightning is almost instantaneous, but sound travels at a speed of 340 m/s in air.

If you hear thunder half a minute after seeing lightning, how far away is the storm?   *AO3* [3 marks]

The electromagnetic spectrum

Visible light

G–E

- White light is made up of different colours. These colours can be dispersed into a spectrum using a **prism**.

- The colours in the visible spectrum are red, orange, yellow, green, blue, indigo and violet.

- Each colour of light has a different wavelength. Red light has a longer wavelength than violet light.

The visible spectrum.

Infrared and ultraviolet

D–C

- William Herschel discovered **infrared** waves in 1800, while investigating the temperature of the visible spectrum.

- He found that the hottest temperatures were beyond the red end of the spectrum, where there is no visible light. This is known as the infrared region.

- Infrared waves have a longer wavelength than red light.

- Johann Ritter discovered **ultraviolet** waves in 1801, while experimenting with silver chloride used in photography. The rate of reaction was highest beyond the violet end of the visible spectrum.

- Ultraviolet waves have a shorter wavelength than violet light.

B–A*

- Ultraviolet waves from the Sun are harmful. They will damage skin cells and eyes. The most common form of this damage is sunburn, but long exposure can cause skin cancer.

- There are three types of ultraviolet waves, known as UV-A, UV-B and UV-C. UV-C waves have the shortest wavelength and cause the most damage.

EM waves

G–E

- Ultraviolet, visible light and infrared are all part of the family of waves known as the **electromagnetic spectrum**.

- All electromagnetic waves are **transverse** and travel at the same speed in a **vacuum** (300 000 000 m/s).

- As with all other waves, EM waves:
  - can transfer energy
  - can be reflected, refracted and **diffracted**
  - obey the wave equation: wave speed = wavelength × frequency.

D–C

- All EM waves are essentially the same, but they have different wavelengths and frequencies.

- In order of decreasing wavelength and increasing frequency, the EM waves are: radio waves, microwaves, infrared, visible light, ultraviolet, X-rays and gamma rays.

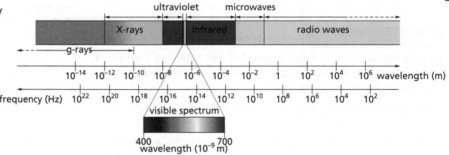

The electromagnetic spectrum. Note how the gamma rays merge into the X-rays.

B–A*

- Electromagnetic waves consist of an oscillating electric field combined with an oscillating magnetic field.

- The two oscillating fields are at right angles to each other and to the direction of wave travel, so EM waves are transverse.

- The wavelength of EM waves varies from about the size of the nucleus of an atom (10^{-15} m) up to several kilometres (10^3 m).

> **Remember!**
> As the wavelength of the EM waves increases, the frequency decreases.

⬤ Improve your grade

Electromagnetic waves

Higher: The speed of electromagnetic waves in air is about 3×10^8 m/s. Microwaves have a wavelength of 3 cm.

Use the wave equation wave speed = wavelength × frequency to calculate the frequency of microwaves.

AO2 [3 marks]

Uses of EM waves

Radio waves and microwaves

- Radio waves are produced and detected by aerials.
- Radio waves are used to broadcast television and radio programmes. They are also used by emergency services for communication.
- Microwaves are very high frequency (short wavelength) radio waves.
- Microwaves are used to cook food in a microwave oven. The microwaves are absorbed by water and fat in the food, and the energy becomes heat.
- Microwaves are also used for mobile-phone communication.

- Medium-frequency radio waves (MW) can reflect off the **ionosphere** to communicate long distances. These reflected waves are called sky waves.
- High-frequency radio waves (and microwaves) travel in straight lines to satellites.
- Mobile-phone networks use satellites to transmit communications around the world at high speeds.

Remember!
Satellite TV is broadcast using very short wavelength radio waves, so satellite dishes need to be pointed directly towards the satellite.

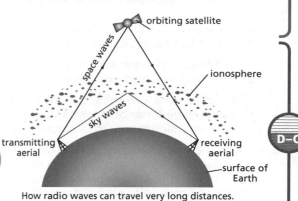

How radio waves can travel very long distances.

Infrared, visible and ultraviolet waves

- All objects with a temperature above absolute zero (–273 °C) emit infrared waves. The higher the temperature, the more infrared waves are emitted.
- Infrared waves are used in cooking, in the grill, toaster and oven.
- Rescue services use infrared cameras to find bodies under rubble.
- Visible light is the only part of the EM spectrum that can be seen with the naked eye.
- Visible light is used in photography and in lighting our homes and streets.
- Ultraviolet lamps produce ultraviolet light when mercury vapour conducts electricity.
- Some security markings can only be seen under UV light, such as the markings on bank notes.
- UV is also used to sterilise water, as it kills bacteria.

- Photographs taken with infrared are called **thermographs**.
- Infrared waves are also used for short-range communication, such as in a remote control or a cordless computer mouse.
- Security systems detect the infrared waves emitted by an intruder.

- Both visible light and infrared waves are transmitted along optical fibres for communications.
- Glass fibres are much lighter and cheaper than copper wires. They can also transmit the information much faster.
- If you increase the angle of incidence as light rays leave a more dense material, the angle of refraction away from the normal increases.
- Eventually it will become refracted along the boundary and there will be some internal reflection. This is the critical angle.
- At even greater angles, the light undergoes **total internal reflection**.

The path of a ray inside an optical fibre.

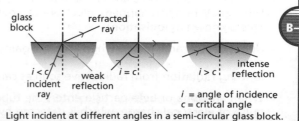

Light incident at different angles in a semi-circular glass block.

Improve your grade

Infrared waves

Foundation: All objects emit infrared waves, but the hotter the object is, the more infrared radiation is given off.

Use this information to explain how the police could use infrared radiation to find fugitives.

AO2 [3 marks]

Gamma rays, X-rays, ionising radiation

Gamma and X-rays

G–E
- X-rays are produced when fast-moving electrons hit a metal plate. They can be used to detect an object's internal structure.
- Uses of X-rays include:
  - airport security scanners
  - medical X-rays to detect the condition of bones and teeth
  - to detect unwanted pieces of metal in machinery.

D–C
- Gamma rays are similar to X-rays, but they are produced by radioactive materials.
- Gamma rays kill bacteria, so they are used to sterilise food and medical instruments.
- Radioactive **tracers** emitting gamma rays are used to detect some forms of cancer, and gamma rays are used in **radiotherapy** to treat cancers.
- The higher the frequency of electromagnetic radiation, the higher its energy and the more harmful it is. The table opposite shows the effects of exposure to electromagnetic waves.

Waves	What they do
Radio waves	They are safe because they do not produce ionisation.
Microwaves	Cause internal heating of the body cells.
Infrared	Can cause skin burns.
Visible light	Intense light can cause permanent damage to the retina.
Ultraviolet	Intense ultraviolet light damages skin (surface) cells; can trigger skin cancer; can cause eye conditions such as cataracts; can destroy proteins in the eye lens.
X-rays and gamma rays	Can damage the DNA of cells; can mutate cells; can trigger cancer.

EXAM TIP
When discussing the use of ionising radiation, make sure you understand both the risks and the benefits.

B–A*
- In an X-ray machine, a very high voltage is applied between the electrodes.
- The electrons are accelerated and hit a tungsten target.
- The fast-moving electrons collide with tungsten atoms, causing them to emit X-rays.

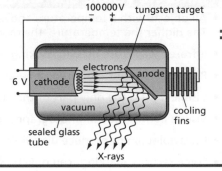
An X-ray machine.

Radioactivity

G–E
- Atoms consist of a tiny, positively charged **nucleus** surrounded by negatively charged electrons.
- Some atoms have unstable nuclei, which break down, releasing energy in the form of alpha (α) particles, beta (β) particles or gamma (γ) rays. This is called radioactivity.
- Radioactivity occurs naturally and randomly, and cannot be switched off. It is not affected by temperature or other physical or chemical conditions.
- X-rays and radioactivity were discovered at the end of the 19th century.

D–C
- To a physicist, radiation can mean either electromagnetic waves, or alpha or beta particles from radioactive sources.
- All types of radiation transfer some of their energy to the atoms of the material they are passing through. When this happens, electrons can get knocked off the atom to form a positive **ion**. This is known as **ionisation**.
- Negative ions are formed when an atom gains extra electrons by ionisation.

B–A*
- Ionising radiation from radioactive sources can be detected using a Geiger–Müller tube.
- When an alpha or beta particle enters the tube, it ionises the gas inside the tube. This produces a pulse of electrical charge, which is detected by the counter attached to the GM tube. Each click or count represents one particle being detected.

Improve your grade

Radioactivity
Foundation: Radioactive materials emit ionising radiation.
Explain what is meant by the term 'ionising radiation'.

AO1 [2 marks]

The Universe

Our place in the Universe

- The **Universe** is made up of billions of **galaxies**, which are collections of stars held together by the force of gravity.
- Our **Solar System** is part of the **Milky Way** galaxy.
- Our Sun is an average star in the Milky Way.
- Distances in the Solar System are measured in Astronomical Units (AU). One AU is the average distance from the Earth to the Sun: 1 AU = 1.5×10^{11} m

Object	Average distance from the Sun (AU)	Diameter relative to Earth
Sun	–	110
Mercury	0.39	0.38
Venus	0.72	0.95
Earth	1.00	1.00
Moon	1.00	0.27
Mars	1.52	0.53
Jupiter	5.20	11.20
Saturn	9.54	9.44
Uranus	19.20	3.69
Neptune	30.10	3.48

The scale of things

- The Universe is believed to be about 10^{27} m across.
- It contains about 10^{11} galaxies, which each contain about 10^{11} stars.
- Each star may have its own system of planets, just like our Solar System.

Remember!
Using standard notation, the power of ten gives the number of zeros you need – for example, 10^3 is 1000 (3 zeros).

Object	Approximate distance (m)	Comments
Earth and Moon	3×10^8	Our nearest neighbour is about 30 Earth diameters away.
Sun and Earth	1.5×10^{11}	It takes about 8 minutes for light from the Sun to reach us.
Solar System	4×10^{12}	The outermost planet Neptune is about 30 times further from the Sun than Earth.
Nearest star	4×10^{16}	Proxima Centauri is the nearest star beyond the Sun. No planets have been detected around it.
Closest galaxy	2×10^{22}	Andromeda is the nearest galaxy and is very similar to the Milky Way. It is about 10^{21} m across.
Size of the Universe	1×10^{27}	The Universe contains all the galaxies in space.

- Very large distances in space are measured in **light-years**. A light-year is the distance travelled by light (through a vacuum) in a year.
- The size of the Universe is about 100 billion light-years.

Exploring the Universe

- Modern telescopes use all parts of the electromagnetic spectrum to observe and record the Universe.
- Larger magnification telescopes allow astronomers to view stars and galaxies in deep space.
- Stars and galaxies (some of which cannot be seen using visible light) also emit other types of electromagnetic waves such as X-rays.

- Space probes are sent out on flyby missions to detect the presence of water and minerals on other planets. They send data and photographs back to Earth using radio waves.
- Robotic **landers** can be sent out to planets. They collect soil samples and search for microscopic life forms, such as bacteria.
- The Search for **Extraterrestrial** Intelligence (SETI) is looking for radio broadcasts from alien civilisations, which may exist on other planets orbiting distant stars.

- Photographs can be taken using parts of the EM spectrum other than visible light, such as X-rays. These reveal different information about the structure of space bodies, such as their magnetic fields.
- Images like the one of Jupiter opposite are taken with X-rays. They are analysed by computers and given false colours.

Jupiter in X-rays.

⊙ Improve your grade

Studying the Universe
Foundation: Humans have always been fascinated with space. Ancient civilisations relied on the naked eye to study the stars. In medieval times, telescopes were used to study objects in space.

Explain **two** advantages of the technology used to study the Universe today. *AO2* [4 marks]

Analysing light

The spectrometer

- White light can be split up into its component colours by **refraction** using a glass prism. It can also be split up by reflecting it off a shiny surface such as the surface of a DVD.
- A **spectrometer** is a device used for looking at the spectrum of light.
- You can make a simple model spectrometer using a cardboard box and an old DVD, as shown in the diagram opposite.
- To get a clearer image, make the slit narrower and observe the light source in a darkened room.

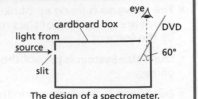

The design of a spectrometer.

Light from different sources

- Different light sources produce different spectra in the spectroscope.
- Hot stars emit more blue light and cool stars emit more red light.
- The gases in the Earth's atmosphere absorb some electromagnetic waves more than others.
- Visible light is not well absorbed, but it is better to site large telescopes at the top of high mountains, where the air is thinner.
- Scientists can identify the chemical composition of stars by examining the light they emit with a spectrometer.
- Modern telescopes have sophisticated built-in spectrometers.

Analysing data

- When electromagnetic waves pass through the atmosphere, some wavelengths are absorbed much more than others.
- Radio waves are not absorbed at all by the atmosphere, so they can be used for communications.
- Gamma and X-rays are almost all absorbed by the atmosphere.

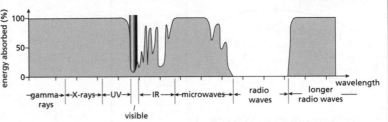

A graph showing the effect of the Earth's atmosphere on electromagnetic waves.

The Doppler effect

- As a siren on a police car passes, you will notice that the sound changes from high pitch as it approaches you to lower pitch as it moves away again.
- The sound waves are being squashed together as the car comes towards you and stretched as the car moves away. This is known as the **Doppler effect**.

- The Doppler effect can be observed for all waves, not just sound waves. For example, it can be observed in light emitted from a star.
- The wavelength of light decreases (and frequency increases) as a light source moves towards an observer. This is known as blue-shift.
- The wavelength of light increases (and frequency decreases) as a light source moves away from an observer. This is known as red-shift.

Remember!

The faster a source is moving, the greater the Doppler shift will be.

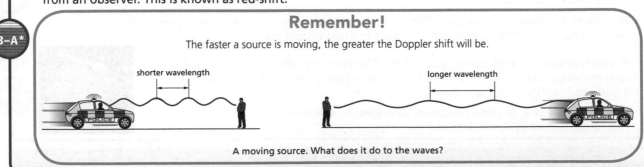

A moving source. What does it do to the waves?

Improve your grade

Positioning telescopes

Higher: Explain why it is best to site optical telescopes at the top of high mountains.   AO1 [2 marks]

The life of stars

How stars are formed

- The Sun is our local star. It was formed about 4.6 billion years ago from a thin cloud of dust, gas (mostly hydrogen and helium) and ice, called a **nebula**.
- All the material in the nebula was pulled together by the **force of gravity**. It eventually started to spin and contract, and the temperature increased.
- At a few million °C, **nuclear fusion** reactions started to occur, the temperature rose even higher, and the new star began to emit its own light.

Fusion reactions

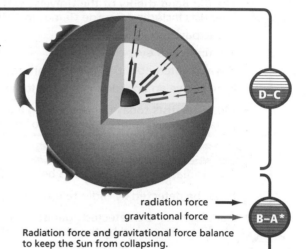

- The Sun is an average star, known as a **main sequence star**.
- Its surface temperature is about 5500 °C, which makes it appear yellow. The core temperature is about 14 million °C.
- The Sun's energy comes from nuclear fusion reactions. The vast amount of energy produced is emitted as electromagnetic radiation.
- The Sun is in a delicate balance between the force of gravity trying to pull all matter towards its centre and the force produced by the EM radiation pushing its way out.

- The nuclear fusion reaction in the Sun occurs when hydrogen nuclei fuse together to form a helium nucleus.
- The mass of the helium nucleus is slightly less than the total mass of the hydrogen nuclei. The lost mass becomes energy: hydrogen → helium + energy

radiation force →
gravitational force →
Radiation force and gravitational force balance to keep the Sun from collapsing.

Evolution of stars similar to the Sun

- Stars are born and eventually, after millions of years, they die. The journey of a star is known as its **life cycle**.
- Our Sun will eventually run out of hydrogen. When this happens, it will start to use helium as fuel to keep burning. This will produce different elements, such as carbon and oxygen.

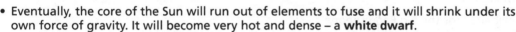

nebula   Sun (main sequence)   red giant   white dwarf   black dwarf

The evolution of the Sun.

- The outer layers of the Sun (and similar-sized stars) will expand and cool, and it will become a **red giant**.
- Eventually, the core of the Sun will run out of elements to fuse and it will shrink under its own force of gravity. It will become very hot and dense – a **white dwarf**.
- The white dwarf will slowly cool. The star ends its life as a **black dwarf**.

Evolution of massive stars

- Larger, bluer stars were created from larger nebulae. Their evolution is different to that of stars like the Sun.
- The outer layers will expand and cool to form a **super red giant**.

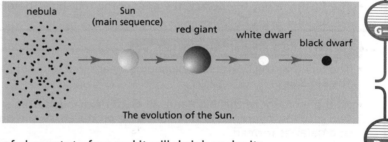

large nebula   star   super red giant   supernova   black hole   neutron star

The evolution of a massive star.

- Nuclear fusion reactions in the star form heavier elements, such as iron and magnesium.
- When all the elements run out, the core will suddenly shrink and all the hot material will be ejected into space as a **supernova**. Very heavy elements will be formed.
- The core shrinks to become a very dense **neutron star**. In extremely massive stars, a **black hole** is formed, which has such strong gravitational force that not even light can escape.

Improve your grade

Energy from the Sun

Higher: Explain where the Sun's energy comes from.

AO1 [4 marks]

Theories of the Universe

Steady State or Big Bang?

- The **Steady State theory** of the Universe was proposed by a group of scientists including Fred Hoyle in 1946. In this model the Universe:
  - is expanding
  - has unchanging density
  - spontaneously created matter, especially hydrogen, from empty space to maintain the same density
  - had no beginning and will never end.

- The **Big Bang** theory of the Universe was proposed by a group of scientists including George Gamov in 1948. In this model the Universe:
  - is expanding
  - is finite and ever-changing
  - was created about 14 billion years ago from an event called the Big Bang
  - may have an end, depending on its density.

- Most scientists today believe the Big Bang theory is correct, because it explains the red-shift of light from distant galaxies and the existence of **cosmic microwave background** (CMB) **radiation**.

> **Remember!**
> Both theories agree with the evidence that the Universe is expanding.

The expanding Universe

- CMB radiation is considered to be 'left-over radiation' from the Big Bang.

- It can be detected by radio telescopes and is the same strength in all directions.

- Dark lines can be detected against a continuous spectrum of colour in the Sun's spectrum. The lines are caused by the absorption of certain frequencies of light.

- The spectrum of light from distant stars show the same pattern, but the dark lines are shifted towards the red end of the spectrum – this is red-shift.

- The light from all distant galaxies is red-shifted. This means that galaxies are all moving away from us and from each other. The Universe is expanding.

- The further away the galaxy, the greater the red-shift. Therefore, the more distant galaxies are moving away more quickly.

Evidence for the Big Bang

- The fact that the Universe is expanding means that it must have started from the same point – the Big Bang.

- At the moment of this Big Bang, all matter expanded from a hot, dense singularity. After the first few seconds, hydrogen and helium were produced. As the Universe expanded, it cooled. Stars and galaxies formed.

- There is strong evidence for the Big Bang theory:
  - All galaxies show red-shift, which means the Universe must be expanding.
  - The temperature of the Universe was predicted to be −270 °C. This was confirmed by the COBE satellite in the 1990s.
  - The original radiation created in the Big Bang would have been gamma rays. Due to the expansion of the Universe, the Doppler effect has stretched the wavelength to become microwave radiation, with a wavelength of about 1 mm.

- The Steady State theory also accounts for the red-shift of galaxies, but it is rejected because it cannot explain the presence of CMB radiation or the abundance of light elements such as helium in the Universe.

How science works

You should be able to:

- understand that nobody really knows how the Universe began – we only have theories and models

- explain how scientists are currently collecting more data to provide evidence for their theories

- describe how scientists may interpret the data in different ways.

Improve your grade

Steady State or Big Bang?
Foundation: The Steady State theory and the Big Bang theory were two opposing theories of the Universe in the 20th century.
Give **one** similarity between the two theories and **one** difference between them.   *AO1* [3 marks]

Ultrasound and infrasound

What you can and cannot hear

- Humans can hear sounds from 20 Hz up to about 20 000 Hz. Younger people can hear much higher-pitched sounds than older people.
- Sound with frequencies above 20 000 Hz is called **ultrasound**. Humans cannot hear ultrasound, but some animals can.
- Some animals, such as dolphins, use ultrasound for communication.
- Sound with frequencies lower than 20 Hz is called **infrasound**. Humans cannot hear infrasound, but they can feel the slow vibrations.
- Some animals, such as whales, use infrasound for communication.

Sound waves

- The speed of sound varies in different materials. The closer the particles in the material, the faster sound travels, so sound travels fastest in solids and slowest in gases.
- Ultrasound is used for foetal scanning, to safely 'see' inside the body.
- Different body tissues reflect ultrasound by differing amounts, and the echoes are used to create an image.
- Very high frequency ultrasound (about 1.5 MHz) is used to see fine detail.
- **Sonar** is a technique used by ships to determine the depth of water:
  - Pulses of ultrasound are reflected off the bottom of the seabed.
  - The time delay between sending the pulse and receiving the echo is used to calculate the distance travelled.

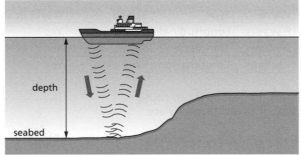

Ships use sonar to find the depth of water.

- Dolphins and bats use the same method to find their prey, but it is called **echolocation**.
- Infrasound is produced naturally by volcanoes, avalanches, ocean waves, hurricanes, earthquakes, meteorite explosions and animal movements.
- Infrasound is also produced by human activity, such as drilling for oil, and nuclear and chemical explosions.
- Infrasound can travel hundreds of kilometres through the Earth and atmosphere.

- Humans use infrasound to detect and monitor:
  - animal movements in remote locations
  - volcanic activity and meteorite strikes.

Remember!
Ultrasound is high-frequency sound. Infrasound is low-frequency sound.

Deep water

- With sonar, the speed of sound and the time delay can be used to calculate the depth of water:

  distance travelled by ultrasound pulse = speed of ultrasound × time delay

- Look at the diagram of the ship above. Remember – the ultrasound pulse has travelled from the ship to the seabed and back, so the depth is half the distance calculated.

Improve your grade

Measuring depth

Higher: A ship uses sonar to measure the depth of water as it approaches a harbour. A short pulse of ultrasound is sent out, and the time for it to return is measured as 20 milliseconds (a millisecond is a thousandth of a second).

If the speed of ultrasound is 1500 m/s, what is the depth of water? *AO2* [2 marks]

Earthquakes and seismic waves

The Earth and earthquakes

- The Earth is not a solid ball. It has several layers:
  - The inner core is very hot and dense solid iron.
  - The outer core is very hot, dense and liquid, mostly iron.
  - The mantle is hot, less dense and a mixture of solid and molten rock.
  - The crust is a very thin layer of solid rock.
- The crust floats on a hot liquid called **magma**. When magma comes to the surface from erupting volcanoes it is known as lava.

The layered Earth.

- The Earth's crust is split into sections known as **tectonic plates**, which move very slowly due to convection currents in the magma.
- Sudden movement of tectonic plates causes the shaking of the ground we call earthquakes:
  - As two plates try to slide past one another, friction between them at first prevents them from moving.
  - Eventually, the frictional forces cannot keep the plates still and they suddenly slip.
  - The immense energy stored in the compressed plates is released in the form of **seismic waves**.

- Alfred Wegener first suggested the movement of the Earth's crust in 1915. Since then, evidence has supported the idea:
  - Ancient rocks found in East Africa are identical to those found in South America.
  - Fossils found in both regions were from the same species of ancient aquatic reptile.

Seismic waves

- There are three types of seismic waves: P, S and L waves.

Wave	Type	Speed	Travel in solids	Travel in liquids
P waves	Longitudinal	Fast	Yes	No
S waves	Transverse	Slow	Yes	Yes
L waves	Combination	Very slow	Yes	Yes

- When an earthquake occurs on the seabed it can cause a tsunami (tidal wave). Earthquakes and tsunamis are unpredictable because:
  - Scientists cannot measure the pressure between the tectonic plates.
  - The fault lines are often deep within the Earth's crust.

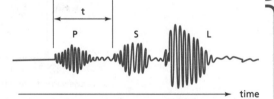
A seismograph showing S, P and L waves.

- Scientists make predictions of the risk of earthquakes happening in a specific area, based on its earthquake history.
- Seismic waves are monitored around the world using **seismometers**. The trace from a seismometer is called a seismograph.
- The distance from the seismometer to the epicentre of the earthquake can be calculated from the time delay between receiving the P and S waves and their speeds.

EXAM TIP

You may be asked to investigate the unpredictability of earthquakes by measuring the force needed to move a block on a horizontal surface or the angle of tilt needed for a block to move on a slope. Learn these experiments and practise plotting the results on a graph.

- Transverse S waves cannot travel through the liquid outer core of the Earth, but they can travel in the solid crust and semi-solid mantle.
- Longitudinal P waves can travel through all the different layers of the Earth, but at differing speeds, so they **refract** at the boundaries.
- Both P and S waves travel faster in more dense materials, so the deeper inside the Earth –where there is very high pressure – the more dense the material and the faster both waves travel. This causes the waves to refract and follow curved paths.
- Both P and S waves will **reflect** at the boundaries between the layers.

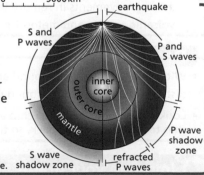
S and P waves through the Earth. Note that there are no S waves at the core.

Improve your grade

S waves

Foundation: The Earth is made up of four layers – crust, mantle, outer core and inner core.
Through which layers can S waves travel? Explain your answer.   AO1 [3 marks]

Electrical circuits

Charge and current

- Circuit diagrams are used to show electrical circuits. When there is a complete closed circuit, an electric **current** will flow.

- Electric current is a *flow* of **charge**. In copper wires, the charge is due to negatively charged **electrons**.

- Electric current is measured in **amperes** (A), using an **ammeter** connected in **series**.

- The electrons in the circuit are attracted to the positive terminal of the battery, but **conventional current** flows from positive to negative.

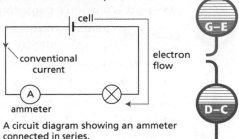

A circuit diagram showing an ammeter connected in series.

Types of circuits

- In series circuits, components are connected end to end in a single loop. The current is the same all the way round the series circuit. An ammeter placed at points A and B will show the same reading.

- In **parallel** circuits, the components are connected across each other. The current will split at the junction X and re-join at junction Y.

- The unit of electric charge is the **coulomb** (C). An electron carries a tiny amount of charge – only -1.6×10^{-19} C.

$$\text{electric current (A)} = \frac{\text{electric charge (C)}}{\text{time (s)}}$$

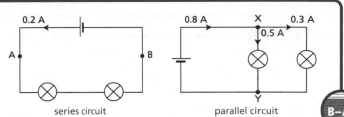

series circuit   parallel circuit

In the series circuit, the current is 0.2 A in each lamp. In the parallel circuit, the currents in the lamps add up to 0.8 A.

Remember!

Electric current is the rate of flow of charge. A current of 1 A is 1 C of charge per second.

Voltage

- The cell or battery in a circuit gives energy to the electrons so that they can transfer energy to the components in the circuit.

- The chemical energy in the cell is converted to electrical energy of the electrons.

- A **voltmeter** measures the amount of energy transferred in a component. It is connected in parallel across the component.

- The **voltage**, or **potential difference**, is measure in **volts**.

How is the voltmeter connected to the lamp?

- In series circuits, the voltage across individual components adds up to the voltage across the power supply: $V = V_1 + V_2$

- In parallel circuits, the voltage across each component is the same as the voltage across the power supply: $V = V_1 = V_2$

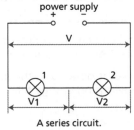

A series circuit.

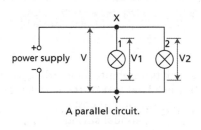

A parallel circuit.

- Potential difference is the amount of energy per unit charge. A p.d. of 1 volt is 1 joule of energy per coulomb of charge.

$$\text{potential difference (volt)} = \frac{\text{energy (joule)}}{\text{charge (coulomb)}}$$

Improve your grade

Potential difference

Higher: A cell has a potential difference of 1.5 V.

What is meant by the term 'potential difference'?

AO1 [3 marks]

Electrical power

Calculating power

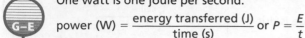

G–E

- **Electrical power** is the rate of energy transfer. It is measured in **watts** (W). One watt is one joule per second.

$$\text{power (W)} = \frac{\text{energy transferred (J)}}{\text{time (s)}} \quad \text{or} \quad P = \frac{E}{t}$$

- All electrical appliances are marked with a power rating in watts or kilowatts.

D–C

- In the circuit opposite, the voltmeter reads 12 V and the ammeter reads 2.0 A. The power of the lamp can be worked out as: 12 V × 2.0 A = 24 W

  electrical power (W) = current (A) × potential difference (V)
  or $P = I \times V$

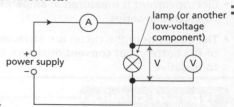

A circuit for determining electrical power.

Fuses

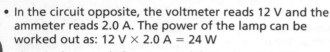

B–A*

- A **fuse** is a very thin piece of wire that acts as a safety device in a mains plug.

- If a fault occurs and the current becomes too high, the fuse wire will melt and break the circuit, preventing further damage or a fire.

- The fuse rating is the maximum current the fuse can carry without melting. Fuses are available at 1 A, 3 A, 5 A and 13 A.

- To find the correct fuse to use, calculate the normal operating current for an appliance. For example, a toaster has a power of 1100 W and is connected to 230 V mains supply:
  - power = current × potential difference
  - current = power / potential difference = 1100 / 230 = 4.8 A
  - The fuse rating must be higher than 4.8 A, so choose a 5 A fuse.

Kilowatt-hours

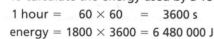

G–E

- Lower-power appliances use less energy in a given time, so they cost less to run.

- The energy in joules used by an appliance can be calculated: energy transferred (J) = power (W) × time (s)

- To calculate the energy used by a 1800 W heater in one hour:
  1 hour = 60 × 60 = 3600 s
  energy = 1800 × 3600 = 6 480 000 J

- This number is very large, so we prefer to use the unit **kilowatt-hour** (kW h) for mains electricity bills. One kilowatt-hour is the energy used by a 1 kW appliance in 1 hour.
  energy (kW h) = power (kW) × time (h)

- The cost of electricity is calculated using the equation below:
  cost (p) = power (kW) × time (h) × cost per kW h (p/kW h)

Remember!
The kilowatt-hour is a unit of energy, not power.

Saving energy and money

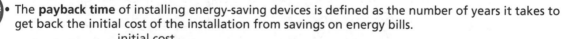

D–C

- Using energy-saving devices in the home has financial and environmental benefits. Examples include: compact fluorescent bulbs, motion-activated lights, home insulation and standby detection devices.

- The **payback time** of installing energy-saving devices is defined as the number of years it takes to get back the initial cost of the installation from savings on energy bills.

$$\text{payback time} = \frac{\text{initial cost}}{\text{annual saving}}$$

- For example, if installing double glazing costs £5000, and in one year saves £250 on heating bills, the payback time is 20 years ($\frac{5000}{250}$).

B–A*

- Sometimes other factors influence decisions about installing energy-saving devices. For example, installing double glazing will provide better soundproofing or increase the saleability of a home.

⦿ Improve your grade

Kilowatt-hours

Foundation: An oven has a power rating of 2000 W and it takes 45 minutes to cook a cake. The cost of 1 kW h of electrical energy is 22 p.

What is the cost of the electricity to cook the cake?

AO2 [2 marks]

Energy resources

Non-renewable energy resources

- Coal, oil and natural gas are called fossil fuels because they formed from the remains of prehistoric plants and animals.

- Fossil fuels and nuclear fuel are **non-renewable resources** because they cannot be replaced and will eventually run out.

Resource	Advantages	Disadvantages
Coal	Fuel is cheap; coal-burning power stations have a quick start-up time; coal will last at least 200 years.	Burning coal releases CO_2 and SO_2; SO_2 produces acid rain; mining can be dangerous; stockpiles are needed to meet demand.
Natural gas	Gas-fired power stations are efficient and have the quickest start-up time, so they are flexible at meeting demand for power; gas will last another 50 years; gas does not produce SO_2.	Burning gas releases CO_2 (although less than coal and oil); pipelines necessary for transporting gas are expensive.
Oil	Oil-burning power stations have a quick start-up time; there is enough oil left to last 50 years.	Burning oil releases CO_2 and SO_2; oil prices are extremely variable; there is a danger of spillage and pollution during transport of oil by road, rail or sea.
Nuclear	Nuclear power stations are located away from population centres; it does not produce CO_2 or SO_2.	Building and decommissioning a power station is expensive; start-up time is the longest; radioactive waste remains dangerous for thousands of years.

The greenhouse effect

- Infrared radiation emitted by the surface of the Earth is absorbed by **greenhouse gases** in the atmosphere. This causes the atmosphere to warm up in a process called the **greenhouse effect**.

- Burning fossil fuels increases the levels of carbon dioxide in the atmosphere. Some scientists believe that this will cause the average temperature of the Earth to rise in the future.

Renewable energy resources

- **Renewable energy resources** are also used to generate electricity:
  - Solar power turns light energy from the Sun into electrical energy, using solar cells.
  - Wind power is used to rotate huge propeller blades on a turbine to generate electricity.
  - Wave power is generated when large floats containing coils and magnets move up and down with ocean waves.
  - Hydroelectric power is when fast-flowing water, stored in a reservoir above a power station, is used to generate electricity.
  - Tidal power uses seawater from incoming and outgoing tides to create electricity.
  - Biomass is organic material from decaying plant or animal waste that can be used as a fuel in a power station, in the same way as fossil fuels are. Wood can be used in a power station in a similar way.

- Most renewable sources cost nothing to use, and produce no greenhouse gases, but the cost of building renewable power stations is substantial. Other advantages and disadvantages of renewable resources are listed in the table below.

Resource	Advantages	Disadvantages
Solar	Useful in remote areas; single homes can have their own electricity supply.	No power at night or when cloudy.
Wind	Can be built offshore.	Can cause noise and visual pollution; amount of electricity depends on the weather.
Wave and tidal	Ideal for island countries.	May be opposed by local or environmental groups.
Hydroelectric	Creates water reserves as well as electricity supplies.	Can cause flooding of surrounding communities and landscapes.
Biomass and wood	Cheap and readily available source of energy; if replaced, biomass can be a long-term, sustainable energy source.	Gives off CO_2 and SO_2 when burnt; biomass and wood are only renewable resources if crops and trees are replanted.

Improve your grade

Understanding resources

Foundation: Fuels are burnt in power stations. Fossil fuels are non-renewable energy sources, and biomass is a renewable energy source.

Explain what is meant by the terms 'non-renewable' and 'renewable'. *AO1* [2 marks]

Generating and transmitting electricity

Inducing a current

- A voltage is **induced** when a wire is moved in a magnetic field. If the piece of wire is part of a circuit, a current will flow. This is called electromagnetic induction.

- The direction of the current is reversed when the motion of the wire is reversed, or when the magnet is turned round.

- A voltage is always induced when there is relative movement between a magnet and a coil of wire. The induced voltage is larger when the magnet is moved more quickly.

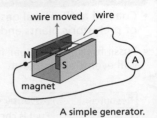

A simple generator.

Generators

- In an electrical **generator**, a coil is rotated in a magnetic field. As the coil rotates, it cuts the magnetic field lines to induce a voltage across the coil.

- The coil has slip rings, which are connected to a circuit via brushes. This causes an alternating current to flow in the circuit.

- The current will increase if the speed of motion increases, if a stronger magnet is used, or if there are more turns of wire in the coil.

- **Direct current** (d.c.) always flows in the same direction. A cell provides direct current.

- **Alternating current** (a.c.) changes direction at a frequency determined by the rotating coil.

- Mains electricity is generated at a frequency of 50 Hz.

- The variation of current with time follows a sinusoidal curve.

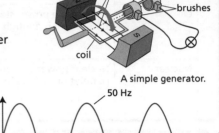

A simple generator.

Typical alternating current from a mains generator.

Transformers

- A **transformer** is a device that changes the size of an alternating voltage.

- Two coils of wire are wrapped round a soft iron core. The alternating voltage supply is connected to the **primary coil**, and the output alternating voltage is induced across the **secondary coil**.

- A **step-up transformer** converts a low voltage input to a higher voltage output. The primary coil will have fewer turns than the secondary coil.

- A **step-down transformer** converts a high voltage input to a lower voltage output. The primary coil will have more turns than the secondary coil.

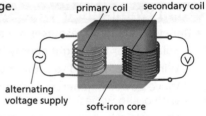

A simple transformer.

- The **National Grid** is the network of pylons and cables that transmit electrical energy from power stations to users across the UK.

- Transformers are used in the National Grid:
  - Step-up transformers increase the 23 kV power-station output to higher voltages (275 kV or 400 kV). This reduces heat loss in the cables and improves efficiency of transmission.
  - Step-down transformers lower the voltage to 230 V, which is safer for homes.

- Electricity transmission is hazardous. Overhead cables are not insulated, so can cause serious harm or even death if touched.

- Risks can be minimised by burying cables underground, but this is expensive and makes it difficult to maintain the cables. Suspending cables from pylons is a cheaper option.

Remember!
Transformers only work with a.c. electricity.

- The a.c. voltage in the primary coil of a transformer creates an ever-changing magnetic field around it.

- The magnetic soft-iron core channels the magnetic field through the secondary coil.

- The alternating magnetic field will continuously cut through the wires in the secondary coil and an a.c. voltage will be induced.

- The turns ratio is equal to the voltage ratio:

$$\frac{\text{output voltage at secondary coil}}{\text{input voltage at primary coil}} = \frac{\text{number of turns on secondary coil}}{\text{number of turns on primary coil}}$$

Improve your grade

Transformers
Foundation: Describe how step-up transformers are used in the National Grid.

AO1 [4 marks]

Topic 5: 5.6, 5.7, 5.8, 5.9, 5.10, 5.11, 5.12, 5.13, 5.14, 5.15

Energy and efficiency

Types of energy

- Energy exists in many forms, as summarised in the table below.

Name	Description	Examples
Thermal (heat) energy	An object at a higher temperature has greater thermal energy	Heater; the Sun; hot water
Light energy	Light is a wave that is emitted from anything at a very high temperature	Lamp; stars; fire
Electrical energy	This is usually associated with electric current	Mains supply; overhead cables; output from a transformer
Sound energy	An object vibrating will emit sound	Buzzer; bell; siren; person talking
Kinetic energy	This is energy due to movement	Person running; high-speed train; planet orbiting the Sun
Chemical energy	This is energy stored by atoms	Food; chemical cell; coal
Nuclear energy	This is energy stored by the nuclei of atoms	Nuclear power station; radioactivity; nuclear bombs
Elastic potential energy	An object that is pulled or squashed has this type of energy	A stretched rubber band; cables supporting a bridge
Gravitational potential energy	This is energy due to an object's position in the Earth's gravitational field. An object lifted higher will have greater gravitational potential energy	Aeroplane in the sky; person up a ladder

Energy transfers

- Whenever anything moves or changes, an energy transfer must happen.

- For example, in a firework, the chemical energy stored in the firework becomes heat, light and sound when the firework explodes.

  chemical energy → thermal energy + light energy + sound energy

- Energy cannot be created or destroyed. It can only be transferred from one form to another. This is known as the **principle of conservation of energy**.

Remember!
The total energy input must equal the total energy output.

Efficiency of devices

- In all energy transfers, some of the energy output is useful and some is unwanted or wasted energy. For example, in a light bulb the useful energy output is the light, and the wasted energy output is the heat.

- In most devices heat energy is wasted. The heat energy produced may be due to electric currents flowing in wires or due to friction between moving surfaces.

- **Efficiency** is a measure of how well a device transfers energy in the form we want. Efficiency is calculated using the equation:

  $$\text{efficiency} = \frac{\text{(useful energy transferred by the device)}}{\text{(total energy supplied by the device)}} \times 100\%$$

- **Sankey diagrams** are energy-transfer diagrams that show the different forms energy takes during a transfer. The thickness of the arrow in the diagram is drawn to scale to show the amount of energy transferred.

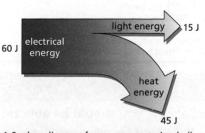

A Sankey diagram for an energy-saving bulb.

Improve your grade

Calculating efficiency

Higher: An electric drill transfers every 50 J of electricity into 20 J of useful kinetic energy.

Calculate the percentage efficiency of the drill.

What forms does the wasted energy in the transfer take?

AO2 [4 marks]

Radiated and absorbed energy

Infrared radiation

- All objects can emit infrared radiation, absorb infrared radiation and reflect infrared radiation.
- The amount of infrared radiation emitted by an object depends on its temperature and its surface.
- All objects at temperatures greater than absolute zero emit (or radiate) infrared radiation. The hotter an object, the more power it radiates.
- A dull black surface loses energy more quickly than a bright shiny surface because:
  - a dull black surface is a good radiator of heat
  - a bright shiny surface is a poor radiator of heat.
- A dull black surface is also a good absorber of heat radiation.
- A bright shiny surface is a poor absorber, as it reflects the heat radiation away.
- Objects that are warmer than their surroundings will emit more energy per unit time than they absorb.
- Objects that are cooler than their surroundings will absorb more energy per unit time than they emit.

G–E

Remember!
Dull black surfaces are good emitters and good absorbers.

Staying at the same temperature

- All objects are continually absorbing and emitting heat radiation.
- When the rate of absorption is greater than the rate of radiation, the temperature will rise (and vice versa).
- Eventually, the rate of heat absorption will equal the rate of heat radiation and the temperature remains steady. This is called thermal equilibrium.
- For example, the hot filament in a bulb will reach a steady temperature when the input of electrical power is equal to the power radiated away from the filament.

D–C

Experimenting with radiation

- All atoms have both **kinetic energy** (due to their vibrations) and **potential energy** (due to their position).
- When substances are heated, the temperature increases and the atoms gain more kinetic energy.
- The heat energy of a substance is the total kinetic energy and potential energy of all the atoms in the substance.
- To investigate the different rates of heat radiation of different surfaces, you can compare the temperature over time of a silvered beaker and a blackened beaker (see below).
- To investigate the different rates of heat absorption, you can use a radiant heater between two plates – one silvered and one blackened. Then time how long it takes for each plate to get hot enough to melt some wax.

B–A*

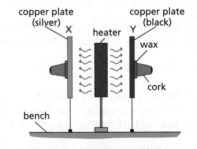

An absorption experiment with shiny and black metal plates. The melting wax is an indicator of the amount of thermal energy absorbed.

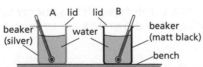

A cooling experiment with shiny and black beakers. The thermometers in each beaker will indicate the amount of thermal energy radiated.

⊙ How science works

In the investigations mentioned above, you should be able to:
- ensure that there is a valid test – there must be only one independent variable
- repeat readings to increase the reliability of the investigation.

⊙ Improve your grade

Staying cool

Foundation: In many Mediterranean countries, such as Greece, the houses are all painted white. Explain why this is, using ideas about heat radiation.

AO2 [3 marks]

P1 Summary

Light is refracted and reflected at surfaces and boundaries between different materials.

Refracting telescopes use an objective lens to form a real image, which is then magnified. Reflecting telescopes use a concave mirror, which gives a clearer image and allows larger telescopes to be made.

Visible light and the solar system

The model of the Universe changed from a geocentric to a heliocentric model.

Observations of the Universe in the past used light, but nowadays many different types of wave are used to gain more evidence.

Waves (transverse or longitudinal) transmit energy but not matter. They are described in terms of wavelength, frequency and amplitude.

The EM spectrum is a family of waves that all travel at the same speed in a vacuum. They are: radio waves, microwaves, infrared, visible light, ultraviolet, X-rays and gamma rays.

The electromagnetic spectrum

Ionising alpha, beta and gamma radiation comes from the nucleus of radioactive elements. The higher the frequency of the EM waves, the higher the ionising power.

The Solar System is our group of planets, moons, asteroids and comets that orbit the Sun.

Scientists study the Universe with all parts of the EM spectrum.

Waves and the Universe

Nuclear reactions take place within all stars, which give out vast amounts of energy in the form of EM waves.

The human hearing range is approximately 20 Hz to 20 000 Hz.

Frequencies lower than 20 Hz are called infrasound.

Frequencies higher than 20 000 Hz are called ultrasound.

Waves and the Earth

Seismic waves are generated by earthquakes. S waves and P waves have different properties.

The Earth's crust is made up of tectonic plates, which are constantly moving due to convection currents in the mantle. Earthquakes are caused by the relative movement of the plates at plate boundaries.

Power is the rate at which energy is used. Power (W) = energy (J)/time (s).

Electrical power (W) = current (A) x potential difference (V)

Fossil fuels are non-renewable resources. Burning them and contributes to the greenhouse effect.

Renewable energy resources include: wind, waves, tidal, hydroelectric and solar.

Generation and transmission of electricity

Electromagnetic induction occurs when there is relative motion between a wire and a magnetic field. An electric current is generated.

Electricity generated in power stations is alternating current (a.c.). Electricity from a battery is direct current (d.c.).

Transformers are used to change the voltage of a.c. electricity.

Energy cannot be created or destroyed – only converted from one form to another, as depicted in Sankey diagrams.

In all energy transfers, some energy is wasted as heat.

Energy and the future

All objects at temperatures above absolute zero emit infrared radiation.

Electrostatics

- All matter is made up of tiny **atoms**. Each atom is about 0.0000002 mm in diameter.
- Scientists believe in the nuclear model of the atom:
  - The **nucleus** is about 100 000 times smaller than the diameter of the atom.
  - The nucleus contains particles called **protons** and **neutrons**.
  - Protons are positively charged and neutrons have no charge, so the nucleus is positive.
  - Negative **electrons** occupy the space around the nucleus.
- The table below shows the properties of atomic particles. The mass and charge are shown relative to that of the proton.

Particle	Where found	Relative mass	Relative charge
Neutron	Inside the nucleus	1	0
Proton	Inside the nucleus	1	+1
Electron	Outside the nucleus	0	−1

- If an atom loses electrons it will have an overall positive charge; if it gains electrons it will have an overall negative charge.
- A charged atom is called an **ion**.

G–E

Static electricity

- An **insulator** can be charged by **friction**. For example, if you rub a balloon against your clothes and then hold it against a wall, the balloon sticks to the wall.
  - The friction transfers electrons to the balloon, making it negatively charged.
  - The charged balloon repels some of the electrons away from the surface of the wall, leaving the surface of the wall with a positive charge.
  - Opposite charges attract, so the balloon is attracted to the wall.
- The separated charges in the wall are called induced charges.
- Some other examples of electrostatic phenomena are:
  - A stream of water can be bent towards a charged insulator.
  - Synthetic clothing clings to your body.
  - A comb becomes charged when you comb your hair or rub it with a cloth. The comb will attract your hair or small pieces of paper.
- Simply rubbing a glass rod with a woollen cloth will cause a transfer of charge. The wool gains electrons to become negatively charged. The glass loses electrons and is left with an equal positive charge. This is called static charge.
- All insulators can be charged by friction:
  - Some insulators lose electrons by friction. They become positively charged.
  - Other insulators gain electrons by friction and become negatively charged.
  - Each insulator acquires an equal but opposite charge.

D–C

> **Remember!**
> Charged objects exert a force on each other. The force depends on the type of charge each one has: like charges repel and unlike charges attract.

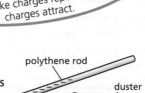

polythene rod

duster

What is the net charge on the rod and cloth?

Gold-leaf electroscope

- A gold-leaf electroscope is used to find the charge on an insulator. A metal rod connects a metal cap to gold leaf.
- Bringing a negatively charged rod close to the cap will repel the electrons down the metal rod. The gold leaf moves away from the metal rod because they both have the same charge. The greater the charge, the further away the leaf moves.
- Bringing a positively charged rod close to the cap will attract the electrons towards the cap, leaving the gold leaf and the metal rod with a positive charge. Again, the gold leaf moves away from the metal rod because they both have the same charge.

B–A*

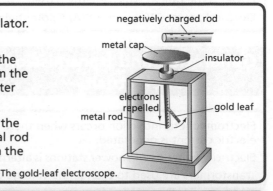

negatively charged rod

metal cap

insulator

electrons repelled

gold leaf

metal rod

The gold-leaf electroscope.

⊙ Improve your grade

Forces and attraction

Foundation: Daisy combs her hair with a plastic comb. She can then use the plastic comb to pick up small pieces of paper.

Explain why the paper is attracted to the comb. *AO1* [3 marks]

Uses and dangers of electrostatics

Uses of electrostatics

- Electrostatic paint sprayers are used on many metal objects:
  - The object to be sprayed, such as a car, is connected to a negative supply.
  - The sprayer charges the tiny droplets of paint as they emerge from the nozzle.
  - The charged droplets of paint repel each other and spread out to form a dispersed cloud.
  - The positive droplets are attracted to the negatively charged object being sprayed.

- The advantages to using this technique are that:
  - less paint is used
  - the object is given an even coating of paint
  - every part of the object attracts the paint, even the underside.

- The same technique is used to spray plants with **insecticides**. The insecticide is given a positive charge and the plants acquire a negative charge by induction.

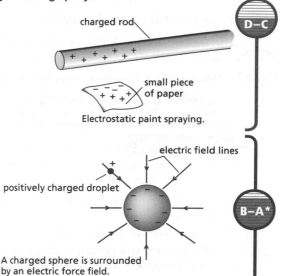

Electrostatic paint spraying.

- The diagram opposite shows a positively charged droplet near a negatively charged metal sphere. The sphere creates an **electric force field** around it, represented by the purple lines known as electric field lines:
  - Closer field lines represent a stronger field.
  - The arrows in the field lines show the direction of the field.
  - Any charged object in this electric field will experience a force. A positive charge would tend to move in the direction shown by the arrows.

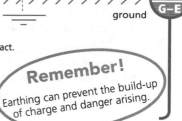

A charged sphere is surrounded by an electric force field.

Dangers of electrostatics

- Certain clothing fibres can rub together and cause charges to separate. The crackling noise when you undress, for example, comes from tiny electrical sparks.

- If you bring your finger close to a charged insulator, the electrons on the surface will jump the tiny distance of air and travel through you to the Earth.

- Electrical shocks can come from everyday objects:
  - Walking on a synthetic (nylon) carpet will charge the carpet and you.
  - You may get a shock when you are about to touch a metal tap or radiator.
  - Your clothing can become electrically charged when rubbing against the synthetic material of a car seat. You will get a shock as you get out of the car.

Like charges repel and unlike charges attract.

Remember!
Earthing can prevent the build-up of charge and danger arising.

- Most objects can be made safe by **earthing**. Earthing involves placing a metal conductor between the object and the Earth to channel the charges safely to Earth.

- Fuelling aircraft and tankers causes static as the fuel rubs against the pipe. The charges can build up and create a spark, igniting the fuel. This is why aircraft and tankers are earthed when fuelling.

- Lightning during a thunderstorm is dangerous. It can destroy buildings and start forest fires.

- Lightning occurs when charge builds up within a cloud. When the charge is large enough, it leaps to another part of the cloud, or to the ground.

- To protect tall buildings from lightning strikes, a lightning conductor is often installed on top of the building:
  - The top of the conductor has a pointed spike and the bottom is embedded in the ground.
  - When a negatively charged cloud passes overhead, it induces positive charges at the top of the lightning conductor.
  - The spike repels positive ions in the air towards the cloud. This neutralises the charge so there is no lightning.

Improve your grade

Painting bicycles

Higher: A factory uses an electrostatic paint gun to paint new bicycles.

Explain why the paint coats the surface of each bicycle evenly, even the back.   *AO1* [2 marks]

Current, voltage and resistance

Charge and current

- **Electric current** is the rate of flow of charge. It is measured with an **ammeter** (connected in **series**) and its unit is **amperes** or 'amps' (A).
- A cell is a chemical device with its own positive and negative terminals which push electrons around a circuit. A battery is a collection of cells, often joined together in series.
- A cell or a battery provides a **direct current** (d.c.). This means that the electrons travel in one direction only.

- Charge is measured in **coulombs** (C).
- In metals, current is due to the flow of electrons. Each electron has a tiny negative charge of -1.6×10^{-19} C.
- The current is 1 ampere when the rate of flow of charge is 1 coulomb per second.
- Current and charge are linked by the following equation, where Q is the charge in coulombs, I is the current in amperes and t is the time in seconds: $Q = I \times t$

Resistance

- Components have difference resistances, and can be used to control the size and current in circuits.
- The diagram opposite shows components (lamps) connected in series.
  - The current in a series circuit is the same all the way round.
  - The current will be larger when more cells are used, but the ammeters will still show the same current at all points around the circuit.

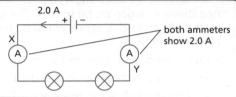

both ammeters show 2.0 A

The current in each lamp is the same.

- The diagram opposite shows components connected in **parallel**.
  - The current splits at the junction J.
  - Electric current is conserved at the junction.
- Current is conserved at a junction because the total number of electrons entering a junction must be equal to the total number of electrons leaving the junction.

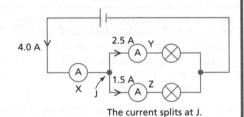

The current splits at J.

Potential difference

- A **voltmeter** in a circuit measures the **potential difference** (p.d.) V in volts (V) across a component. It is connected in parallel with the component.
- The **resistance** of the component can be determined from the ammeter and voltmeter readings. The resistance R in ohms (Ω) of a component is given by the equation $R = V / I$, which can also be written:

$$\text{resistance (ohm, } \Omega) = \frac{\text{potential difference (volt, V)}}{\text{current (ampere, A)}}$$

Remember!
If you know the resistance of a component and the current in it, then you can use the equation V = I R to calculate the potential difference across it.

- As electrons travel round a circuit, they transfer some of their electrical energy to other forms such as heat and light.
- Potential difference across a component is defined as the energy transferred per unit charge:

$$\text{potential difference (volt, V)} = \frac{\text{energy (joule, J)}}{\text{charge (coulomb, C)}}$$

- In the diagram opposite, a charge of 1 C going round the circuit will transfer 1.2 J of its electrical energy into heat in the resistor, and 0.3 J of its electrical energy into heat and light in the lamp.

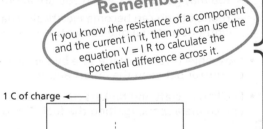

1 C of charge
1.2 V   0.3 V
1.2 J of heat per coulomb
0.3 J of heat and light per coulomb
Potential difference is energy transferred per unit charge.

Improve your grade

Calculating current

Foundation: Look at the circuit opposite.

What is the current at points Z and X in the circuit?

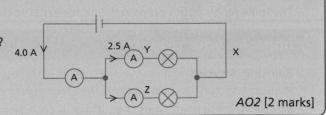

AO2 [2 marks]

Lamps, resistors and diodes

Changing currents in circuits

- The potential difference *V* across the wire and the current *I* in the wire can be used to determine its resistance.

- The resistance of the wire is directly proportional to its length. For example, the resistance doubles when the length is doubled.

- The resistance of a variable resistor (or **rheostat**) can be changed manually.

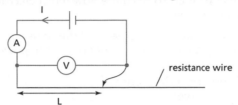

The meters help you to determine the resistance of the wire.

G–E

- The diagram below shows a rheostat used in a circuit to change the brightness of a lamp. When the resistance of the variable resistor is set:
  - to its maximum value, the current in the circuit is low and the lamp is dimly lit
  - to its lowest value (often zero), the current in the circuit is high and the lamp is fully lit.

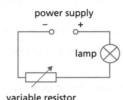

A variable resistor in a circuit.

Current against potential difference graphs

- Circuits can be used to investigate how a current *I* varies with potential difference *V* for devices such as a **filament lamp**, a **diode** and a **fixed resistor**. In the diagram opposite, the current is altered using the variable resistor.

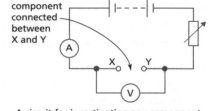

A circuit for investigating any component.

- The graphs below show the *I* against *V* for a filament lamp, a 100 W fixed resistor and a diode.

  - In the filament lamp the current increases with p.d. As the current increases, the filament gets hotter and the resistance increases.

  - In the fixed resistor, the current is directly proportional to the p.d. Its resistance is constant.

  - The diode only conducts in one direction. It has an infinite resistance when the current is zero. The diode has low resistance when it conducts.

D–C

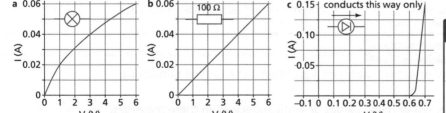

Current against voltage graphs for **a** a filament lamp, **b** a fixed resistor and **c** a (silicon) diode.

EXAM TIP

You may come across these graphs with current on the *x*-axis and voltage on the *y*-axis. The gradient of the graph for the filament lamp would increase rather than decrease. Remember to check the axis.

Practical application of a diode

- A diode only conducts in one direction, when its anode is connected to a positive terminal and its cathode is connected to a negative terminal.

B–A*

- A diode and a lamp can be used to work out the polarities of an unmarked cell or power supply.

Improve your grade

Variable resistors

Foundation: Explain how a circuit with a variable resistor can be used to control the brightness of a lamp in the same circuit.

AO2 [3 marks]

Heating effects, LDRs and thermistors

Heating effect of electric current

- The current in a resistor causes it to heat up. This heat is made use of in electric heaters and the heating element of an electric kettle.
- Excessive heat produced in electrical devices can cause fires if there are no safety features. This is why you need a cooling fan in a laptop. An overloaded mains socket is a potential fire hazard.

- The **electrical power** of a device can be used to calculate how much electrical energy it will transfer and how much it would cost to use the device.
- The power of a device is related to the current it carries and the potential difference (p.d.) across it. Power can be calculated using the equation $P = I \times V$, where P is the power, I is the current and V is the p.d. You can also write this equation as:

$$\underset{\text{(watt, W)}}{\text{electrical power}} = \underset{\text{(ampere, A)}}{\text{current}} \times \underset{\text{(volt, V)}}{\text{potential difference}}$$

- The energy transferred by any component can be found by multiplying the power by the time:

$$\underset{\text{(joule, J)}}{\text{energy transferred}} = \underset{\text{(watt, W)}}{\text{power}} \times \underset{\text{(second, s)}}{\text{time}}$$

or:

$$\underset{\text{(joule, J)}}{\text{energy transferred}} = \underset{\text{(ampere, A)}}{\text{current}} \times \underset{\text{(volt, V)}}{\text{potential difference}} \times \underset{\text{(second, s)}}{\text{time}}$$

> **Remember!**
> You can find the current and the voltage from the ammeter and voltmeter readings of a device in a circuit, and use the values to calculate the power and the energy transfer.

- When current flows, electrons travel through the **lattice** of a metal.
- Electric current causes heating. The heating is the result of collisions between electrons and the ions in the lattice.
- The collisions cause increased vibrations (around fixed positions) of the ions. This is what we mean by heat energy.

LDRs and thermistors

- A **light-dependent resistor** (LDR) is a special type of resistor made from a **semiconductor** material.
  - The resistance of an LDR decreases as the intensity of light incident on it increases.
  - More light means less resistance.

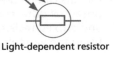

Light-dependent resistor

- A **thermistor** is a special type of resistor. Its resistance depends on its temperature.
  - The resistance of a thermistor decreases as its temperature increases.
  - Greater temperature means less resistance.

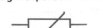

Thermistor

- Circuits with LDRs can be used to detect changes in light **intensity**. Examples of where LDRs are used include simple light-meters, light-sensitive circuits in burglar alarms, and circuits for automatically controlling how much light enters a camera.

- Circuits with thermistors can be used to detect changes in temperature. Examples of where thermistors are used include a simple electrical thermometer, a circuit to switch on a fan inside your laptop when it overheats, and a central-heating controller that senses the temperature using a thermistor.

The variation of resistance

- Increasing the temperature of a metal does not significantly change the number of conduction electrons.
- Thermistors are made from either metal oxides or semiconductors. In these materials, an increase in temperature frees up more electrons from the atoms, and this causes the resistance to decrease.
- LDRs are made from semiconductors that are responsive to light. As the intensity of light falling on an LDR increases, this frees up more electrons from the atoms and this causes the resistance to decrease.

⬤ Improve your grade

Kettle calculations

Higher: Noah buys an electric kettle that supplies 2000 W of power. A full kettle of water requires 200 000 J of energy to bring it to boiling point.

a Assuming that the kettle wastes no energy, how long will it take Noah to boil a full kettle of water?

b The current in the heating element produces a heating effect. Describe how this occurs.   AO2 [3 marks]

Scalar and vector quantities

Scalars and vectors

- Quantities in physics can be divided into two groups:
  - A **scalar** quantity only has size (or magnitude), e.g. distance, mass, volume, temperature, speed and energy.
  - A **vector** quantity has both magnitude and direction, e.g. displacement, velocity, acceleration and force.

- **Displacement** has a size that is equal to the distance from a specified point. It also has direction. For example, in the graph opposite, the aeroplane has travelled a distance of 40 km and has a displacement of 28 km (its size) at a bearing of 45° (its direction) from A.

- **Speed** is defined as the rate of change of distance. Most journeys are not covered at a constant speed. The average speed of a journey can be calculated using the equations:

$$\text{speed (metre per second, m/s)} = \frac{\text{distance (metre, m)}}{\text{time taken (second, s)}} \text{ or } v = \frac{x}{t}$$

- The **velocity** of an object is its speed in a specified direction. It is also defined as the rate of change of displacement.

- The car in the diagram opposite is travelling at a constant speed, but its velocity is changing continuously.

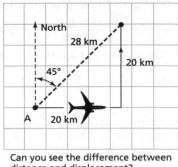

Can you see the difference between distance and displacement? **G–E**

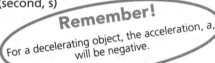

A car going round a roundabout at a constant speed of 5 m/s. Its velocity is different at all points. The velocity at A is 5 m/s due south and at B the velocity is 5 m/s due north.

Acceleration

- If the velocity of an object increases or decreases, or its direction changes, it is **accelerating**. The car in the diagram above is accelerating because its velocity is changing.

- An object whose velocity decreases with time is said to have **deceleration**. Deceleration means 'slowing down'.

- Acceleration is defined as the rate of change of velocity. It is given by the word equation:

$$\text{acceleration (metre per second squared, m/s}^2) = \frac{\text{changing velocity (metre per second, m/s)}}{\text{time taken (second, s)}}$$

- Acceleration can also be given by the equation (where u is the initial velocity, v is the final velocity and t is the time taken):

$$a = \frac{v - u}{t}$$

D–C

Remember!
For a decelerating object, the acceleration, a, will be negative.

- If you know the acceleration of an object, then you can calculate other quantities relating to its motion using the equations in the table below.

Quantity	Equation
Time t taken	$t = \dfrac{v - u}{a}$
Final velocity v	$v = u + at$
Initial velocity u	$u = v - at$
Average velocity v_{av}	$v_{av} = \dfrac{u + v}{2}$
Displacement s	$s = \left(\dfrac{u + v}{2}\right)t$

B–A*

Equations relating to the motion of an object.

Improve your grade

Understanding acceleration
Foundation: Syamala is training for the 1200-m race by running round a 200-m track six times. She runs at a constant speed.
- Explain why she is accelerating.

AO1 [2 marks]

Distance–time and velocity–time graphs

Distance–time graphs

- The **gradient** (or slope) of a line can be determined by drawing a large triangle and carrying out the following calculation:

$$\text{gradient} = \frac{\text{change in } y}{\text{change in } x} \text{ or gradient} = \frac{\Delta y}{\Delta x}$$

- The gradient of a distance–time graph is equal to the speed of the object.

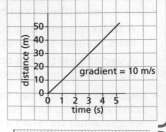

Car travelling at a constant speed of 10 m/s.

- The graph opposite is a distance–time graph for a falling apple. The graph is a curve. It shows that the apple falls longer distances in each succeeding 0.1 s. This means:
  - The apple is accelerating.
  - The gradient of the graph increases with time.

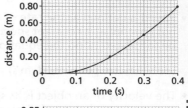

A distance–time graph for a falling apple.

- The graph opposite is a distance–time graph for a ball rolling down a straight slope.
  - The speed at the start is zero.
  - The speed of the ball at 0.2 s can be found from the **tangent** drawn.
  - The speed of the ball at 0.2 s is 0.80 m/s.
  - Since the ball is moving in a straight line, its final velocity at 0.2 s is 0.80 m/s.
- Work out the acceleration of the ball using the equation: $a = \dfrac{v - u}{t}$

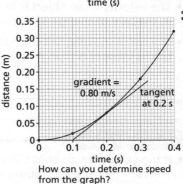

How can you determine speed from the graph?

Velocity–time graphs

- The graph opposite is a velocity–time graph for a car travelling at a constant velocity:
  - The velocity of the car is 20 m/s.
  - The line is horizontal and has zero gradient.

- In a velocity–time graph for an accelerating car:
  - A straight line shows constant (or uniform) acceleration.
  - The line will have a positive (upward) gradient or slope.
  - The steeper the line, the greater the acceleration.

- A velocity–time graph for a decelerating (slowing down) object at constant deceleration is a straight line with a negative gradient (slope).

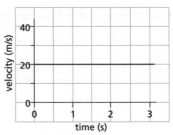

A velocity–time graph showing constant velocity.

- The acceleration is equal to the gradient of a velocity–time graph.

- The distance travelled is equal to the area under a velocity–time graph.

Remember!
A horizontal line on a velocity–time graph means the object is travelling at constant velocity. A horizontal line on a distance–time graph means the object is stationary.

Improve your grade

Interpreting graphs

Higher: Explain how the graph opposite shows that:

a The acceleration of car A is greater than the acceleration of car B.

b Car B has travelled further than car A.

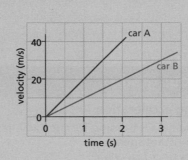

AO3 [2 marks]

Understanding forces

The basics about forces

- A force is a push or a pull exerted by one object on another.

- Force is a **vector** quantity: it has both size (magnitude) and direction.

- Force is measured in units called newtons, N.

- The direction of forces must be taken into account when adding them. For example, in the diagram below, the resultant force of the two 20 N forces can be either 40 N or zero.

resultant force = 40 N   resultant force = 0
Why are the resultant forces different?

- A **free-body force diagram** shows all the forces acting on an object. Some examples are shown opposite.

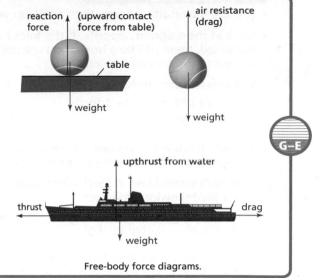

Free-body force diagrams.

Newton's first and third laws

- Sir Isaac Newton's first law of motion states that if there is no resultant force acting on an object, then:
  - if stationary, the object will remain at rest
  - if moving, the object will keep moving at a constant speed in a straight line.

- In the examples of free-body force diagrams above, the forces balance out. Each object is either stationary or moving at a constant speed in a straight line.

- Newton's third law states that when two objects interact, each object exerts an equal but opposite force on the other. We call these equal and opposite forces action and reaction forces.

- A car travelling along a straight road at a constant velocity has the following forces acting on it:
  - the total forward force F between the tyres and the road
  - frictional forces D, including the air resistance or drag opposing the motion of the car
  - the weight W of the car
  - the total upward contact force N provided by the road.

- Since the car is not moving vertically up or down, N is equal to W.

- The resultant force on the car in the horizontal direction is zero. The forces F and D are balanced. The car has no acceleration.

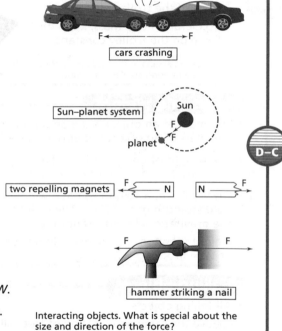

cars crashing

Sun–planet system

two repelling magnets

hammer striking a nail

Interacting objects. What is special about the size and direction of the force?

Action at a distance

- The gravitational force between the interacting Sun and the Earth acts over a long distance. There is action at a distance.

- According to Newton's third law, the force provided by the Sun on the Earth is equal in size but opposite in direction to the force provided by the Earth on the Sun.

- The gravitational force acting on an object on the Earth is called its **weight**. Every object has weight.

- The Earth interacts with every object on it, including you. You are pulling the Earth towards you with a force equal to your weight.

◐ Improve your grade

Forces and velocity

Foundation: An aeroplane experiences the forces of gravity, upthrust, thrust from the engine and friction from the air.

Explain how it can be travelling at a constant velocity.

AO1 [3 marks]

Force, mass and acceleration

Investigating forces

G–E

- If there is a resultant force, then an object will have acceleration.

- Look at the diagram opposite. If the force F on the car is suddenly increased, there will be a horizontal unbalanced force and the car will accelerate.

- The acceleration of the car depends on:
  - the size of the resultant force
  - the mass of the car.

Remember!
No resultant force means either the object is stationary or it is moving with a constant speed in a straight line.

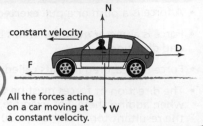

All the forces acting on a car moving at a constant velocity.

D–C

- The acceleration of an object is greater when the resultant force F is greater and when its mass m is smaller.

- **Newton's second law** of motion shows the link between force and mass, using the following equation:

$$\underset{\text{(newton, N)}}{\text{force}} = \underset{\text{(kilogram, kg)}}{\text{mass}} \times \underset{\text{(metre per second squared, m/s}^2\text{)}}{\text{acceleration}} \quad \text{or } F = ma$$

EXAM TIP
You may be asked to calculate the acceleration of an object with more than one force acting on it. Remember – it is the resultant force on the object that determines the acceleration.

B–A*

- You can investigate the link between resultant force, mass and acceleration using trolleys and a motion sensor connected to a computer.

- To investigate the relationship between:
  - force and acceleration, keep the mass of the trolley constant
  - acceleration and mass, keep the force on the trolley constant.

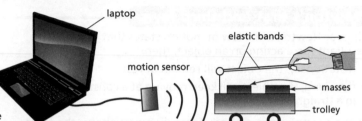

Investigating $F = ma$ in the laboratory. What is the advantage of using an electronic motion sensor?

Mass and weight

G–E

- The **mass** of an object is the amount of matter it contains. It is measured in kilograms (kg). The mass of an object remains a constant. Mass is a scalar property.

- The **weight** of an object is the Earth's gravitational force acting on it. It is measured in newtons (N). The weight of an object depends on where it is. Weight is a vector quantity.

- On the Earth, a 1 kg mass has a weight of 9.8 N or about 10 N. The Earth's **gravitational field strength** g is 10 N/kg. The weight W of an object is equal to its mass m multiplied by the gravitational field strength g:

  weight (newton, N) = mass (kilogram, kg) × gravitational field strength (newton per kilogram, N/kg)
  or $W = m \times g$ ($g = 10$ N/kg)

Falling objects

D–C

- All falling objects:
  - have the same acceleration of free-fall, 9.8 m/s² (about 10 m/s²)
  - have an acceleration that is independent of their mass.

- In a vacuum, all objects fall at the same rate.

- All objects moving in air experience air resistance.

- The drag on an object depends on its speed (drag increases as speed increases), shape and area.

- Air resistance increases as speed increases, until the air resistance is equal to the weight of the object. At this point, the resultant force on the object is zero. The object has reached a constant velocity known as **terminal velocity**.

Improve your grade

Falling objects

Foundation: Kerri is a professional skydiver. When she jumps out of an aeroplane, she initially accelerates, then falls at a steady speed. Explain why. *AO1* [3 marks]

Stopping distance

Thinking, braking and stopping

- Cars rely on friction to stop, using their brakes.
- The **stopping distance** of a car is made up of two parts:
  - **Thinking distance** is the distance travelled by the car as the driver reacts to apply the brakes. It can be calculated using the equation:

  thinking distance (m) = speed of car (m/s) × reaction time (s)
  - **Braking distance** is the distance travelled by the car while the brakes are applied before the car comes to a stop:

  stopping distance = thinking distance + braking distance

| | thinking distance 7 m | braking distance 8 m | total stopping distance 15 m |

At 10 m/s (22 mph)

thinking distance 14 m | braking distance 32 m | total stopping distance 46 m

At 20 m/s (45 mph)

thinking distance 21 m | braking distance 72 m | total stopping distance 93 m

At 30 m/s (70 mph)

Minimum stopping distances for a car.

Remember!
The car continues to move at a constant speed during the thinking distance.

G–E

Factors affecting stopping distance

- The thinking distance increases when:
  - the speed of the car increases
  - the driver's reaction time increases.
- The driver's reaction time increases when he or she:
  - is tired   – is under the influence of drugs or alcohol
  - is distracted (for example, by passengers or mobile-phone calls).
- The braking distance increases when:
  - the mass of the car increases   – the speed of the car increases.
- Braking distance also increases when there is reduced friction:
  - between tyres and road because of worn tyres
  - between tyres and road because of wet or icy road surface
  - because of worn brakes.
- Friction opposes motion. It can be a nuisance because it wastes energy. Friction can also be helpful. It helps us to walk and it also helps vehicles to slow down.
- Polished surfaces have less friction than rough surfaces. Lubricating two surfaces with oil will reduce the friction.

D–C

Explaining stopping distance

- Total stopping distance can be shown on a velocity–time graph.
- On the graph opposite, the car has constant velocity before the brakes are applied and a constant deceleration when the brakes are applied.
- The area under a velocity–time graph is equal to the distance travelled. Therefore:
  - area A = thinking distance
  - area B = braking distance
  - area A + area B = stopping distance.

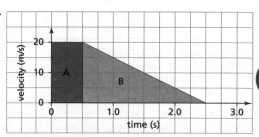

Thinking and braking distances can be found from a velocity–time graph.

B–A*

Improve your grade

Assessing stopping distances

Higher: Some roads have markings that indicate how far apart drivers should be from the car in front. On a particular road the markings are 30 m apart, the speed limit is 60 mph and drivers are advised to keep at least two markings apart.

Use this data and your understanding of stopping distances to evaluate whether a separation of two markings is safe in good weather.

AO3 [3 marks]

Momentum

Linear momentum

- All moving objects have **momentum**.
- The linear momentum of an object moving in a straight line is defined as:
  momentum (kg m/s) = mass (kg) × velocity (m/s) or momentum = $m \times v$
- A minus sign implies that an object is moving in the opposite direction (momentum is a vector quantity).

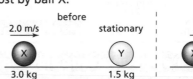

Remember!
The momentum of an object will be big if the object is travelling quickly and it has a large mass.

Collisions and conservation of momentum

- The diagram below shows a moving ball, X, about to collide with a stationary ball, Y.
- As they collide, each ball exerts an equal but opposite force on the other:
  - The force from ball Y slows down ball X.
  - The force from ball X makes ball Y move.
- When two objects collide, their total momentum remains constant as long as no external forces are acting. This is the **principle of conservation of momentum**:

  total momentum before collision = total momentum after collision

- In the diagram, ball Y gains the same amount of momentum lost by ball X.
- The balls stick together after the collision and move with a common velocity v: (3.0 x 2.0) + (1.5 × 0) = (4.5 × v)
  v = 1.33 m/s
- After the collision, the balls have a velocity of 1.33 m/s towards the right.

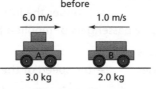
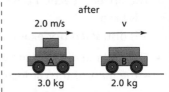

The balls before and after the collision.

- When objects travelling in opposite directions collide, the vector nature of momentum before and after a collision must be taken into account.
- In the example opposite, the 2.0 kg trolley changes direction and has a final velocity of 5.0 m/s to the right:
  (3.0 × 6.0) + (2.0 × −1.0) = (3.0 × 2.0) + (2.0 × v)
  v = 5.0 m/s

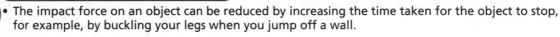

The trolleys before and after the collision.

Car safety and momentum

- The impact force on an object can be reduced by increasing the time taken for the object to stop, for example, by buckling your legs when you jump off a wall.

- During a car crash, seat belts, crumple zones and air bags help to reduce the rate of change of momentum of the driver or passenger.
  - Seat belts stretch slightly during a crash. This increases the time taken for the driver or passenger to stop, reducing impact force to a safe level.
  - Cars are designed to crumple. The car, and hence the driver, takes a longer time to stop. This reduces the impact force on the driver.
  - Air bags inflate suddenly during a collision. The stopping time is longer and the impact force on the driver is reduced.

Force and momentum

- If a force F acts on an object of mass m for a time t, the velocity of the object changes from u to v.

  force = mass × acceleration $\quad F = ma = m \times \dfrac{(v - u)}{t} \quad F = \dfrac{mv - mu}{t}$

- This can be described in words as: force (N) = $\dfrac{\text{change in momentum (kg m/s)}}{\text{time (s)}}$
  or force = rate of change of momentum

Improve your grade

Conservation of momentum

Higher: Caitlin likes to play snooker. She is learning to hit a red ball with the cue ball in such a way that the cue ball stops when it hits the red. Both the cue ball and the red ball have the same mass.

Use the principle of conservation of momentum to describe the velocity of the balls before and after the collision.

AO2 [3 marks]

Work, energy and power

Work and energy

- There is **work done** whenever a force is applied on an object and it moves:

  work done = force (newton, N) × distance moved in the direction of the force (metre, m)

  or $E = F \times d$

- Work done is measured in newton metres or **joules** (J). 1 joule is the work done when a force of 1 newton moves through a distance of 1 metre in the direction of the force.

  work done by a force = energy transferred

- For example, if a force of 40 N is exerted on a box as it moves a distance of 5.0 m along the floor, the work done by the force is calculated:

  $F \times d = 40 \times 5.0 = 200$ J

- In the above example, the work done on the box is transferred to heat between the box and the floor.

- In another example, a person of weight 400 N on an escalator climbs a vertical height of 6.0 m. Here, the work done against the weight is calculated:

  $F \times d = 400 \times 6.0 = 2400$ J

- In this example, the work done is transferred to **gravitational potential energy**.

Power

- **Power** is the rate at which work is done, or the rate at which energy is transferred.

- Power is measured in watts (W). 1 watt = 1 joule per second (J/s).

- Power is also measured in kW (1000 W) and MW (1 000 000 W).

- Power can be calculated using the equation:

  $$\text{power} = \frac{\text{work done (joule, J)}}{\text{time taken (second, s)}}$$

  or (where E is the work done and t is the time taken):

  $$P = \frac{E}{t}$$

- For example, if a crane lifts a weight of 1200 N through a vertical height of 40 m in 5 minutes (300 s), the power can be calculated:

  work done E by the crane = $F \times d = 1200 \times 40 = 48\,000$ J

  power $P = \dfrac{E}{t} = \dfrac{48\,000}{300} = 160$ W

Power at a constant speed

- The diagram below shows a car travelling at a constant speed v. The forward force provided by the car is equal to the drag F.

- All the work done by the car engine is transferred into heat.

- The output power of the car can be calculated:

  $$\text{output power} = \frac{\text{work done}}{\text{time}} = \frac{F \times d}{t} = F \times \left(\frac{d}{t}\right)$$

  speed $v = \left(\dfrac{d}{t}\right)$ therefore output power $P = Fv$

- The equation $P = Fv$ will only work if the car is travelling at a constant velocity.

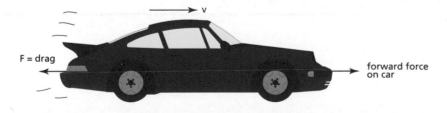

F = drag

forward force on car

A car travelling at a constant speed.

Improve your grade

Work done

Foundation: Explain what work is done when a window cleaner climbs a ladder. *AO1* [2 marks]

KE, GPE and conservation of energy

Kinetic energy

- **Kinetic energy** (KE) is energy of movement. All moving objects have kinetic energy.
- An object's kinetic energy is determined by its mass and speed (or velocity), using the equation:

  KE = ½ × mass × velocity²   or $KE = \frac{1}{2} \times m \times v^2$

- KE is measured in joules (J).
- Mass is measured in kilograms (kg).
- Velocity is measured in metres per second (m/s).

Gravitational potential energy

- The diagram below right shows an object of mass m moved from A to B through a vertical height h.
- The object gains **gravitational potential energy** (GPE). Potential energy is stored energy.
- GPE is calculated:

  GPE = work done

  GPE = weight × vertical height

  GPE (joule, J) = mass (kg) × gravitational field strength (newton/kilogram, N/kg) × vertical height (metre, m)

  or GPE = $m \times g \times h$

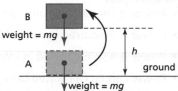

- The principle of conservation of energy states that energy can neither be created nor destroyed, it can only be transformed into different forms. For example:
  - When a cyclist brakes, kinetic energy is transformed into heat in the brakes and sound.
  - In a petrol car, the **chemical energy** of the fuel is changed into kinetic energy, heat and sound.
  - When a parachutist falls, his or her GPE is transformed into KE, heat and a bit of sound.

The object at B has gravitational (stored) potential energy.

- At the top of a diving board, a diver has GPE. As he or she falls, the GPE is converted into KE.
- In the previous example, the principle of conservation of energy can be used to determine the speed v of the diver just before entering the pool.

EXAM TIP

When working out speed in examples like that of the diver, you do not need to know the mass of the diver to complete the calculation. Since the mass appears on both sides of the equation, it cancels. The final speed v is independent of the mass of the object.

Braking distance and velocity of a vehicle

- A car of mass m is travelling at a velocity v. When the brakes are applied, the car decelerates and comes to rest over a braking distance d with a braking force F.
- The kinetic energy of the car is transferred into heat by the brakes:

  work done by the brakes = initial KE of the car

  $Fd = \frac{1}{2} mv^2$

- For a given car, the braking force and mass are constants. This means that braking distance is directly proportional to velocity².
- Doubling the velocity of the car quadruples the braking distance.

v (m/s)	5	10	20	40
Braking distance (m)	2	8	32	128

Improve your grade

Energy transfers

Foundation: Abeni enjoys slides at the playground.

Explain the energy transfers that occur as Abeni climbs up the slide and then slides down it. *AO2* [3 marks]

Atomic nuclei and radioactivity

Atoms and ions

- The nucleus of an atom contains neutrons and protons, together referred to as **nucleons**. Around the nucleus are electrons.
- A neutral atom has the same number of electrons and protons.
- An ion is a charged atom that has lost or gained electrons:
  - Adding electrons makes an atom a negative ion.
  - Removing electrons makes an atom a positive ion.
- Positive ions can be created by:
  - rubbing insulators together (the friction removes electrons from the atoms of one insulator)
  - heating a gas (thermal energy ionises the gas atoms; electrons of the atoms gain energy and fly off).

G–E

- The nucleus of an atom is represented as: $^A_Z X$
- X is the chemical symbol for the element.
- A is the **nucleon number** (or **mass number**), which is the total number of neutrons and protons in the nucleus.
- Z is the **proton number** (or **atomic number**) – the number of protons in the nucleus.
- The number of neutrons N in the nucleus is equal to (A – Z).

D–C

Isotopes

- **Isotopes** of an element are nuclei that have the same number of protons but a different number of neutrons.
- The isotopes of a particular element have the same chemical properties because they all have the same number of electrons.

B–A*

Radioactivity

- Some isotopes are unstable. Over time, the nucleus breaks up and emits a particle or wave in an attempt to become more stable. This is called radioactive decay.
- There are three types of nuclear radiations:
  - **Alpha (α) particles**: each alpha particle is identical to a helium nucleus, with two protons and two neutrons.
  - **Beta (β) particles**: each beta particle is an electron emitted from inside the nucleus.
  - **Gamma (γ) rays**: gamma rays are electromagnetic waves of very short wavelength.
- The radiations from radioactive materials carry energy and can cause **ionisation**.
- Ionisation is the process of removing electrons from atoms. This leaves behind positive ions.

- The table below summarises the main properties of different radiations.

Radiation	What is it?	Charge	Typical speed (m/s)	Mass	Ionising effect	Penetration
α	Helium nucleus ($^4_2$He)	+2	10 million	4	Strong	Stopped by paper, skin or about 6 cm of air
β	Electron ($^0_{-1}$e)	–1	100 million	0.00055	Weak	Stopped by a few millimetres of aluminium
γ	Short-wavelength electromagnetic wave	0	3.0×10^8	0	Very weak	Never completely stopped, but reduced significantly by thick lead or concrete

D–C

- Before it decays, a nucleus is known as the 'parent'. The nucleus left behind after decay is known as the 'daughter'.
- In alpha decay, two protons and two neutrons are removed from the nucleus. The proton number decreases by two and the nucleon number decreases by four.

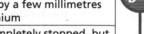

Remember!
In radioactive decay, the daughter nucleus is of a *different* element.

B–A*

Improve your grade

Radioactive decay

Higher: Alice has learnt at school that radioactive decay is random and cannot be predicted. Andrew argues that it must be possible to predict when radiation will occur, otherwise it would not be useable in the treatment and diagnosis of medical conditions.

Who is correct?

AO2 [2 marks]

Nuclear fission

How fission works

- Nuclear reactions produce energy:
  - In radioactive decay, the kinetic energy of the alpha or beta particles emitted from the nuclei can be used to generate electricity on a small scale.
  - **Fusion reactions** cause the energy generated by the Sun and the stars.
  - **Fission reactions** take place in a nuclear reactor to generate electricity on a large scale.
- Fission means 'splitting' the nucleus. In a fission reaction:
  - A slow-moving neutron is absorbed by a $^{235}_{92}U$ nucleus, creating an unstable $^{235}_{92}U$ nucleus.
  - The new nucleus splits into two smaller nuclei (**daughter nuclei**) and two or more fast-moving neutrons.
  - Energy is released as the kinetic energy of the daughter nuclei and the neutrons.

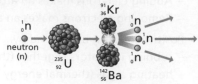

A fission reaction of the uranium-235 nucleus by a slow-moving neutron.

- Radioactive decay happens naturally – we cannot control it. However, scientists can trigger fission reactions by bombarding uranium-235 nuclei with neutrons.
- In a large sample of uranium, the fast-moving neutrons from the fission reactions can go on to split other uranium-235 nuclei. This is called a **chain reaction**.

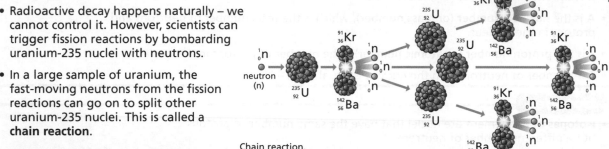

Chain reaction.

Nuclear power stations

- Nuclear power stations get their energy from nuclear fission reactions.
  - Most nuclear power stations use uranium or plutonium as fuel for the reactions.
  - A large amount of energy, in the form of the kinetic energy of the neutrons and the daughter nuclei, is released in fission reactions.
  - The kinetic energy is turned into heat and used to boil water to make steam.
  - The steam powers turbines.
- The chain reaction of the uranium nuclei in a nuclear power station is controlled to maintain a steady output of power.
- The daughter nuclei produced in fission reactions can remain active for thousands of years, so disposal of nuclear waste is a major concern.

- Reactors have several key components, some of which are there to control the chain reaction:
  - **Fuel rods** contain pellets of nuclear fuel in the form of uranium dioxide.
  - A **coolant** removes the thermal energy produced in the fission reactions in the reactor core, so it can be used to heat water to create steam to power generator turbines (in a water-cooled reactor).
  - The **moderator** surrounds the nuclear fuel rods and slows down the fast-moving neutrons. Slow-moving neutrons have a greater chance of reacting with uranium nuclei than fast-moving neutrons.
  - **Control rods** can be lowered into the reactor to absorb the neutrons and so slow down the fission reactions and control the chain reaction.

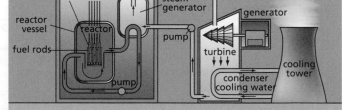

A nuclear power station.

Critical mass

- A chain reaction can only be sustained by large amount of uranium-235. In small amounts, too many neutrons will escape and not take part in fission reactions.
- The minimum mass of a fissile material required to sustain a chain reaction is known as its **critical mass**.

Improve your grade

Controlling neutrons

Foundation: Explain how the neutrons in a nuclear power station are controlled.

AO1 [2 marks]

Topic 5: 5.6, 5.7, 5.8, 5.9, 5.10, 5.11

Fusion on the Earth

Fission and fusion

- In a nuclear fission reaction, a neutron is used to split a uranium-235 nucleus to produce two radioactive daughter nuclei and two or more neutrons.

- The energy released in fission reactions is used by nuclear reactors to produce electricity.

- Nuclear fusion is a nuclear reaction in which two smaller, lighter nuclei join or fuse together to produce one larger nucleus. Fusion reactions produce a vast amount of energy.

- Extremely high temperatures are required for fusion reactions to take place.

- The energy source that keeps our Sun and other stars burning is the fusion of hydrogen and other lighter nuclei.

- The bottom diagram opposite shows the fusion of two isotopes of hydrogen, deuterium and tritium. The reaction produces a stable nucleus of helium and a neutron.

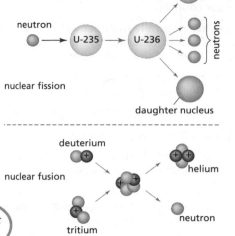

Remember!
Nuclear fission is the splitting of nuclei, nuclear fusion is the joining together of nuclei. Both reactions produce large amounts of energy.

Fission and fusion reactions. Can you spot any differences?

Cold fusion

- In 1989, scientists Stanley Pons and Martin Fleischmann announced that they had produced nuclear fusion at room temperature. This theory became known as **cold fusion**. They claimed their experiment had produced large amounts of thermal energy.

- The announcement gained worldwide publicity. However, Pons and Fleischmann were criticised by many scientists because they had not published enough technical details of their experiment for other scientists to reproduce their results.

- The majority of scientists now reject Pons and Fleischmann's theory of cold fusion because their theory could not be validated by reproducing their experiment.

Bringing fusion to the Earth

- Fusion reactions are more difficult to trigger than fission reactions because hydrogen nuclei (protons) are positively charged, and therefore repel each other.

- Increasing the speed at which the nuclei move improves the chances of a fusion reaction taking place.

- At temperatures around 10 million °C, hydrogen nuclei move rapidly enough to overcome the electrostatic repulsive forces and join together in fusion reactions. Nuclear fusion cannot take place at low temperatures and pressures.

- In order to create fusion, hydrogen nuclei must be heated to temperatures of about 100 million °C and contained by very strong magnetic fields produced by super-cooled electromagnets. It is difficult to create these conditions on Earth.

Remember!
Particles move faster at higher temperatures.

 How science works

You should be able to:
- analyse the reasons that Pons and Fleischmann (and scientists like them) were criticised by the wider scientific community

- understand why the theory of cold fusion has been largely rejected

- explain the reasons why it is necessary for other scientists to be able to reproduce an experiment like theirs.

 Improve your grade

The future of fusion

Foundation: Scientists hope that fusion reactors will soon be able to produce energy for human consumption.

a What fuel will be used in these reactors?
b Explain how nuclear fusion occurs.

AO1/AO2 [3 marks]

Background radiation

Radioactive rocks

- Everything around us is slightly radioactive. This is due to **background radiation**.
- Background radiation comes from a variety of sources and can be detected by a Geiger counter.
- Rocks are naturally radioactive because they contain small traces of radioactive isotopes. Granite is slightly more radioactive than other rocks because it contains higher levels of uranium atoms.
- Uranium nuclei decay naturally over time to produce **radon** nuclei. Radon is a colourless and odourless radioactive gas.
- Houses built over granite can trap radon gas.
- Exposure to radioactive radon can lead to lung cancer.
- Different areas of the UK have different levels of background radiation due to varying amounts of radioactive sources in that region.

G–E

> **Remember!**
> Background radiation is random, low-level radiation that is present everywhere on Earth.

Origins of background radiation

- Most background radiation comes from natural sources. These include:
  - *Cosmic rays:* energetic particles such as electrons, protons and neutrinos that come from the Sun and outer space. They penetrate the Earth's atmosphere to reach the surface. The danger from cosmic rays increases with altitude because there is less atmosphere to stop the radiation.
  - *Rocks:* rocks such as granite contain uranium, which decays to produce radioactive radon gas.
  - *Food:* all foods will have minute traces of radioactive nuclei.
- Humans have also had a small effect on background radiation, through:
  - nuclear power stations
  - fallout from previous nuclear weapons tests, explosions and accidents
  - radiation from equipment or waste from hospitals and industry.

D–C

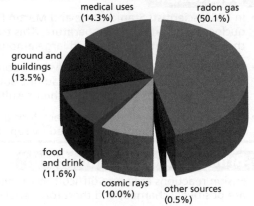

medical uses (14.3%)
radon gas (50.1%)
ground and buildings (13.5%)
food and drink (11.6%)
cosmic rays (10.0%)
other sources (0.5%)

A pie chart showing the origin of background radiation for an average individual in the UK.

Dangers of radon gas

- Radon gas rises from ground that contains granite. Radon gas is particularly dangerous if it remains trapped in the walls of buildings or under floorboards.
- Radon-222 is one of the isotopes of radon gas. It is a non-reactive **noble gas** and itself is not a health hazard.
- Radon-222 nuclei are produced by the decay of radium-226 nuclei.
- The nuclei decay by alpha emission with a short half-life of 3.8 days.
- The daughter nuclei of radon-222 also emit alpha particles.
- It is believed that radon-222 may cause cellular damage in the lungs.

B–A*

◉ Improve your grade

Differences in background radiation

Foundation: Joe and Fred are pen-pals who live in different parts of the UK. They have both been learning about background radiation at school and have discovered that the level of background radiation where they live is different.

Explain why this is.

AO1 [3 marks]

Uses of radioactivity

Industrial uses of radioactivity

- The three types of ionising radiation – alpha particles, beta particles and gamma rays – have practical uses in a number of industries.

- Irradiating food with gamma rays prolongs its shelf-life. The gamma rays kill off **microorganisms** on the food even after it has been packaged.

- In the metal industry, gamma rays are used to check the quality of welding or to detect cracks in metals.

- Beta particles are used in the water industry to detect leaks in underground pipes. A beta-particle emitting radioactive material, known as a **tracer**, is fed into the pipe. Above ground, a radiation detector detects increased levels of radiation.

> **Remember!**
> Radioactive materials emit alpha particles, beta particles and gamma rays.

G–E

- The thickness of paper can be controlled using a source of beta particles (see diagram).

- Pressure is applied by the rollers to control the thickness of the paper. A beta source is placed above the paper and a radiation detector is placed directly below the source and the paper.

- If the paper is thicker than required, the detector shows an increase in the number of beta particles recorded per unit time. A signal is then sent to the rollers to increase the pressure and reduce the thickness.

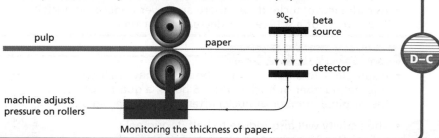

Monitoring the thickness of paper.

D–C

Domestic uses of radioactivity

- Smoke alarms are an example of radioactivity in use in the home. Most smoke alarms use a weak alpha source with a long half-life.

- The alpha particles from the source ionise the air, producing positive ions and electrons.

- The positive ions are attracted towards the negative terminal of the battery. The electrons travel in the opposite direction towards the positive terminal. The ionisation of the air produces a tiny current in the circuit.

- When smoke reaches the smoke alarm, it absorbs the alpha particles, causing a drop in ionisation of the air. The current and potential difference across the resistor drops.

- The electronic circuit detects the decrease in the potential difference and triggers the alarm.

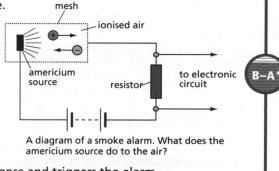

A diagram of a smoke alarm. What does the americium source do to the air?

B–A*

Medical uses of radioactivity

- Gamma rays have enough energy to kill bacteria, so they are used in hospitals to **sterilise** plastic equipment, such as syringes and bandages, that cannot be sterilised by heating (as metal equipment can).

- Sealing syringes in plastic bags then sterilising with gamma rays makes both the package and the content sterile. The sterilisation process minimises the risks of infection.

G–E

- Radioactivity can also be used to detect and treat cancer.

- To detect cancer, the patient is injected with a small amount of a radioactive tracer called technetium-99m, which emits gamma rays. It is carried around the body in the blood and builds up in the cancerous parts of the body. A gamma camera is used to detect and display the gamma rays that pass through the patient.

- To treat cancer, a technique called **radiotherapy** is used. A gamma source of cobalt-60 is used to target cancerous cells. By rotating an intense beam of gamma rays, most of the cancerous cells can be killed off with little damage to healthy cells.

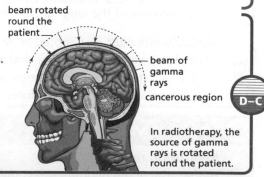

In radiotherapy, the source of gamma rays is rotated round the patient.

D–C

Improve your grade

Using radiation

Higher: Which type of radiation would be most suitable for use in a machine that controls the thickness of thin aluminium sheeting? Give the reasons for your answer.

AO2 [3 marks]

Activity and half-life

Radioactivity and half-life

- Radioactive decay is random and spontaneous, which means it cannot be predicted and that it is not affected by external conditions.
- Some nuclei are very unstable and decay very quickly. Others take a long time to decay.
- The **half-life** of an isotope is the average time it takes for half of the undecayed nuclei in a sample to decay. Half-life can be micro-seconds or thousands of years.

- The rate of decay of a source's nuclei is known as its **activity**.
- Activity is measured in **becquerel** (Bq) – 100 Bq means that 100 nuclei decay per second and that 100 alpha or beta particles are emitted per second.
- Activity is directly proportional to the number of undecayed nuclei in a source, i.e. activity doubles when the number of nuclei is doubled.
- Activity is inversely proportional to the half-life of the isotope.
- As radioactive nuclei decay, there are fewer undecayed nuclei. The activity of a source thus decreases over time.

Remember!
A source with a short half-life will have a large activity.

Exponential decay

- A sample with a half-life of 15 hours will have half the original number of nuclei after 15 hours, a quarter of the original number of nuclei after 30 hours, and so on.
- The activity will also reduce to half its original value after 15 hours, to a quarter of its original value after 30 hours etc.
- This type of decay is known as **exponential decay**.

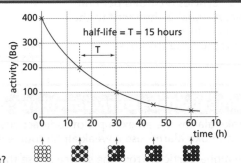

The activity of a source decreases with time. What happens to the activity after one half-life?

Dangers of radioactivity

- Alpha, beta and gamma radiations can damage tissue in the human body. If the damage is severe, the cells cannot repair themselves. They can also damage DNA, which can lead to cancer.
- Radioactive sources used in schools have low activities. However, when using them you should always take precautions such as wearing gloves, washing your hands and keeping the source pointed away from people.

Nuclear power and nuclear waste

- Some people believe that nuclear power is a good alternative to **fossil fuels**, which are fast running out and are non-renewable.
- Advantages of coal are that it is cheap and there are enough reserves to last 100 years. Disadvantages are that it causes pollution, it is costly to control and transportation can be expensive.
- Advantages of nuclear power are that the waste is compact, very little background radiation is produced and there is only a very low risk of a nuclear accident. Disadvantages are that power stations are costly to build and decommission, waste is radioactive for thousands of years and public perception is poor because of the long-term dangers posed by accidents.

- Nuclear waste must be disposed of carefully. Disposal can take place in the following ways:
  - Low-level waste (paper, tools, clothing) can be buried in shallow trenches in steel drums.
  - Intermediate-level waste is buried about 8 m under ground and shielded by water, concrete or lead.
  - High-level waste can only be stored deep underground or in special tunnels made under mountains.

Improve your grade

Measuring radioactive decay

Higher: The graph opposite shows the count rate of a radioactive sample over a period of 40 days.

Using the graph, estimate the half-life of the sample.

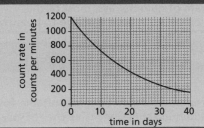

AO3 [2 marks]

Topic 6: 6.4, 6.5, 6.6, 6.7, 6.8, 6.9, 6.10, 6.11, 6.12

P2 Summary

An insulator can be charged with static electricity by friction, which results in the transfer of electrons. Like charges attract and unlike charges repel.

The dangers of static electricity can be minimised by earthing.

Static and current electricity

Electric current is the rate of flow of charge. Cells and batteries supply direct current (d.c.), which is the movement of charge in one direction only.

A variable resistor can be used to change the resistance in a circuit.

The potential difference across a component can be measured using a voltmeter connected in parallel with the component.

The current can be measured using an ammeter connected in series.

Controlling and using electric current

Electrical power = current × potential difference

Energy transferred = current × potential difference × time

Motion and forces

Displacement, velocity, acceleration and force are vector quantities. They all take into account direction as well as magnitude.

Acceleration = change in velocity / time

Acceleration can be determined from the gradient of a velocity–time graph.

If the resultant force on a body is zero it will remain at rest or continue to move at the same velocity.

If the resultant force acting on a body is not zero, it will accelerate in the direction of the resultant force according to the equation force = mass × velocity.

The stopping distance of a vehicle is made up of the thinking distance and the braking distance.

Momentum, energy, work and power

Momentum = mass × velocity

The rate of change of momentum can be used to explain how safety features of vehicles, such as air bags and crumple zones, reduce the force.

Work done = force × distance moved in the direction of the force

Power is the rate of doing work.

Gravitational potential energy = mass × gravitational field strength × vertical height.

Kinetic energy = half mass × velocity squared

Energy is conserved in energy transfers.

An alpha particle is equivalent to a helium nucleus, a beta particle is an electron emitted from the nucleus and a gamma ray is electromagnetic radiation.

Nuclear fission and nuclear fusion

The fission of U-235 produces two daughter nuclei and two or more neutrons.

The chain reaction is controlled in a nuclear reactor by the action of moderators and control rods.

Heat energy from the chain reaction is converted into electrical energy in a nuclear power station.

Nuclear fusion is the creation of larger nuclei from smaller nuclei, accompanied by a release of energy, and is the energy source for stars.

Radioactivity has many uses, including in smoke alarms, sterilisation of equipment, and tracing and gauging thickness. It can also be used to diagnose and treat cancer.

Advantages and disadvantages of using radioactive materials

Issues associated with nuclear power include the lack of carbon dioxide emissions, risks, public perception, waste disposal and safety.

The activity of a radioactive source decreases over time. The half-life of a radioactive isotope is the time taken for half the undecayed nuclei to decay.

Page 4 Naming new species

Higher: Scientists exploring an area of the South American rainforests discovered several species of frog. They gave each a new binomial name.

Explain why classifying organisms in this way is important. *AO1* [3 marks]

It identifies new species.

> **Answer grade: C.** The answer only gives one reason. For full marks, mention two other reasons why classification is important. For example, it enables scientists who speak different languages to communicate about the frogs, it prevents confusion about different names for them, and it shows that this area of rainforest has an increased biodiversity as new species have been found there.

Page 5 Polar bear adaptations

Foundation: Explain how the fur of the polar bear is a good adaptation for surviving in the Arctic. *AO2* [4 marks]

It is thick, which keeps it warm and because it is white it is camouflaged.

> **Answer grade: D/E.** This answer lists the functions of the fur but does not explain *how* they would enable the bear to survive in the Arctic. To raise this to a grade C, explain that the thick fur acts as insulation by trapping air. You should also explain why the polar bear needs to be camouflaged (to hide from its prey so it can hunt effectively). Many students believe that a polar bear's thick fur keeps it warm because it stops the cold from the outside reaching the bear's skin. In fact, the reverse is true – the fur stops heat from the polar bear escaping. Remember – heat always travels from where the temperature is higher to where it is lower.

Page 6 Human evolution

Higher: Scientists believe that humans evolved from ape-like animals. Millions of years ago the Earth became drier and forests were replaced by grasslands. Before this change all apes walked on four feet. Afterwards, populations of apes emerged that walked upright on two feet and were able to see further across the grassland.

Use your knowledge of natural selection to explain why apes evolved in this way. *AO1* [3 marks]

The apes that stood upright could see predators coming and so could run away and not get eaten. They then survived to pass on their genes to offspring which also walked upright.

> **Answer grade: B.** The answer correctly identifies two points about natural selection. However, in order to achieve full marks you should mention that the apes show variation: some were more upright than others. You should also include the fact that the apes passed on their genes to their offspring through reproduction.

Page 7 Hair-colour inheritance

Foundation: Katy has red hair. Both her parents have brown hair. Explain how Katy inherited red hair, even though her parents do not have it. *AO2* [3 marks]

Katy got the gene for red hair from her parents.

> **Answer grade: E.** To achieve full marks, explain why Katy's parents do not have red hair. The red-hair allele must be recessive. Her parents both have the red-hair allele but their other allele (brown hair) masks it. Katy received two red-hair alleles from her parents during fertilisation, so Katy has two red-hair alleles (her genotype) and so therefore has red hair (her phenotype). Notice that red hair is referred to as an allele in these comments, not a gene. The red-hair allele is a type of hair-colour gene. Always use the correct terminology.

Page 8 Mendel's experiments

Foundation: Mendel discovered that the colour of pea flowers was controlled by a single gene. The red allele (R) was dominant, and the white allele (r) was recessive. He crossed a pure-breeding red-flowered plant with a pure-breeding white-flowered plant. What would be the genotype and phenotype of the pea plants that were produced? Use a genetic diagram to help you. *AO2* [4 marks]

Cross: RR × rr		
	R	R
r	Rr	Rr
r	Rr	Rr

> **Answer grade: D.** The Punnett square is correctly drawn and the correct genotypes of both parents and offspring are shown. However, the question asks for the phenotype of the offspring, which the Punnett square alone does not show. For full marks, include a genetic diagram with this information. In answering this type of question, it is important to understand the difference between genotype and phenotype. Here, the genotype of the offspring is Rr. This shows the alleles of the offspring. The phenotype is the physical characteristic. In this case, all the offspring would have red flowers.

Page 9 Sickle cell disease inheritance

Higher: Ben carries the mutated gene for sickle cell disease in his sperm. Janet is neither affected by the disease, nor is she a carrier of it.

What is the probability of their children having sickle cell disease? Show how you worked out your answer. *AO2* [3 marks]

		Ben	
		D	d
Janet	d	Dd	dd
	d	Dd	dd

Probability of children having it is 2 in 4.

> **Answer grade: C.** This would only achieve one of the three available marks. Ben's genotype of Dd is correctly identified, as he is a carrier of the allele, but sickle cell allele is recessive. This means that Janet would have the genotype DD. This would result in none of the children having the disease. Therefore, the probability is zero.

Page 10 Cooling down

Foundation: Sandeep is running a marathon. When he gets too hot he begins to sweat. Explain how sweating cools the body down. *AO2* [2 marks]

It evaporates.

Answer grade: E. Only 1 mark would be awarded here for mentioning evaporation. In order to bring this answer up to a C grade, you need to explain how the sweat evaporating will decrease body temperature. Many students believe that sweating cools down the body simply because having water on the skin makes you cooler. Remember, there is more to it than that: the sweat evaporates from the skin. To do this it needs heat energy, which it takes from the skin. This lowers the temperature of the skin and cools you down.

Page 11 Paralysis

Foundation: An injury that results in a broken spine may cause a person to be paralysed. Explain why. *AO1* [3 marks]

The nerves in the spine might be broken.

Answer grade: E. This answer only makes one suggestion and thus only achieves 1 of the 3 possible marks. To gain full marks, explain that the spinal cord is part of the central nervous system and contains many nerves that send messages from the brain to all parts of the body. When these impulses meet the muscles, they contract. If the nerves in the spinal cord are broken then impulses cannot cross them, so muscles will not respond.

Page 12 Synapses

Higher: Explain how an impulse travels over a synapse. *AO1* [3 marks]

Chemicals are released into the synapse and travel across it.

Answer grade: C. There is not enough detail in this answer to earn more than 1 of the 3 marks available. For an A grade, you need to describe every step in the process, e.g. the impulse reaches the end of the neurone, which triggers the release of chemicals called neurotransmitters into the synapse. This diffuses across the synapse and stimulates a new impulse in the next neurone.

Page 13 Calculating BMI

Higher: Susan is a normal-sized adult woman. She is 1.6 m tall and has a mass of 82 kg. Work out her BMI by using this equation:
BMI (kg/m^2) = mass (kg) / height $(m)^2$

Use the graph to explain how much of a risk Susan has of developing Type 2 diabetes. *AO3* [2 marks]

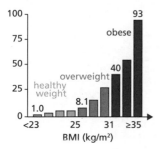

Her BMI is 32. She is obese so therefore has an increased risk of developing Type 2 diabetes.

Answer grade: B. The answer 32 has been calculated correctly, as has the fact that her risk of developing Type 2 diabetes is increased. However, the units for BMI have not been included (kg/m^2) so a mark has been lost here. It is also good practice to include the working for any calculations you do.

Page 14 Weedkiller

Higher: Explain how plant hormones can be used as weedkillers that kill weeds only, without affecting the crops. *AO1* [3 marks]

The herbicide contains plant hormones that stimulate the growth of plant stems. Because the rate of root growth does not keep pace with the stem, the roots are not able to absorb enough water to support the growing plant and it dies.

Answer grade: C. This would fail to achieve anything higher than a C grade, as the answer has not explained why weedkillers only affect the weeds. For full marks, explain that weeds are usually broad-leaved and absorb more herbicide than narrow-leaved crop plants.

Page 15 Drink driving

Foundation: Explain why it is dangerous to drink alcohol and drive. *AO1* [3 marks]

Alcohol slows down your reaction times.

Answer grade: E. This answer would gain only 1 of the 3 available marks because it is incomplete. For full marks, explain why slow reaction times would be dangerous to a driver. Including scientific words in your answer would also show evidence of a C grade level of understanding. For example, mentioning that alcohol is a depressant. Recall also that even a small amount of alcohol can affect reaction times, but how much a person is affected depends on their age, body mass and gender (whether they are male or female).

Page 16 Ethical concerns

Higher: Jon and Margaret are both on the liver transplant waiting list. Jon is a 36-year-old father of three. He has cirrhosis caused by alcohol abuse. The cause of Margaret's liver failure is unknown. She is 83 years old. Doctors have to decide who receives the next available liver.

Choose who this should be and argue your case. *AO2* [5 marks]

Jon because even though he has abused alcohol in the past, he is young and has a family to support. With help he can overcome his alcohol addiction and go on to lead a healthy life.

Answer grade: C. This answer puts forward several good reasons why Jon should receive the liver. However, to achieve the full marks available and an A grade you should demonstrate an awareness of Margaret's case, too. Having outlined both cases, a conclusion should be drawn.

Page 17 Malaria prevention

Higher: The spread of malaria can be prevented cheaply and easily by covering beds with mosquito nets. Explain how this technique works. *AO2 [3 marks]*

Mosquitoes cannot get through the net to bite the person so the malaria cannot be passed on.

> **Answer grade: B.** A bit more detail is needed to achieve full marks. Remember – it is only an *Anopheles* mosquito carrying the malaria pathogen that needs to be prevented from biting the person. Try to include as much scientific vocabulary as you can, for example the terms vector, host and protozoa. Remember, the vector is the animal carrying the pathogen. The vector cannot get the disease. The host is the animal that the pathogen is passed on to and is susceptible to the disease, e.g. a human.

Page 18 Interdependence

Foundation: Plants and humans are interdependent.

Explain what this statement means, giving an example. *AO1 [2 marks]*

Being interdependent means that plants and animals need each other for survival. An example is that animals eat plants.

> **Answer grade: E.** This answer shows an understanding of the term interdependent, but has not applied it to an example. Not all animals eat plants. Lions eat plant-eaters (herbivores), but without plants there would be no prey for them to eat. Plants are at the start of every food chain.

Page 19 Metal recycling

Higher: Evaluate the use of metal recycling as an alternative to landfill. *AO3 [4 marks]*

Metals are extracted from ores, which are non-renewable so our supplies of metal are running out. If we put metal in landfills it is lost forever, but if we recycle metal then we can use it again and we will not run out. Also, we are running out of space to put landfills and metal is non-biodegradable so it will stay in the ground.

> **Answer grade: B.** This answer has covered nearly all the major points so would achieve a B grade. To push it up to an A, mention the chances of water pollution from burying metal in the ground. Also, discuss the drawbacks of recycling – such as the energy required to collect, sort and melt down the recycled metal – to present a balanced evaluation.

Page 20 Pond pollution

Foundation: A farmer grows wheat on his field. He notices that the water in a small pond next to the field has turned green and the fish have all died.

Explain what has happened. *AO2 [5 marks]*

Chemicals from fertiliser have run off into the pond. This has caused the rapid growth of plants and algae which use up all the oxygen in the water, so the fish will die as they have no oxygen.

> **Answer grade: D.** Two correct explanations have been given, but there is also a mistake in the answer – it is not the plants that use up oxygen in the water. To push this up to a C grade, explain that the bacteria use up the oxygen, not the plants. When the plants die they get decomposed by the bacteria. It is the bacteria that use up the oxygen in the water and cause other organisms in the pond to die.

Page 21 Carbon dioxide levels

Foundation: Scientists around the world are monitoring the amount of carbon dioxide in the atmosphere and collecting data such as that shown in the graph below.

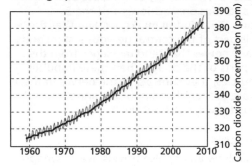

Describe the trend in the graph and suggest a reason for it. *AO3 [3 marks]*

The graph shows that the amount of carbon dioxide in the atmosphere has increased since 1960. This is because we have more cars on the roads.

> **Answer grade: D.** The trend has been described correctly, but the reason is not fully explained. The answer should explain that cars burn fossil fuels (petrol) in the engine, which releases carbon dioxide. You could also mention that it is not just cars that release carbon dioxide. Other vehicles such as lorries, trains and aeroplanes do, too. Also the burning of fossil fuels in power stations and deforestation by burning forests releases a lot of carbon dioxide.

Page 23 Looking at cells

Foundation: The image below shows some cells as seen down a microscope.

Are these animal or plant cells? Explain how you decided. *AO2* [2 marks]

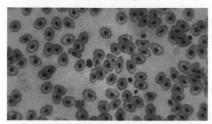

Animal cells because they have a nucleus.

Answer grade: E. The correct cell has been identified but the reason given is not sufficient – all cells have a nucleus. For full marks the answer should mention that this cell has no chloroplasts, vacuole or cell wall so therefore cannot be a plant cell. Remember, plant cells have both a cell membrane and a cell wall.

Page 24 Cell division

Higher: Why must the two strands in the DNA molecule separate during cell division? *AO2* [2 marks]

So each cell has one DNA strand.

Answer grade: U. This answer is incorrect and would not achieve any marks. To gain both marks available, explain that the strands need to separate in order to create two new DNA molecules. The DNA in cells consists of two complementary strands with their nucleotides held together by weak hydrogen bonds. When a cell is about to divide, these bonds break and the strands separate to enable new nucleotides to join up with each strand. This results in two new DNA molecules being formed – one for each cell.

Page 25 GM protest

Foundation: Fifty people turned up to protest on a piece of land that was growing GM wheat.

Outline the concerns they have. *AO1* [3 marks]

They don't think it is natural and are worried that it could harm the health of the people who eat it.

Answer grade: D. This answer gives two reasons and would therefore achieve only 2 of the 3 available marks. To improve to a C grade, give another reason why the people might be protesting. Remember, one of the things that concerns people the most about GM crops is that they will be able to breed with normal crops that are nearby (cross-pollination). This could cause changes – for example, the crop could produce a toxin in its leaves, which could spread to other plants, where it may disrupt food chains.

Page 26 Zygote to foetus

Foundation: Explain why mitosis is an essential process in the formation of a baby from a fertilised egg. *AO1* [3 marks]

Mitosis is needed to produce the egg and sperm cells that join to form the fertilised egg.

Answer grade: U. This answer is incorrect as the terms mitosis and meiosis have been mixed up, and would thus gain no marks. The correct answer should state that a fertilised egg is one cell (called a zygote), which will divide by mitosis to form the many cells that make up a baby. Remember that meiosis is the process that produces gametes (sex cells). An easy way to remember is that meiosis contains an 'e' for egg and an 's' for sperm.

Page 27 Cloning Daisy

Foundation: Daisy the cow produces the most milk in her herd. Her farmer is considering using her eggs for embryo transplants.

Explain to him why cloning her might be an even better idea. *AO2* [3 marks]

If you clone her then you will get another cow exactly like her who will produce lots of milk.

Answer grade: D. The answer is correct but is not detailed enough. To push this up to a C grade, explain why in this case cloning is better than embryo transplants. Embryo transplants involve using the eggs from the animal with desired characteristics and fertilising them with sperm from a male animal. The resulting embryos will have a mix of genes from both parents, and not necessarily the desired ones that you were trying to achieve. Cloning animals will produce offspring that are *identical* to the parent.

Page 28 Future applications of the HGP

Higher: In the future we may be able to use an individual's genome to calculate the likelihood of them developing diseases such as cancer.

Evaluate this potential application of the Human Genome Project. *AO2* [4 marks]

If a person's genome shows they might get a disease, they might not be able to get a job because the employer might be worried about them getting ill and taking time off work. But, it might be useful to know this as you can be told what symptoms to look out for and be able to get treatment quickly before the disease gets too bad.

Answer grade: B. This is a good answer, which discusses both the potential benefits and drawbacks to this technology. However, a little more detail would help to push this up to an A grade – for example, another benefit might be that doctors may be able to help with treatment such as stem cell therapy before the disease develops at all. Remember that when a question asks you to 'evaluate' something, it is asking you to provide both sides of the story – in this case both the benefits and drawbacks of this new technology. Try to get the word 'but' into the answer.

Page 29 Sickle cell mutation

Higher: Haemoglobin is a protein that is found in red blood cells. Sickle cell disease is a genetic illness where the red blood cells have a distorted shape. It is caused by a mutation in the gene for haemoglobin that converts a GAG codon into a GTG.

Explain how this causes a change in the shape of the hameoglobin molecule. *AO1* [4 marks]

The change in the DNA will result in the wrong amino acids being put into the haemoglobin protein, which will affect its shape.

> **Answer grade: C.** Although this answer is correct, there is not enough detail to gain higher than a C grade. To improve this, explain your understanding of the process of protein synthesis. For example, how could the change in the base bring about the wrong order of amino acids in the protein?

Page 30 Sex cell mutation

Foundation: A mutation occurs in the DNA of an organism's sperm cells.

Explain how this mutation will be passed to its offspring. *AO1* [2 marks]

The mutation will be passed onto offspring by fertilisation.

> **Answer grade: D.** Fertilisation has been mentioned as the process by which the mutation is passed on, but the answer lacks detail. For full marks, explain that if there is a mutation present in the DNA inside the nucleus of a sex cell then it will be present in the DNA of the zygote formed. As this one cell will undergo mitosis to form the organism, this mutation will be present in the nucleus of every cell in the adult organism.

Page 31 Understanding enzymes

Higher: Biological washing tablets contain enzymes that help break down stains found on clothes. Mark's clothes were very dirty so he decided to wash them at a much higher temperature than recommended on the instructions. The stains did not come off his clothes.

Use your understanding of enzymes to explain why. *AO2* [3 marks]

The high temperature killed the enzymes so they did not work and did not break down the stains on the clothes.

> **Answer grade: U.** This answer would not get any marks, as the science is incorrect. Enzymes cannot die as they are not living! At high temperatures, they change shape, or denature. The answer should have included this information and explained why denaturing makes enzymes less effective.

Page 32 The race

Foundation: You take part in a race. At first you sprint off feeling full of energy. However, half way through your legs start to ache and you have to stop. Explain why this happened. *AO1* [3 marks]

My heart could not beat fast enough and my lungs could not breathe fast enough to get oxygen to my leg muscles, so they could not carry out respiration and make energy for my leg muscles to work.

> **Answer grade: E.** This answer correctly states that the lungs and heart could not work fast enough to get the required levels of oxygen to the leg muscles, but respiration would not stop – anaerobic respiration would take over instead. The answer should explain this and then go on to link this to why the leg muscles would start to feel sore. Some of the energy stored in glucose is released without oxygen, a product of which is lactic acid, which is the chemical that makes the legs ache.

Page 33 Stroke volume

Higher: Ben is a professional runner. Through training he has been able to increase the stroke volume of his heart.

Explain how this enables Ben to run faster. *AO2* [5 marks]

Increasing stroke volume will also increase cardiac output. This means that more blood is being pumped by the heart and so more oxygen and glucose are able to get to his respiring muscle cells in his legs. This will increase the rate of respiration, enabling him to release more energy.

> **Answer grade: B.** This is a good answer that would achieve 4 marks out of a possible 5. What is missing is an explanation of why an increased rate of respiration will allow Ben to run faster. If muscle cells are able to release energy quickly, they will not just be able to contract more quickly, but more strongly as well.

Page 34 Weed removal

Foundation: Suzanne has weeds growing in her flowerbeds and notices that her plants are not growing. Explain why this is. *AO2* [3 marks]

The weeds are taking water from the plants which they need to grow.

> **Answer grade: E.** To improve the grade, this answer needs to explain how a lack of water leads to a low rate of photosynthesis, which will lead to poor growth. It should also mention that weeds will cover the plants, which reduces the amount of light reaching their leaves. This will also affect the rate of photosynthesis in the plants.

Page 35 Preserving fish

Foundation: Covering fish with salt and leaving it causes the fish to dry out and become hard.

Use osmosis to explain why this happens.
AO2 [3 marks]

The salt sucks the water out of the fish by osmosis so it dries out.

Answer grade: E. This answer would gain 1 mark for offering a basic explanation. However, it does not use the process of osmosis to explain what is happening. You must be able to apply osmosis to many different situations. In this case, the salt surrounding the fish will have a much lower water concentration than the fish cells. This means that there is a net movement of water out of the fish cells into the salt by osmosis. This will cause the fish to lose water and become dry.

Page 36 Growth chart

Higher: Adam is four years old and has a weight of 13 kg.

Use the growth chart below to comment on his weight. *AO3 [3 marks]*

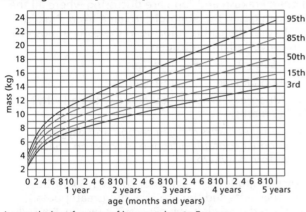

A growth chart for mass of boys aged up to 5 years
(Source: WHO Child Growth Standards)

He is underweight.

Answer grade: C. The answer is correct, but does not give enough information to score higher than a C grade. To bring it up to A* standard, percentiles need to be explained. A four-year-old with a mass of 13 kg is on the 3rd percentile. This means that their mass is equal to or greater than 3% of children of the same age; 97% of children at that age are heavier than that child, so Adam is below average weight for a four-year-old.

Page 37 The heart

Foundation: Why is the heart classed as an organ?
AO1 [2 marks]

It is made up of lots of different types of cell.

Answer grade D. In order to achieve full marks, the answer needs to explain what we mean by an 'organ'. Organs are made up of different types of tissue and have a particular function. For example, the function of the heart is to pump blood. It contains cardiac muscle tissue that can contract and relax, as well as nervous tissue and ligaments to hold the different tissues in place.

Page 38 Capillaries

Higher: Capillaries are blood vessels that are very thin (0.005 mm in diameter) and whose walls are only one cell thick.

Explain how these features enable them to carry out their function. *AO2 [4 marks]*

Being so thin they can reach all parts of the body, so all cells are close to a capillary. This also increases the pressure in a capillary bed so tissue fluid is forced out. Because they are one cell thick, the substances they carry can easily pass out.

Answer grade: B. This answer gives functions for each of the features described in the question, but it lacks detail and does not show a thorough understanding of the science. For full marks, explain that the pressure in the capillary bed is a result of the blood being at high pressure in the artery that supplies the bed with blood, and the narrow diameter of the capillary. Remember, at capillary beds there is an exchange of substances from the blood to the tissues, and waste products from the tissues to the blood.

Page 39 Enzyme experiment

Foundation: Dipesh mixed some starch with amylase in a beaker and left the mixture in a water bath at 37 °C. After 30 minutes he tested the mixture to see if there was any starch present.

Predict what Dipesh will find and give a reason for your answer. *AO3 [3 marks]*

There would be no starch because the amylase is a carbohydrase enzyme, which helps breaks down starch.

Answer grade: C/D. This is almost a complete answer, but it would have better to add that the starch would have been broken down into maltose and glucose. Remember that enzymes are only *catalysts* in the breaking down of food. The starch would eventually break down by itself, it would just take a lot longer.

Page 40 Probiotics vs prebiotics

Higher: Explain why some people believe that prebiotics are more likely to affect the health of the gut than probiotics. *AO1 [5 marks]*

Probiotics contain live bacteria, which need to get to the intestines to have an effect. This means they have to survive storage and their journey through the digestive system. They also have to compete with the bacteria already present in the gut. Prebiotics contain sugar (food) for the 'good' bacteria in the gut, so this is more likely to get to the bacteria and have an effect.

Answer grade: A. This answer contains all the information needed so would achieve an A grade. To bring it up to an A* standard, more detail could be added, for example, stating that acid in the stomach may kill the bacteria in probiotics.

Page 42 The changing atmosphere

Foundation: Explain how the early atmosphere changed to become the atmosphere we have today. *AO1* [4 marks]

The atmosphere used to be made by volcanoes but now there is much more oxygen.

Answer grade: E. The answer gains a mark for noting that there is more oxygen in the modern atmosphere, but there is no explanation of how changes occurred. For full marks, list the gases in the early atmosphere, explain that water vapour condensed to make seas and carbon dioxide dissolved into it, describe how photosynthesis increased the oxygen and show your understanding by giving the composition of today's atmosphere.

Page 43 Measuring gases

Higher: The percentage of gases in the atmosphere has remained the same for thousands of years, but recently there have been small changes.

Explain how we can tell when the atmosphere changes and suggest some reasons for the changes. *AO1* [4 marks]

Scientists have instruments like satellites that can measure accurately the amount of gases in the atmosphere and detect when they change. The gases are changing because of burning fossil fuels and volcanoes.

Answer grade: C. For full marks, specific detail and more examples are needed. Explain that burning fossil fuels for energy releases carbon dioxide. There is more deforestation to provide land for agriculture, which means that less carbon dioxide is absorbed by trees for photosynthesis. Some types of farming release nitrogen oxides or methane. Remember – always give specific examples rather than making general statements.

Page 44 Igneous rock formation

Foundation: Scientists can gain a great deal of information by looking at the properties of rocks.

What is igneous rock, and what does the appearance of igneous rock tell scientists about the way it has formed? *AO1* [5 marks]

Igneous rock used to be lava. There are two different types with different sized crystals.

Answer grade: E. Two marks are awarded for describing igneous rock, but there is no explanation of what this tells scientists. Improve to grade C by explaining that igneous rock forms when molten lava or magma cool and solidify. Say what the different sizes of crystals tell us about where the rock formed.

Page 45 Uses of limestone

Foundation: The limestone industry plays an important part in the British economy but it also causes environmental problems.

Explain what limestone is and how it is used. Discuss whether you think that the damage to the environment means we should no longer use it. *AO2* [4 marks]

Limestone is a sedimentary rock. We should still use it because you have to have it to make concrete and other things. You have to take more care about quarrying it so it doesn't do so much damage.

Answer grade: E. This explains what limestone is but does not fully describe its uses or how quarrying damages the environment. To improve to grade C, give specific examples of environmental damage, like dust or road damage. Give more than one example of the uses of limestone. Make a suggestion of how more care could be taken in quarrying, such as repairing the damage to animal habitats after quarrying is finished.

Page 46 Rearranging atoms

Higher: Prakash and John are discussing where all the nitrogen gas in the atmosphere came from. Prakash thinks that the carbon dioxide in the early atmosphere was converted into nitrogen.

$$CO_2 \text{ (g)} \longrightarrow N_2 \text{ (g)}$$
carbon dioxide   nitrogen

John disagrees. Who do you agree with and why? *AO2* [3 marks]

John. The reactants contain carbon and oxygen. The product contains nitrogen so they can't have come from nowhere.

Answer grade: B. The answer correctly identifies the reactants and the product, but greater detail is needed to explain why Prakash is wrong. For full marks, explain that in a chemical reaction the atoms are rearranged to form the products. Atoms cannot change from one type to another, so carbon and oxygen atoms cannot become nitrogen atoms.

Page 47 Limestone reactions

Foundation: Explain the reaction that occurs when limestone is heated and why this makes it more useful. *AO2* [4 marks]

Limestone breaks down when you heat it. It makes calcium oxide, which is used for lots of things.

Answer grade: E. The answer gains 2 marks for explaining what happens when you heat it and naming one of the products. However, it lacks specific examples. To improve to grade C, give the chemical name of limestone and named examples of its uses such as in making glass. You could also gain marks by writing an equation for the reaction.

Page 48 Neutralising acid spills

Higher: Commercial spill kits are available to help deal with acid spills in the workplace. They often contain sodium carbonate.

Explain how such kits would help to prevent damage from a spill of sulfuric acid. Include any reactions that occur in your answer. AO2 [4 marks]

Sodium carbonate can neutralise the acid so it isn't dangerous. It reacts acid + base ⟶ salt + water.

Answer grade: C/D. Two marks would be gained for mentioning neutralisation and giving a basic equation. However, the answer does not fully explain the properties of sodium carbonate that enable it to neutralise the acid. For full marks, explain that sodium carbonate is a base/alkali that will neutralise the sulfuric acid, give the full names in the reaction equation, describe the product (sodium sulfate) and explain why it is safe (not harmful/corrosive). If a question uses specifically named chemicals (sodium carbonate, sulfuric acid), use these names in your answer rather than making general statements.

Page 49 Creating chlorine

Foundation: Describe how you could produce a test tube full of chlorine from hydrochloric acid. AO2 [4 marks]

You would make electricity go through it and it would bubble up.

Answer grade: E. The answer gains 1 mark for explaining the need for electricity, but it does not give enough detail on the process of the experiment. You should explain what apparatus you would use to pass electricity through the acid (a beaker with two electrodes) and how you would collect the chlorine gas (from the positive electrode in an upturned test tube filled with water). You could draw a diagram to show you fully understand the process.

Page 50 Pricing metals

Higher: Aluminium is the most common metal in the Earth's crust, yet 100 g of aluminium costs about ten times more than 100 g of iron. Explain the reasons for this difference in value. AO1 [5 marks]

Aluminium is more expensive to extract than iron. You have to use electrolysis.

Answer grade: C. This answer does not explain why aluminium must be extracted by electrolysis or what the method of extraction for iron is. Explain that aluminium is more reactive than carbon and so cannot be extracted by heating with carbon. Iron is less reactive than carbon and so you can use this method, which is much cheaper than electrolysis. Compare the two metals in the answer.

Page 51 The right metal

Higher: A town council is trying to decide on the best metal to choose as a building material for a bridge over the local river.

Explain what factors they should take into account and suggest a suitable metal. Give reasons for your choice. AO2 [5 marks]

They should consider economic and environmental factors. It would be best to use a recycled metal. Aluminium is a good choice because it does not corrode.

Answer grade: B/C. This answer addresses the main points but does not explain or give examples. The answer should include the cost of the metal and the cost of upkeep of the bridge. Aluminium is more expensive because it is costly to extract, but has lower maintenance costs because it does not need to be painted. Recycled metal is best because it reduces damage to the environment from mining and from energy use in extracting the metal. Make sure that you always give specific examples of the type of damage.

Page 52 Making crude oil

Foundation: Explain how and why kerosene is made from crude oil. AO2 [4 marks]

Kerosene is made by fractional distillation. You heat up the oil and it turns into vapour, then it separates into fractions and one of them is kerosene.

Answer grade: D. Two marks would be earned for mentioning fractional distillation and how this works. However, the answer does not include any explanation of how the fractions separate and has not offered any uses for kerosene. For full marks, include the condensation of the fractions at different temperatures because of their different sizes and mention why it is made, such as its use as an aircraft fuel.

Page 53 Causes of climate change

Higher: Discuss the evidence that the activities of humans have resulted in climate change. AO2 [5 marks]

Humans have burned fossil fuels to provide energy. This has meant that a lot of carbon dioxide has been produced. Carbon dioxide is a greenhouse gas which traps energy on the Earth. So this has caused global warming.

Answer grade: B. This answer includes some important points about global warming but it does not fully discuss the evidence. To improve to grade A, discuss the correlation between the carbon dioxide levels in the atmosphere and the global temperatures over a number of years. Alternative ideas about why global temperatures have risen should be mentioned. It is a common misconception that all the carbon dioxide produced by burning hydrocarbons remains in the atmosphere. In fact, most is absorbed by the oceans or used in photosynthesis.

C1 Improve your grade Chemistry in our world

Page 54 Energy from ethanol

Foundation: Describe an experiment that you could do to decide whether ethanol or hexane releases the most energy. Ethanol and hexane are both liquids.
AO2 [4 marks]

You could make the ethanol and hexane heat up some water. The one that heated it up the most would be the one that released the most energy.

Answer grade: G. The student probably had a clear idea of the experiment in their head but they have not written enough on paper. To improve, describe how the apparatus would be set up and say how you measure the temperature of the water before and afterwards. Explain how you will get valid results by controlling the volume of water and measuring the mass of fuel burned. A diagram would help.

Page 55 Ethane and ethene

Higher: Most of the ethane obtained from natural gas is converted into ethene and made into polymers.

How could you distinguish between ethane and ethane? Illustrate your answer with drawings of the molecules concerned. *AO2 [5 marks]*

You can tell the difference because ethene has a double bond and it will react with bromine water.

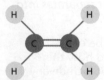 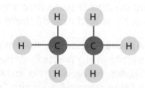

Answer grade: C. Marks are gained here for outlining the characteristics of ethane. The molecules are drawn correctly but they are not labelled. There is no comparison between ethane and ethane. For full marks, explain that ethane does not have a double bond. Give the outcome of the reaction with bromine water and the results with both ethane and ethane.

Page 56 Polymers and energy

Higher: Explain why making bin-liners from new polymers uses so much more energy than using recycled polymers. Suggest why only a small proportion of polymers are recycled. *AO2 [4 marks]*

When you make bin-liners from new polymer you have to use up oil. This takes a lot of energy. Recycled polymers don't need oil – they are just made from old polymers melted together. But many people don't recycle their polymers because it is too much trouble.

Answer grade: C: The first sentence mentions the need for oil in the process, gaining a mark, and the last sentence gives one reason why polymers are not recycled. However, the answer does not contain enough specific information. Explain how the energy is used in making polymers (extracting oil, distilling/refining, manufacturing polymer etc.). Identify that some energy is required to recycle polymers (collecting, transporting, sorting and manufacturing) and indicate why it is likely to be less than making new polymer. Demonstrate understanding that mixed polymers cannot be recycled together but must be sorted. Some cannot be recycled. This increases the costs of recycling, making it less economic to recycle.

Page 58 Atomic structure

Foundation: Describe the structure of an oxygen atom.
AO1 [4 marks]

*Atoms are made from protons and neutrons.
They have electrons going round the nucleus.*

> **Answer grade: E.** This answer lists the particles that make up an atom, but it does not make clear how these particles are arranged in an oxygen atom. To gain full marks, state that protons and neutrons are in the nucleus. Explain that an oxygen atom has 8 protons, 8 electrons and 8 neutrons.

Page 59 Similar reactions

Higher: Lithium and potassium both react in a very similar way when they are added to water.

Use your knowledge of atomic structure to explain why lithium and potassium react in a similar way, even though they are different elements.
AO2 [4 marks]

Lithium and potassium are both in the same group of the periodic table. Elements in the same group have similar properties.

> **Answer grade: B/C.** Although the information given is correct, the student has not fully answered the question because they have not explained in terms of atomic structure. To gain full marks, include an explanation of the similarities in the outer-shell electrons of the two elements. Including the electronic structure of both elements would show this clearly.

Page 60 Ionic compounds

Higher: Explain the meaning of 'ionic compound'. Illustrate your answer with a diagram that shows an example of how ionic compounds form. *AO1* [4 marks]

Ionic compounds are compounds formed from ions.

transfer of electrons

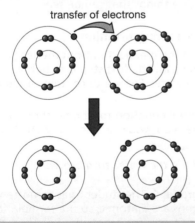

> **Answer grade: B.** The student has forgotten to include the fact that ions are held together by electrostatic attraction. A very good diagram is spoilt because the positive and negative charges on the ions have not been included. Remember – it's not sufficient just to describe an ion as positive or negative, you should always state how many positive or negative charges an ion carries.

Page 61 Formulae equations

Higher: Potassium iodide and lead nitrate react together to form lead iodide. Lead forms a Pb^{2+} ion.

Decide the name of the other product and write a balanced formula equation for this reaction.
AO2 [4 marks]

The other product is potassium nitrate.

$$KI + PbNO_3 \longrightarrow PbI + KNO_3$$

> **Answer grade: B.** The name of the product is correct. However, the formula for lead nitrate is wrong, because lead ions have a 2+ charge while nitrate ions only have a 1+ charge. There should be two nitrate ions for every lead ion. The equation is balanced for this wrong formula, but would not be balanced with the correct formula.

Page 62 Ionic compounds

Foundation: Jon has a test tube containing white crystals. He wonders if the white crystals are ionic.

What properties will the crystals have if they are ionic? Suggest a simple experiment Jon could carry out to test for one of these properties. *AO2* [5 marks]

It will have ionic bonds. It will have a high melting point and conduct electricity. He could see if it melts easily.

> **Answer grade: E.** This answer does not give enough information to gain all the marks available. Saying that a substance has ionic bonds is not naming properties. Ionic substances only conduct electricity if they are dissolved or molten, so Jon's white crystals will not conduct electricity. Give more detail about the test and explain what he would expect to see. For example, Jon should heat the crystals in a strong Bunsen flame. If they do not melt, they have a high melting point.

Page 63 Preparing copper carbonate

Higher: Describe how a pure sample of copper carbonate could be made. *AO2* [5 marks]

A soluble copper compound should be mixed with a soluble carbonate compound. Then you filter it and you have copper carbonate.

> **Answer grade: C.** Marks have been lost because the student did not give enough detail. Suggest the name of a soluble copper and carbonate, state that they must be in solution at the start. Remember to wash and dry the precipitate. Many students think that when an ionic compound dissolves the ions remain in pairs. In fact, each ion is completely separate from all the others and behaves independently.

Page 64 Bonding in water

Foundation: Describe the type of bonding that holds the atoms together in a molecule of water. Illustrate your answer with a diagram. *AO2* [4 marks]

Water molecules are covalent.

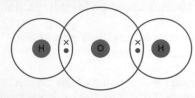

Answer grade: E. This answer names the bonding, but it does not describe it as a shared pair of electrons. The diagram shows the shared pairs of electrons, but the student has forgotten to draw the other outer-shell electron of oxygen. Many students think that it is sufficient to show only the bonding electrons in dot and cross diagrams. It is important to show *all* outer-shell electrons on your diagram.

Page 65 Using graphite

Higher: Sodium is extracted from sodium chloride by electrolysis. A graphite electrode is inserted into the molten compound and then electricity is passed through it.

Describe the properties of graphite that make it particularly useful as an electrode for this purpose, and explain why graphite has these properties. *AO2* [5 marks]

Graphite can conduct electricity. This is because it has free electrons that can move when a voltage is applied.

Answer grade: C. This answer mentions one property of graphite. However, the student has failed to mention a second important property – the high melting point of graphite. The description of the structure of graphite does not contain enough detail to gain full marks. The number of marks available is a clue that more is needed. Describe the layered structure of graphite. A labelled diagram would be helpful.

Page 66 Identifying colours

Foundation: A sweet manufacturer wants to know what colours have been used in a chewy sweet.

Describe a method that they could use to find this out. *AO2* [5 marks]

You could use paper chromatography. You put the colour on the bottom of the paper, and when it is ready you can tell which colours are in it.

Answer grade: E. This answer does not describe the method clearly enough to gain more than 2 of the 5 marks available. To improve, describe what the sweet manufacturer should do with the paper once the mixture has been spotted onto it. Explain what will happen to the different colours in the mixture and how the manufacturer can tell what the colours are.

Page 67 Electrical conductivity

Higher: Use ideas about structure and bonding to explain why solid copper oxide cannot conduct electricity while solid copper can. *AO2* [5 marks]

Copper oxide is ionic, because it is made from copper ions and oxide ions in a giant lattice. This means that it cannot conduct electricity unless it is molten or dissolved. Solid copper is metallic, so it can conduct electricity as a solid.

Answer grade: C. This answer has described the structure and bonding in copper oxide accurately, but has not explained why this means that it cannot conduct electricity. The description of copper only repeats the information in the question so it does not gain any marks. For full marks, describe the structure and explain that the delocalised electrons can move to carry the electricity.

Page 68 Observing a reaction

Foundation: Sarah's teacher demonstrates the reaction of potassium with water. She carefully cuts a small piece of potassium and drops it into a large trough of water.

Describe what Sarah would see. *AO2* [5 marks]

The potassium would fizz and shoot around on the water. It would turn into potassium hydroxide.

Answer grade: E. This answer would gain no more than 2 marks and a grade E, because the description is incomplete. Remember that the potassium would melt and form a ball and would get smaller. There might be a flame. Although the chemistry is good, there are no marks for saying that potassium turns into potassium hydroxide because the question asks what Sarah would *see*. Always read the question carefully. It is not always necessary to explain the chemistry behind what you are describing. If the question asks what would you see then you should only describe things that can be seen.

Page 69 Extracting bromine

Higher: Bromine is extracted from the sea on a commercial basis. The first step is to bubble chlorine through sea water. Sea water contains sodium bromide.

Write a symbol equation for the reaction that occurs. Explain what you would see and why. *AO2* [4 marks]

$NaBr + Cl \rightarrow NaCl + Br$

You would see the water turn brown.

Answer grade: C. The student has forgotten that the halogens are diatomic, so the formula for chlorine is Cl_2 and for bromine is Br_2. Thus, two parts of the formula equation are incorrect. Although the answer correctly identifies that the water would turn brown, there is no explanation of *why* the brown colour appears.

Page 70 Proving exothermic reactions

Foundation: Andy says that adding iron filings to copper sulfate solution is an exothermic reaction.

Describe an experiment he could do to show this. *AO2* [4 marks]

He could put some iron filings in copper sulfate and see if the temperature goes up.

> **Answer grade: D/E.** This answer would gain 2 marks for mentioning iron filings and the use of temperature as a measure. However, the answer is not detailed enough to achieve full marks. To improve, describe the experiment including the apparatus needed. Remember that the temperature must be measured before and after. The student has not said whether a rise in temperature confirms that the reaction is exothermic or demonstrates that it is not.

Page 71 Collision theory

Higher: In glow sticks, two chemicals react to form a product that glows for a short time.

Use collision theory to explain why manufacturers recommend keeping glow sticks in the fridge to make them last longer. *AO2* [4 marks]

Keeping the sticks in the cold slows down the rate of reaction. The reactants have less energy so there will be less collisions.

> **Answer grade: C.** This answer would receive half marks at best because collision theory is not fully explained. The important fact is that there are fewer effective collisions per second. Explain that lowering the temperature means that the reactants have less energy so a lower percentage of collisions will be effective. For this reason, the reaction will take longer to complete.

Page 72 Calculating mass

Foundation: Which compound contains the highest percentage by mass of nitrogen, KNO_3 or NH_4Cl? Show your working. *AO1* [4 marks]

$$M_r\ KNO_3\ 39 + 14 + 16 = 69\ \frac{14}{69} \times 100\% = 20.3\%$$

$$M_r\ NH_4Br\ 14 + (4 \times 1) + 80 = 98\ \frac{98}{14} \times 100\% = 700\%$$

KNO_3 has the highest percentage nitrogen.

> **Answer grade: E.** The student has shown their working very clearly, and this means that they would gain some marks, even though there are mistakes in the calculation.
>
> KNO_3 – the student has forgotten to multiply the oxygen by three so has the wrong M_r but still gets credit for doing the correct calculation for the percentage mass.
>
> NH_4Br – the student has the numbers the wrong way up for the percentage mass – the clue was that the answer is higher than 100%.
>
> Many students panic when they see a calculation, but they can be one of the easiest ways to pick up marks. Stay calm, follow the method you have learned and always show your working!

Page 74 Jupiter's moons

Higher: Galileo observed four moons orbiting Jupiter. Explain how this observation did not fit with the geocentric model of the Solar System. *AO2* [3 marks]

If moons are orbiting Jupiter, then not everything is orbiting Earth.

> **Answer grade: A/B.** This answer is essentially correct, but it is not detailed enough to gain full marks. To improve, introduce the fact that in the geocentric model, everything in the Solar System orbited Earth.

Page 75 Refraction

Higher: Light travels more slowly in water than it does in air.

Explain why this makes it difficult to pick up an object from the bottom of a swimming pool. *AO2* [4 marks]

Light gets refracted in water so it distorts the image you see. The object at the bottom of the pool isn't where it seems to be.

> **Answer grade: B/C.** This answer would gain 2 marks for mentioning refraction and distortion, and for accurately stating that this changes the apparent position of the object. To get full marks, however, you should say that the object appears to be closer than it really is, because the light travels more slowly in water, and your brain assumes it is travelling at the faster speed of light.

Page 76 Advantages of modern telescopes

Foundation: Early telescopes used two lenses. Modern astronomical telescopes are reflecting telescopes.

Describe **two** advantages of reflecting telescopes over refracting telescopes. *AO1* [4 marks]

The reflecting telescope uses a curved mirror instead of a lens, so it is much lighter. It is also easier to make a bigger mirror, so you can get a larger telescope.

> **Answer grade: C/D.** This answer mentions two advantages of reflecting telescopes, but needs to include more detail. It is easier to move a lighter telescope around, and you can see more distant objects with a larger aperture telescope. Another advantage that could be mentioned is that colours refract slightly differently in a lens, so a clearer image can be obtained with a reflecting telescope.

Page 77 Thunder and lightning

Higher: During a thunderstorm you always see the lightning before you hear the thunder. Light travels so fast that the lightning is almost instantaneous, but sound travels at a speed of 340 m/s in air.

If you hear thunder half a minute after seeing lightning, how far away is the storm? *AO3* [3 marks]

$Speed = \dfrac{distance}{time}$, so $distance = speed \times time$.

$Distance = 340 \times 0.5 = 170 \text{ m away}$.

> **Answer grade: C.** The equation has been rearranged correctly, which would gain 1 mark, but the student has forgotten to change the time into seconds. Half a minute is 30 seconds, so the answer should be $340 \times 30 = 10\,200$ m or 10 km away.

Page 78 Electromagnetic waves

Higher: The speed of electromagnetic waves in air is about 3×10^8 m/s. Microwaves have a wavelength of 3 cm.

Use the wave equation wave speed = wave length × frequency to calculate the frequency of microwaves. *AO2* [3 marks]

Wave speed = frequency × wavelength, so frequency = speed ÷ wavelength.

$Frequency = 3 \times 108 \div 3 = 1 \times 10^8 \text{ Hz}$.

> **Answer grade: B.** The equation has been correctly rearranged, so one mark would be gained here. However, the units of the wavelength need to be converted to metres, not centimetres, so the calculation is incorrect. The answer should be 1×10^{10} Hz.

Page 79 Infrared waves

Foundation: All objects emit infrared waves, but the hotter the object is, the more infrared radiation is given off.

Use this information to explain how the police could use infrared radiation to find fugitives. *AO2* [3 marks]

They use infrared cameras in helicopters to find criminals on the run, because the criminals will emit infrared waves and show up in a photo.

> **Answer grade: D.** This answer is correct, but it needs a bit more detail to gain full marks. Explain that the fugitives are warmer than their surroundings, so they will emit more infrared radiation, and that this shows up on a thermograph.

Page 80 Radioactivity

Foundation: Radioactive materials emit ionising radiation.

Explain what is meant by the term 'ionising radiation'. *AO1* [2 marks]

Ionising radiation means it turns atoms into ions.

> **Answer grade: E.** This answer is essentially correct, but needs to explain what an ion is. For example 'when the radiation collides with an atom, it knocks off an electron to form an ion'.

P1 Improve your grade Universal physics

Page 81 Studying the Universe

Foundation: Humans have always been fascinated with space. Ancient civilisations relied on the naked eye to study the stars. In medieval times, telescopes were used to study objects in space.

Explain **two** advantages of the technology used to study the Universe today. *AO2* [4 marks]

Nowadays we send space probes to orbit other planets, and we use the Hubble space telescope.

Answer grade: E/F. Two modern methods are identified, but no advantages are discussed. To improve, mention that the space probes can find out the elements present on other planets, and the Hubble space telescope can obtain much more detailed images of space because there is no interference from the atmosphere.

Page 82 Positioning telescopes

Higher: Explain why it is better to site optical telescopes at the top of high mountains. *AO1* [2 marks]

The air is thinner, and the light waves do not have so far to travel.

Answer grade: C/D. This answer would gain 1 mark for the first part, stating that the air is thinner. However, the student should also explain that when there is less air, the light waves will not get absorbed as much. This means that more light will arrive at the telescope, so you will be able to see further into space.

Page 83 Energy from the Sun

Higher: Explain where the Sun's energy comes from. *AO1* [4 marks]

There is a nuclear reaction inside the Sun, which gives off a lot of energy in the form of electromagnetic radiation.

Answer grade: B/C. Marks would be awarded here for correctly identifying that the energy comes from nuclear reactions in the Sun. However, to gain full marks the answer should say that the nuclear reaction is a *fusion* reaction that turns hydrogen into helium. It should also explain that some of the mass of the hydrogen is converted to energy.

Page 84 Steady State or Big Bang?

Foundation: The Steady State theory and the Big Bang theory were two opposing theories of the Universe in the 20th century.

Give one similarity between the two theories and one difference between them. *AO1* [3 marks]

They both said that the Universe is expanding. The Big Bang theory suggested that the Universe began at a certain point and has been expanding ever since.

Answer grade: C/D. This answer correctly states that both theories support the idea of an expanding Universe. However, to gain full marks, the answer should include how the Steady State theory differs from the Big Bang theory with regard to the beginning of the Universe – i.e. that according to the Steady State theory, some matter was being created all the time, not just at the beginning.

Page 85 Measuring depth

Higher: A ship uses sonar to measure the depth of water as it approaches a harbour. A short pulse of ultrasound is sent out, and the time for it to return is measured as 20 milliseconds (a millisecond is a thousandth of a second).

If the speed of ultrasound is 1500 m/s, what is the depth of water? *AO2* [2 marks]

$$distance = speed \times time = 1500 \times \frac{20}{1000} = 30\ m$$

Answer grade: C. The time has been correctly converted into seconds, so 1 mark would be gained here. However, the student has forgotten that the distance needs to be halved, as the ultrasound has to travel to the seabed and back again. The answer should therefore be 15 m.

Page 86 S waves

Foundation: The Earth is made up of four layers – crust, mantle, outer core and inner core.

Through which layers can S waves travel? Explain your answer. *AO1* [3 marks]

S waves can travel through the crust and the mantle because they are transverse.

Answer grade: D. This answer correctly identifies the layers through which S waves can travel and also that they are transverse. However, to gain the additional mark available for this question, the student should also have explained that transverse waves cannot travel in liquids, which is why S waves cannot travel through the liquid outer core.

Page 87 Potential difference

Higher: A cell has a potential difference of 1.5 V.

What is meant by the term 'potential difference'? *AO1* [3 marks]

Potential difference is the scientific word for voltage. It pushes the electrons around the circuit.

Answer grade: B. This answer is correct and would gain 2 marks, but to achieve the additional mark available, the answer should also explain that potential difference is a measure of how much *energy* each electron gets from a battery or cell.

Page 88 Kilowatt-hours

Foundation: An oven has a power rating of 2000 W and it takes 45 minutes to cook a cake. The cost of 1 kW h of electrical energy is 22 p.

What is the cost of the electricity to cook the cake? *AO2* [2 marks]

No. of kW h = 2 × 45 = 90 kW h.

Cost = no. of kW h × 22 p = 1980 p = £19.80

Answer grade: D/E. The formulas used here are correct, but the answer is wrong. Although the power has been correctly converted to kW, the student has failed to convert 45 minutes into ¾ hour – or 0.75 hours, so the subsequent calculations are incorrect. The final answer should be 2 × 0.75 × 22 p = 33 p.

Page 89 Understanding resources

Foundation: Fuels are burnt in power stations. Fossil fuels are non-renewable energy sources, and biomass is a renewable energy source.

Explain what is meant by the terms 'non-renewable' and 'renewable'. *AO1* [2 marks]

Non-renewable means that once you have used it, it can't be used again. Renewable means it cannot be used up.

> **Answer grade: D.** The first sentence is correct, but the answer should also include the fact that non-renewable energy sources will eventually run out. The second sentence is not strictly true. For example, biomass and wood can be used up. They are only renewable resources if crops and trees are replanted to make more. The definition of a renewable energy source is that it will not run out, not that it cannot be used up.

Page 90 Transformers

Foundation: Describe how step-up transformers are used in the National Grid. *AO1* [4 marks]

A step-up transformer is used to make voltage bigger so it can be carried across cables to people's homes.

> **Answer grade: E.** The answer correctly identifies that a step-up transformer increases the voltage, but it does not fully explain how it is used in the National Grid. For full marks, state that each transformer from the power-station generator increases the voltage from 23 000 V to 400 000 V. Include the fact that the current in transmission cables is smaller when higher voltages are used and that this reduces heat loss in the cables, improving efficiency. Less wastage of energy also saves money.

Page 91 Calculating efficiency

Higher: An electric drill transfers every 50 J of electricity into 20 J of useful kinetic energy.

Calculate the percentage efficiency of the drill.

What forms does the wasted energy in the transfer take? *AO2* [4 marks]

$$\frac{20}{50} = 0.4 \times 100 = 40\%$$

The wasted energy is sound.

> **Answer grade: B.** The student has used the efficiency equation correctly, and has remembered to multiply the total by 100 to reach the correct percentage. The answer also correctly identifies sound as one of the products in the energy transfer. However, for full marks, the answer should mention that heat energy is also created in the transfer.

Page 92 Staying cool

Foundation: In many Mediterranean countries, such as Greece, the houses are all painted white.

Explain why this is, using ideas about heat radiation. *AO2* [3 marks]

The white surfaces are good at reflecting the heat radiation, and poor at absorbing heat radiation.

> **Answer grade: D.** This is scientifically correct, but needs more detail to achieve full marks. Explain that in hot countries such as those in the Mediterranean, it is useful to have a good reflector and poor absorber of heat radiation, to keep the house cool in hot weather.

Page 94 Forces and attraction

Foundation: Daisy combs her hair with a plastic comb. She can then use the plastic comb to pick up small pieces of paper.

Explain why the paper is attracted to the comb. *AO1* [3 marks]

Because the comb and the paper have opposite charges and opposite charges attract.

> **Answer grade: E.** To achieve full marks in this question, the answer needs to include the fact that the comb has been charged by friction, as Daisy was combing her hair. It should also explain that the charged comb repels some of the electrons away from the surface of the paper, leaving the surface of the paper with a positive charge.

Page 95 Painting bicycles

Higher: A factory uses an electrostatic paint gun to paint new bicycles. Explain why the paint coats the surface of each bicycle evenly, even the back. *AO1* [2 marks]

Because the paint is charged positive and the bicycles are charged negative. Opposite charges attract.

> **Answer grade: B.** Although this answer is correct, only one reason is given so only 1 mark would be awarded. To achieve full marks, also explain that the paint is spread out evenly. Each of the droplets of paint has a positive charge, so they repel one another, causing them to spread out evenly as they leave the nozzle.

Page 96 Calculating current

Foundation: Look at the circuit below.

What is the current at points Z and X in the circuit? *AO2* [2 marks]

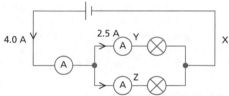

At Z the current is 2.5 A and at X it is 4.0 A.

> **Answer grade: E.** The current at X is 4.0 A, so the answer is correct here. However, the current at Z is not 2.5 A, it is 1.5 A. Notice that 2.5 A + 1.5 A = 4.0 A. The current in a circuit is always conserved.

Page 97 Variable resistors

Foundation: Explain how a circuit with a variable resistor can be used to control the brightness of a lamp in the same circuit. *AO2* [3 marks]

When the resistance of the variable resistor is increased, the lamp is dimly lit.

> **Answer grade: D.** To improve the grade, the answer should include the value of the current. To achieve full marks, you also need to explain that the brightness of the lamp when the resistance of the variable resistor is set to its maximum value should also be explained. When the resistance of the variable resistor is set to a high value, the current in the circuit is low and the lamp is dimly lit. When the resistance of the variable resistor is set to a low value, the current in the circuit is high and the lamp shines brightly.

Page 98 Kettle calculations

Higher: Noah buys an electric kettle that supplies 2000 W of power. A full kettle of water requires 200 000 J of energy to bring it to boiling point.

a Assuming that the kettle wastes no energy, how long will it take Noah to boil a full kettle of water?

b The current in the heating element produces a heating effect. Describe how this occurs. *AO2* [3 marks]

$$Time = \frac{energy}{power} = \frac{200\ 000\ J}{2000\ W} = 100\ s$$

The heat is produced by the movement of the electrons in the element.

> **Answer grade: B.** The calculation is correct, gaining 1 mark, but the explanation is too brief. To achieve full marks, a more detailed explanation of the heating effect must be given. For example, the movement of the electrons in the element causes energy transfer from the electrons to the atoms in the element, causing it to heat up.

Page 99 Understanding acceleration

Foundation: Syamala is training for the 1200-m race by running round a 200-m track six times. She runs at a constant speed.

Explain why she is accelerating. *AO1* [2 marks]

Because she is changing direction.

> **Answer grade: D.** The answer is correct, but too brief. To achieve full marks, a fuller explanation is needed. Explain that velocity is a vector quantity – speed in a specific direction. As Syamala runs, her velocity is changing because her direction is constantly changing; a change in velocity is an acceleration.

Page 100 Interpreting graphs

Higher: Explain how the graph below shows that:

a The acceleration of car A is greater than the acceleration of car B.

b Car B has travelled further than car A. *AO3* [2 marks]

Car A has a greater acceleration because its line is steeper. Car B has travelled further because its line is longer.

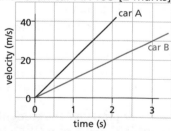

> **Answer grade: B.** Car A has greater acceleration because the line is steeper/the gradient of the line is greater. This is correct and would achieve 1 of the 2 available marks. The gradient of a velocity–time graph is equal to the acceleration. The distance each car has travelled is equal to the area under the line, so car B has travelled further than car A not simply because its line is longer, but because the area under the line for car B is greater than the area under the line for car A.

Page 101 Forces and velocity

Foundation: An aeroplane experiences the forces of gravity, upthrust, thrust from the engine and friction from the air.

Explain how it can be travelling at a constant velocity. *AO1* [3 marks]

If the forces add up to zero and there is no resultant force.

> **Answer grade: E.** This explanation would gain 1 mark for stating that the forces add up to zero. However, with 3 marks available, more explanation is needed. If the forces on an object add up to zero, there is no resultant force and the object will either remain stationary or continue to move with constant velocity. If the upward force on the aeroplane (upthrust) equals the downward force of gravity, and the forward thrust of the engine equals the backward force of air resistance/friction/drag, then the forces add up to zero and there is no resultant force.

Page 102 Falling objects

Foundation: Kerri is a professional skydiver. When she jumps out of an aeroplane, she initially accelerates, then falls at a steady speed. Explain why.
AO1 [3 marks]

Air resistance increases until the force of gravity and the air resistance are balanced. Therefore, the resultant force is zero and she falls at a steady speed.

> **Answer grade: D.** This explanation would achieve 2 of the 3 available marks. However, the answer has left out an important link between air resistance/friction and speed. For full marks, explain that as Kerri's speed increases, the air resistance/friction increases.

Page 103 Assessing stopping distances

Higher: Some roads have markings that indicate how far apart drivers should be from the car in front. On a particular road the markings are 30 m apart, the speed limit is 60 mph and drivers are advised to keep at least two markings apart.

Use this data and your understanding of stopping distances to evaluate whether a separation of two markings is safe in good weather. *AO3* [3 marks]

At 70 mph the total stopping distance is 93 m, so a separation of two markings is not enough.

> **Answer grade: C.** The student has correctly identified that a separation of two markings is not enough, but the way in which he or she has reached this conclusion is not fully explained. To attain full marks, the answer should include the fact that if the distance between markings is 30 m, the distance between two markings would be 60 m. To estimate the stopping distance at 60 mph quantitatively, compare the distances for 45 mph and 70 mph. For example, at 70 mph the total stopping distance is 93 m and at 45 mph it is 46 m. At 60 mph the stopping distance would be between these values, at about 70 m. So a separation of 60 m is not a safe distance.

Page 104 Conservation of momentum

Higher: Caitlin likes to play snooker. She is learning to hit a red ball with the cue ball in such a way that the cue ball stops when it hits the red. Both the cue ball and the red ball have the same mass.

Use the principle of conservation of momentum to describe the velocity of the balls before and after the collision. *AO2* [3 marks]

Since both the balls have the same mass, the red ball will have the same velocity as the cue ball had before it hit the red ball.

> **Answer grade: C.** This answer would only gain 1 of the available marks. For full marks, the principle of momentum should be stated – that the total momentum before the collision is equal to the total momentum after the collision – and used to explain the answer. Momentum = mass × velocity. The velocity of the red ball is zero before the collision, so the red ball has no momentum before the collision. The velocity of the cue ball is zero after the collision, so it has no momentum after the collision, as all its momentum is transferred to the red ball. Since the balls have the same mass, the velocity of the red ball after the collision is the same as the velocity of the cue ball before the collision.

Page 105 Work done

Foundation: Explain what work is done when a window cleaner climbs a ladder. *AO1* [2 marks]

The window cleaner does work climbing the ladder because he has weight.

> **Answer grade: E.** This does not fully answer the question – it describes why the window cleaner is doing work but not what the work is. To attain full marks in this question, explain that the window cleaner is doing work against the force of gravity. Since work done = force × distance, the window cleaner will do an amount of work equal to his weight (the force) × distance (the height he climbs).

Page 106 Energy transfers

Foundation: Abeni enjoys slides at the playground.

Explain the energy transfers that occur as Abeni climbs up the slide and then slides down it. *AO2* [3 marks]

As she climbs up she gains gravitational potential energy and as she slides down she gains kinetic energy.

> **Answer grade: D.** This answer would attain 2 of the possible 3 marks. To attain full marks, the student should expand on the answer to explain that the gravitational potential energy Abeni gains becomes kinetic energy as she slides down the slide, and that the total energy remains constant. Some of this energy will be transferred to heat through friction with the slide and the air.

Page 107 Radioactive decay

Higher: Alice has learnt at school that radioactive decay is random and cannot be predicted. Andrew argues that it must be possible to predict when radiation will occur, otherwise it would not be useable in the treatment and diagnosis of medical conditions. Who is correct? *AO2* [2 marks]

Andrew is correct. It is possible to predict when radioactive decay occurs when it is used in medicine because the samples of radioactive material have many millions of atoms in them.

Answer grade: B. While this answer is correct, to attain full marks in this question the reasons need to be explained in more detail. For example, although it is not possible to predict when an individual radioactive nucleus will decay, in a large sample of nuclei it is possible to predict roughly how many will decay in a given time.

Page 108 Controlling neutrons

Foundation: Explain how the neutrons in a nuclear power station are controlled. *AO1* [2 marks]

The control rods can be lowered into the reactor to absorb the neutrons and slow down the fission reactions.

Answer grade: D. The student has identified one way in which the neutrons in a power station are controlled, but two methods of controlling the reactions need to be given to gain both marks. As well as the control rods, the moderator in a reactor slows down the fast-moving neutrons, making them more likely to react with the uranium nuclei. Always look at the number of marks available to get an idea of how much information you should give in your answers.

Page 109 The future of fusion

Foundation: Scientists hope that fusion reactors will soon be able to produce energy for human consumption.

a What fuel will be used in these reactors?

b Explain how nuclear fusion occurs. *AO1/AO2* [3 marks]

Hydrogen is the fuel for these reactors. Hydrogen nuclei join together to produce helium.

Answer grade: D. This answer correctly identifies hydrogen as the fuel, but the answer to part **b** of the question is not complete. To push this up to a grade C, include the fact that the hydrogen nuclei collide at high speeds in order to fuse/join together to form helium nuclei.

Page 110 Differences in background radiation

Foundation: Joe and Fred are pen-pals who live in different parts of the UK. They have both been learning about background radiation at school and have discovered that the level of background radiation where they live is different.

Explain why this is. *AO1* [3 marks]

Because the rocks that give off radiation are different in each location.

Answer grade: E. This answer is too superficial to earn more than 1 mark. To achieve full marks, explain that the rock granite contains uranium. Uranium decays to produce radioactive radon gas. Different parts of the UK have different amounts of granite in the landscape, so the amount of background radiation varies.

Page 111 Using radiation

Higher: Which type of radiation would be most suitable for use in a machine that controls the thickness of thin aluminium sheeting? Give the reasons for your answer. *AO2* [3 marks]

Beta radiation would be most useful because it can pass through the aluminium. Alpha radiation could not be used because it cannot pass through aluminium.

Answer grade: B. This answer mentions both alpha and beta radiation. However, the student has overlooked gamma radiation. It is correct that beta radiation can pass through a thin sheet of aluminium, but gamma radiation can also penetrate aluminium. The advantage of beta radiation is that it cannot penetrate through a thick sheet of aluminium, so the amount of beta radiation passing through the sheet would change as the thickness of the sheet changes. It would not be possible to measure any change in the amount of gamma radiation passing through a sheet of aluminium.

Page 112 Measuring radioactive decay

Higher: The graph below shows the count rate of a radioactive sample over a period of 40 days. Using the graph, estimate the half-life of the sample. *AO3* [2 marks]

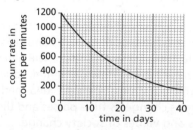

13.5 days.

Answer grade: B. The answer is correct, but in order to achieve both of the available marks, some working should be shown. For example, reference lines could be drawn on the graph. Remember, you should always show your working out. Sometimes marks can be earned for correct parts of working, even if the final calculation is wrong.

How Science Works

Data, evidence, theories and explanations

As part of your Science and Additional Science assessment, you will need to show that you have an understanding of the scientific process – How Science Works.

This involves examining how scientific data is collected and analysed. You will need to evaluate the data by providing evidence to test ideas and develop theories. Some explanations are developed using scientific theories, models and ideas. You should be aware that there are some questions that science cannot answer and some that science cannot address.

Practical and enquiry skills

You should be able to devise a plan that will answer a scientific question or solve a scientific problem. In doing so, you will need to collect data from both primary and secondary sources. Primary data will come from your own findings – often from an experimental procedure or investigation. While working with primary data, you will need to show that you can work safely and accurately, not only on your own but also with others.

Secondary data is found by research, often using ICT – but do not forget books, journals, magazines and newspapers are also sources. The data you collect will need to be evaluated for its validity and reliability as evidence.

Communication skills

You should be able to present your information in an appropriate, scientific manner. This may involve the use of mathematical language as well as using the correct scientific terminology and conventions. You should be able to develop an argument and come to a conclusion based on recall and analysis of scientific information. It is important to use both quantitative and qualitative arguments.

Applications and implications of science

Many of today's scientific and technological developments have both benefits and risks. The decisions that scientists make will almost certainly raise ethical, environmental, social or economic questions. Scientific ideas and explanations change as time passes and the standards and values of society change. It is the job of scientists to validate these changing ideas.

How Science Works

How science ideas change

From the information you have learnt, you will know that science is a process of developing, then testing theories and models. Scientists have been carrying out this work for many centuries and it is the results of their ideas and trials that has provided us with the knowledge we have today.

However, in the process of developing this knowledge, many ideas were put forward that seem quite absurd to us today.

In 1692, the British astronomer Edmund Halley (after whom Halley's Comet was named) suggested that the Earth consisted of four concentric spheres. He was trying to explain the magnetic field that surrounds the Earth and suggested that there was a shell of about 500 miles thick, two inner concentric shells and an inner core. Halley believed that these shells were separated by atmospheres, and each shell had magnetic poles with the spheres rotating at different speeds. The theory was an attempt to explain why unusual compass readings occurred. He also believed that each of these inner spheres, which was constantly lit by a luminous atmosphere, supported life.

Reliability of information

It is important to be able to spot when data or information is presented accurately and just because you see something online or in a newspaper, does not mean that it is accurate or true.

Think about what is wrong in this example from an online shopping catalogue. Look at the answer at the bottom of the page to check that your observations are correct.

From box to air in under two minutes!

Simply unroll the airship and, as the black surface attracts heat, watch it magically inflate.

Seal one end with the cord provided and fly your 8-metre, sausage-shaped kite.

- Good for all year round use.
- Folds away into box provided.
- A unique product – not for the faint hearted.
- Educational as well as fun!

Once the airship is filled with air, it is warmed by the heat of the sun.

The warm air inside the airship makes it float, like a full-sized hot-air balloon.

Answer

Black absorbs heat, it does not attract it.

Glossary

A

abuse non-medical use of drugs legally available only with a doctor's prescription 15

acceleration the rate of change of the velocity of an object 99, 100, 101, 102, 104

acid rain rain with a pH of 2–5, usually caused by the emission of pollutants such as the oxides of sulfur and nitrogen 20, 47, 53

acquired characteristics non-hereditary or environmental changes in an organism 7

acquired immune deficiency syndrome (AIDS) the collection of diseases associated with HIV infection 17

active site part of an enzyme to which a substrate bonds 31

active transport movement of molecules through a cell membrane against the concentration gradient; this process requires energy 35

activity the activity of a source is the rate of decay of nuclei 112

adaptation refers to characteristics that best suit the survival of an individual 5

adapted suited to surviving in a particular environment 6

addiction being dependent (hooked) on drugs or any other habit-forming substance 15

addition polymerisation the process of making polymers by adding monomers to the end of chains of monomers 56

aerobic respiration respiration that involves oxygen 32, 33, 35

algae aquatic organisms capable of photosynthesis 19, 20, 53

alkali a substance that makes a solution that turns red litmus paper blue 48

alkali metals very reactive metals in group 1 of the periodic table, e.g. sodium 67, 68

alkanes a family of hydrocarbons found in crude oil containing only hydrogen and carbon, with single covalent bonds 55, 56

alkenes a family of hydrocarbons found in crude oil with double carbon to carbon covalent bonds 55, 56

alleles different versions of a gene which control a particular characteristic 7, 8, 9

alpha particle a helium nucleus emitted from an unstable nucleus 80, 107, 108, 111, 112

alternating current a current that repeatedly changes direction 90

ammeter meter used in an electric circuit for measuring current 87, 88, 96, 98

amperes (amps) units used to measure electrical current 87, 96, 98

amplitude the maximum displacement of a wave measured from the mean position 77

amylase an enzyme that catalyses reactions that break down starch to the sugar maltose 31, 39

anaerobic respiration respiration without using oxygen 32

animal vector an organism that transmits pathogens from host to host – insects are common disease vectors 17

animalia the animal kingdom 4

anion a negatively charged ion 60, 61, 63

anode a positively charged electrode 49

Anopheles the mosquito genus that transmits the malarial protozoan; *Anopheles gambiae* is the most common species in Africa 17

antacid a base or basic salt used to treat indigestion 48

antibacterials substances that prevent bacteria from multiplying but are toxic to ingest, e.g. antibacterial soap 17, 18

antibiotic therapeutic drug acting to kill bacteria or prevent them from multiplying, which is taken into the body 17, 18, 40

antibodies proteins produced by a particular type of white blood cell (B-lymphocytes) that bind to substances on the surface of pathogens and destroy them 18

antifungals substances that kill fungi 17

antiseptics substances that prevent bacteria from multiplying on the body and other surfaces 17

arteries blood vessels that carry blood away from the heart 38

atmosphere the layer of gases surrounding a planet 21, 42, 43, 53, 76, 82, 110

atom the basic 'building block' of an element which cannot be chemically broken down 43, 46, 51, 55, 58, 59, 60, 61, 64, 67, 70, 80, 92, 94, 107

atomic mass unit a unit of mass for expressing the mass of atoms or molecules 58

atomic number the number of protons inside the nucleus of an atom (same as the proton number) 58, 59

atria chambers of the heart that receive blood from the veins 38

auto-immune disease a disease caused by the body's immune system attacking its own tissues and organs, e.g. Type I diabetes 13

auxin a plant hormone involved in plant growth 14

axon a long projection from a nerve fibre that conducts impulses away from the body of a nerve cell 11, 12

B

backbone a flexible rod running along the length of the body near to its upper surface; it supports the body 4

bacteria single-celled microorganisms which can either be free-living or parasites (they sometimes invade the body and cause disease); the cell does not contain a distinct nucleus 5, 17, 18, 19, 21, 23, 25, 37, 40

barium meal barium sulfate swallowed just before an X-ray 63

base (1) part of a nucleotide unit of DNA which is adenine (A), thymine (T), guanine (G) or cytosine (C); (2) any substance that neutralises an acid 24, 28, 29, 30 48

becquerel the unit for activity: one becquerel is equal to one nucleus decaying per second 112

beta particle an electron emitted from the inside of an unstable nucleus 80, 107, 108, 111, 112

beta-carotene an orange-yellow pigment that human cells convert into vitamin A 25

Big Bang an explosion some 14 billion years ago that created both space and time 84

Big Bang theory a theory that proposes the creation of the Universe from the Big Bang 84

bile a chemical produced by the liver which digests fats 39

binge drinking consumption of large amounts of alcohol in a short period of time 15

binomial system the method of giving an organism a two-part name consisting of genus and species 4

biodegradable a biodegradable material can be broken down by microorganisms 56

biodiversity the range of different living organisms in a habitat, e.g. a woodland or pond 4

biofuels fuels made from plants and animal waste 54

biomass the amount of organic material of an organism (usually measured as dry mass); waste wood and other natural material which are burned in power stations 19, 89

black dwarf the final stage of a white dwarf, when it has lost all its energy 83

black hole an extremely dense core of a supermassive star left behind after the supernova stage; light cannot escape its strong gravitational pull 83

body mass index (BMI) a measure of someone's weight in relation to their height, used as a guide to thinness or fatness – values over 30 indicate obesity 13

Glossary

boiling point the temperature at which a substance changes its state from a liquid to a gas 52, 62, 65, 66, 67, 69

braking distance the distance travelled by a car while the brakes are applied and the car comes to a stop 103

C

calorimeter a device used to measure the heat of chemical reactions 54, 70

capillaries small blood vessels that join arteries to veins 32, 38

capillary beds a dense network of capillaries 38

capillary vessels small blood vessels that join arteries to veins 32

carbohydrates Organic molecules composed of carbon, hydrogen and oxygen with the generalised formula $C_nH_{2+n}O_n$. 13, 21, 36, 39, 54

carbon monoxide a toxic gas formed during incomplete combustion 15, 52, 71

carbonate a compound containing carbonate ion, CO_3^{2-} 42, 45, 47, 48, 53, 62, 63

cardiac muscle the muscle found in the heart that squeezes and relaxes continuously 38

catalyst a substance added to a chemical reaction to alter the speed of the reaction; it is effective in small amounts and is unchanged at the end of the reaction 31, 54, 56, 71

cathode a negatively charged electrode 49

cation a positively charged ion 60, 61

cell membrane a membrane surrounding the cell and through which substances pass in solution into and out of the cell 23

cell wall surrounds plant cells and some bacterial cells – the cell wall of plants consists of 40% cellulose; the bacterial cell wall does not contain cellulose 4, 23

cellulose large polysaccharides made by plants for cell walls 4, 23, 56

chain reaction a process in which an enormous amount of energy is produced when neutrons from previous fission reactions go on to produce further uncontrolled fission reactions 108

chalk porous, fine-grained rock composed mainly of calcareous shells of microorganisms 21, 44, 45

charge a physical property of particles which causes them to experience a force when near other electrically charged particles 87, 94, 95, 96

chemical energy Energy available from atoms when electron bonds are broken. 70, 87, 91, 106

chemical formula the name of a compound written down as chemical symbols 61

chemical properties properties that cannot be observed just by looking at a substance – a chemical property depends on how that substance reacts chemically with other substances, e.g. flammability, reactions with acids and bases 46, 49, 58, 67

chemosynthesis the chemical reactions in different species of bacteria which utilise hydrogen from sources other than water to reduce carbon dioxide, forming sugars 18

chlorides a group of chemicals all containing the anion chloride 48, 60, 61, 62, 63

chlorophyll a pigment found in plants that is used in photosynthesis (gives green plants their colour) 4, 23, 34

chloroplast a structure in plant cells and algae that absorbs light energy – where photosynthesis takes place 23, 34

cholesterol fatty substances that can block blood vessels 40

chordata a phylum of animals that possess a rod for supporting the body 4

chromatids produced as a result of replication of a chromosome (DNA replication); they appear as a pair joined by a centromere under the high power of a light microscope 26

chromatogram the final result of chromatography: the chromatography paper with the result on 66

chromatography a method of separating substances using solvent passing through paper or a similar medium 66

chromosomes thread-like structures in the cell nucleus that carry genetic information – each chromosome consists of DNA wound round a core of protein 7, 24, 26

clone an organism whose genetic information is identical to that of the organism from which it was created 26, 27

codon triplets of base pairs in DNA that contain the instructions for making protein 24, 29, 30

coefficients numbers used in chemical equations to indicate the number of each reactant and product 43, 61

cold fusion an invalidated theory that proposed nuclear fusion occurring at room temperature 109

collision theory an idea that relates collisions among particles to their reaction rate 71

common ancestor an individual from which organisms are directly descended 4, 6, 36

compaction the process of compressing sediments together and squeezing out water between the grains 45

competition rivalry between competitors for supplies of limited resources 6, 40

complementary base pairing bonding between the bases of each strand of a double-stranded DNA molecule – each base on one strand of DNA bonds with its complementary partner on the other strand; adenine always binds with thymine; guanine always bonds with cytosine 24

compound two or more elements that are chemically joined together, e.g. H_2O 21, 47, 48, 49, 52, 55, 60, 61, 63, 65, 67, 72

compressions regions where particles are pushed together and create a region of higher pressure in a sound wave 77

concentration gradient the difference in concentration of a substance between regions where it is in high concentration and where it is lower concentration 32

condensation the process of vapour molecules forming a liquid 42

conferences meetings where participants exchange and present new ideas for research 7

conservation of matter the principle that states that atoms are neither created nor destroyed in a chemical reaction 46

consumers organisms that feed on food already made 19

continuous variation a characteristic that varies continuously shows a spread of values between extreme values of the characteristic in question, e.g. the height of people shows a range of values between short and tall 5

control rods material used to absorb the neutrons in a nuclear reactor in order to produce a controlled chain reaction 108

conventional current the flow of positive charges 87

converging lens a lens that focuses parallel rays of light to a point 75, 76

coolant gas or liquid used to remove thermal energy from a nuclear reactor 108

cosmic microwave background (CMB) radiation the 'left-over' radiation from the Big Bang – radiation coming very faintly from all directions in space 84, 110

coulomb the unit for charge 87, 96

covalent bonds bonds between atoms where electrons are shared 55, 64, 65, 67

cracking the thermal decomposition of long-chain hydrocarbons into smaller and more useful hydrocarbons 56

critical mass the minimum mass of fissile material that can sustain a chain reaction 108

cross-breeding mating (or plant equivalent) between two individuals, resulting in offspring 8

Glossary

crude oil a complex mixture of hydrocarbons mined from the Earth from which petrol and many other products are made 52, 54, 55, 56

cryogenics the study of substances at low temperatures 49

culture a combination of microorganisms and all the substances they need to live and multiply; the substances may be in solution or part of a jelly-like material (e.g. agar) on which microorganisms grow 25

current the rate of flow of charge 87, 88, 90, 96, 97, 98

cuttings a piece of stem cut from a parent plant, which has the potential to develop roots and grow into a new plant 14, 27

cystic fibrosis a recessive genetic disorder in which thick, sticky mucus is produced, affecting the lungs and digestive tract in particular 9

cytoplasm a jelly-like material that fills the cell, giving it shape 7, 23, 29

D

daughter cells the new cells produced when parent cells divide 26, 37

daughter nuclei the nuclei produced in a fission reaction 108, 109, 110

deceleration negative acceleration 99, 100

decompose the separation of a chemical compound into simpler compounds 20, 21, 36, 47

decomposers fungi and bacteria whose feeding activities cause decomposition 21

decomposition the process resulting from the feeding activities of fungi and bacteria that release nutrients from dead organic matter into the environment 21

deforestation clear cutting of large tracts of land 43

denatured refers to irreversible changes in the structure of proteins (including enzymes); the changes stop the proteins from working properly 31, 34

dendrites the fine branches at the end of axons and dendrons 11

dendron a long projection of a nerve fibre that conducts impulses to the nerve cell 11, 12

denitrifying bacteria bacteria that converts nitrates into nitrogen gas 21

density the density of a substance is found by dividing its mass by its volume 46, 69, 75

deoxygenated blood blood where the oxyhaemoglobin has reverted to haemoglobin 37, 38

depressants substances that slow down responses 15

diabetes a disease where the body cannot control its sugar levels 13, 28

diatomic elements elements that exist as pairs of covalently bonded atoms 64, 69

differentiation the process during which a stem cell (unspecialised) develops into a particular type of cell (specialised) 28, 36, 37

diffusion the spread of particles through random motion from regions of higher concentration to regions of lower concentration 32, 35

diminished an image that is smaller than the object 76

diode a device made from semiconductor material that conducts in one direction only 97

diploid refers to cells with two sets of chromosomes (one set from each parent) – most cells are diploid (except gametes) and the symbol 2n represents the diploid state 26

direct current an electric current that flows in one direction only 49, 90, 96

discontinuous variation a characteristic that only has a limited number of values, e.g. blood group, shoe size 5

disinfectants substances applied to an object to kill microorganisms 49

displacement distance moved in a specific direction 77, 99

displacement reaction a reaction in which one substance displaces another substance from a compound 69, 70

dissolved to make something pass into solution; to turn something into liquid form. 42

diverging lens a lens that makes parallel rays of light spread out rather than focus to a point 75, 76

DNA deoxyribonucleic acid – a molecule found in all body cells in the nucleus; its sequence determines how our bodies are made (e.g. whether we have straight or curly hair), and gives each one of us a unique genetic code 7, 23, 24, 25, 26, 28, 29, 30, 31

DNA polymerase an enzyme that catalyses reactions that join up nucleotides forming DNA 31

dominant refers to an allele which controls the development of a characteristic, even if it is present on only one of the chromosomes of a pair of chromosomes 7, 8

dominant characteristic any characteristic that appears in the heterozygote 8

donor a person who gives (donates) and organ 16

donor eggs eggs taken from a female animal and fertilised in the laboratory 27

Doppler effect the change in wavelength or frequency of a wave as a result of relative motion between the source and an observer 82, 84

dot and cross diagrams a way of drawing the formation of covalent bonds using dots and crosses to represent electrons in the outer shells of the atoms involved 64

double bond a bond formed by two shared pairs of electrons 55, 64

drug a substance from outside the body that affects chemical reactions inside the body 15

ductile capable of being drawn into a wire 51

E

earthing a method used for ensuring the safe discharge of charges to the Earth 95

echolocation a technique similar to sonar used by some animals to navigate and find their prey 75, 85

ecosystem a habitat and all the living things in it 35

ectoparasite a parasite that lives on the external surface of the host 18

efficiency the proportion of the input energy that is transferred to useful form, calculated using the equation: efficiency = (useful energy transferred by the device / total energy supplied to the device) x 100% 91

electric current when electricity flows through a material we say that an electric current flows 49, 87, 91, 96, 98

electric force field a region where electric charges experience a force 95

electrical conductivity how well a substance conducts an electrical current 65, 67

electrical power the rate of energy transfer 88, 92, 98

electrode bars of metal or carbon that carry electric current into a liquid 49, 65

electrolysis a process in which compounds are decomposed by passing an electric current through a solution that conducts electricity 49, 50, 54

electromagnetic spectrum electromagnetic waves ordered according to wavelength and frequency, ranging from low-frequency radio waves to high-frequency gamma rays 78, 79, 80, 81, 82

electron configuration notation indicating the distribution of electrons in electron shells 59, 60, 64

electrons tiny negatively charged particles within an atom that orbit the nucleus – responsible for current in electrical circuits 23, 55, 58, 59, 60, 64, 67, 68, 69, 80, 87, 94, 95, 96, 98, 107, 111

electrostatic forces the very strong forces between positive and negative ions in an ionic substance 60, 62, 94, 95

Glossary

element a substance made out of only one type of atom 46, 58, 59, 60, 61, 63, 65, 67, 68, 69, 107

embryo the early stages in the development of an organism, from the time of first cell division 27, 28, 37

embryonic stem cells undifferentiated cells that are able to develop (differentiate) into any type of body cell 28

emission spectrum the specific frequencies of light an element emits or gives out 63

empirical formula a formula that shows the correct ratio of all of the elements in a compound 72

emulsification the breaking down of fat into more manageable molecules 39

endocrine glands glands that release the substances produced (hormones) directly into the blood 13

endoparasite a parasite that lives inside the host's body 18

endothermic reaction a chemical reaction that takes in heat 70

endotoxin a poison produced by a pathogen within the body 17

energy profile diagram a diagram showing energy taken in or given out during a chemical reaction 70

enzymes biological catalysts (usually proteins) produced by cells that control the rate of chemical reactions in cells 10, 18, 25, 31, 39, 54

erosion where rock is worn away by wind or rain 45

ethene a gas that is an alkene (C_2H_4), which is used to make polymers and is also a plant hormone 14, 55, 56

ethics the actions taken as a result of moral judgements 16, 27

eutrophication the processes that occur when water is enriched with nutrients (from fertilisers) which allow algae to grow and use up all the oxygen 20

evolution the process whereby organisms change through time – present-day living things are descended from organisms that were different from them 6, 36, 42

excess post-exercise oxygen consumption (EPOC) additional oxygen required after a period of anaerobic respiration 32

exothermic reaction a chemical reaction in which heat is given out 70

exponential decay a graph in which the quantity halves after a given interval of time 112

extraterrestrial a term used to describe things 'beyond Earth' 81

extrusive rock igneous rock formed by rapid cooling of lava when exposed to the air or seawater 44

eyepiece the lens at the end of a telescope that you look through 76

F

family a group consisting of several genera (used in classification of living things) 4

feedback regulation of a process by the results (outcomes) of the process 10

fermentation the conversion of carbohydrates to alcohol and carbon dioxide by yeast or bacteria 54

fermenter a large vessel containing a liquid culture of microorganisms and all the substances (nutrients) they need to live and multiply 25

fertile able to reproduce sexually 4

fertilisation the moment when the nucleus of a sperm fuses with the nucleus of an egg 7, 26, 27, 28, 30

fibrin an insoluble protein involved in the clotting of blood 37

filament lamp a lamp that emits light when its thin metal filament gets very hot 97

filtrate the soluble material that passes through a filter paper 63

fineness system a system for denoting the purity of gold, platinum and silver, indicating parts per thousand 51

fission reactions the splitting of a nucleus when it absorbs a neutron 108, 109

fixed resistor a resistor that can only resist a specific amount of current 97

flaccid soft, droopy, lacking turgor 35

flagellum a whip-like extension of a cell that lashes from side to side, driving the cell through liquid 23

flame test the heating of metal ions in a flame to produce a colour as an aid to identifying the material 63

focal length the distance between the centre of the lens and the focal point 75, 76

foetus a stage in the development of an organism when tissues and organs are forming; after the embryo stage 26, 37

food web a flow chart to show how a number of living things get their food (more complicated than a food chain) 19

force of gravity an attractive force between all particles that have mass 81,

fossil fuels fuel (coal, natural gas, oil) formed from the compressed remains of plants and other organisms that lived long ago 20, 21 43, 54, 89, 112

fossils the preserved remains of organisms that lived long ago 36, 45, 86

fractional distillation a method of separating liquid mixtures by evaporation 43, 52, 66

free-body force diagram a diagram showing all the forces acting on an object 101

frequency the number of vibrations per second or number of complete waves passing a set point per second 77, 78, 80, 82, 85, 90

friction energy losses caused by two or more objects rubbing against each other 86, 94, 103

fuel cell a cell that produces energy by combining a fuel and an oxidant 54

fuel rods rods containing nuclear fuel for a fission reactor 108

functional foods any healthy foods claimed to have health-promoting or disease-preventing properties 40

fungi organisms which can break down complex organic substances (some are pathogens and harm the body) 4, 21

fuse a thin wire used in an electrical circuit as a safety device 88

fusion reactions reactions in which lighter nuclei join together (fuse) and produce energy 83, 108, 109

G

galaxy a collection of billions of stars held together by the force of gravity 81, 84

gall bladder a small, sac-like structure connected to the small intestine by the bile duct; it stores bile, which breaks down fats in partly digested food 39

galvanising a method of corrosion protection that uses a layer of zinc to protect the underlying metal 50

gametes the male and female sex cells (sperm and eggs) 8, 26

gamma rays electromagnetic waves of short wavelength emitted from unstable nuclei 80, 107, 111, 112

gene a section of DNA that codes for a particular characteristic, by controlling the production of a particular protein or part of a protein by cells 7, 8, 9, 24, 25, 28, 30

generator a device used for producing electrical energy by moving wires through a magnetic field 90

genetic code all of the base sequences of the genes that enable cells to make proteins 24

genetic disorder an inherited disease that arises as the result of a mutated gene, passed on from parents to children 9

genetic engineering techniques that make it possible to manipulate genes in the cells of organisms 16, 25

Glossary

genetically modified (GM) a GM organism has had its DNA modified by the insertion of DNA from another species 25

genome all of the DNA in each cell of an organism 28

genotype all of the genes of an organism 7

genus a group consisting of more than one species (used in classification of living things) 4

geocentric model Earth-centred model of the Solar System 74

geotropism (or gravitropism) growth movement in response to the stimulus of gravity 14

giant molecular covalent compounds very large molecules consisting of non-metals covalently bonded together 67

glucagon a hormone produced by the pancreas that promotes the conversion of glycogen to glucose 13

glycogen a type of carbohydrate whole molecule consists of many glucose units joined together 13

gradient a quantity determined by dividing the change in y by the change in x; it is the slope of a line 100

gravitational field strength gravitational force acting on an object per unit mass 102, 106

gravitational potential energy the energy associated with the position of an object in the Earth's gravitational field 91, 105, 106

greenhouse effect a process in which the atmosphere is warmed up by infrared radiation; it then re-radiates some of the infrared radiation back towards the Earth's surface, which warms the surface 53, 89

greenhouse gases gases in the atmosphere whose absorption of infrared solar radiation is responsible for the greenhouse effect, e.g. carbon dioxide, methane and water vapour 53, 89

H

haemoglobin the chemical found in red blood cells which carried oxygen 9, 29, 37, 52

half-life the half-life of an isotope is the average time taken for half of the undecayed nuclei in the sample to decay 110, 111, 112

hallucinogens substances that give a false sense of reality 15

halogens reactive non-metals in group 7 of the periodic table, e.g. chlorine 67, 69

haploid refers to cells with only one set of chromosomes – gametes are haploid; the symbol n represents the haploid state 26

heart rate the number of heartbeats every minute 33

heartbeat the two-tone sound of one complete contraction and relaxation of the heart 33

heliocentric model Sun-centred model of the Solar System 74

herbicides chemicals that kill plants – used to remove weeds (unwanted plants) from crops, gardens and public places 14, 25

heterozygous refers to the pair of alleles of a gene where the alleles are different 7, 8

HIV human immunodeficiency virus – the virus that causes AIDS 17

homeostasis self-adjusting mechanisms that allow the body to keep a constant internal environment 10, 13

homeotherms animals that regulate their body temperature 4

homologous series a group of compounds that change in some incremental way 52, 55

homozygous refers to the pair of alleles of a gene where the alleles are the same 7, 8

hormones substances produced by animals and plants that regulate activities; in animals, hormones are produced by and released from endocrine tissue into the blood to act on target organs, and help to coordinate the body's responses to stimuli 10, 13, 14

host the individual infected with a pathogen transmitted by a vector; the organism on which a parasite lives (the parasite takes food from the host) 18

host mother an animal that receives cells that are not its own 27

Human Genome Project an international group of scientists aiming to map the human genome 28

hybrids the offspring of parents which are not the same species 4, 5

hydrocarbons molecules containing only carbon and hydrogen – many fuels are hydrocarbons, e.g. natural gas (methane) and petrol (a complex mixture) 43, 52, 53, 55

hydrothermal vents cracks in the seabed where water is heated as a result of volcanic activity 5, 18

hypothalamus part of the brain that has several functions, the most important being to link the nervous system to the endocrine system 10

hypothesis a possible explanation for an observation 44, 69

I

igneous rock rock formed from magma or lava 44

immiscible two liquids that are completely insoluble in each other 66

incomplete combustion when fuel burns in a small amount of oxygen so that carbon monoxide, particles and water are produced 52

indicator species the presence or absence of these species indicates how polluted (or not) a particular environment is 20

induced a term used to mean 'created' 90, 94

inert an inert substance is one that is not chemically reactive 69

infectious a disease that passes (is transmitted) from person to person 17

infrared radiation part of the electromagnetic spectrum, thermal energy 53, 79, 89, 92

infrared waves non-ionising waves with a wavelength longer than red light that are radiant heat 74, 76, 78, 79, 80

infrasound sound with frequencies less than 20 Hz 85

insecticide chemicals used to kill insects 95

insoluble a substance that will not dissolve (something that will not dissolve in water may dissolve in other liquids) 31, 37, 39, 47, 62, 63

insulators materials that are poor electrical conductors 94, 95, 107

insulin a hormone produced by the pancreas that promotes the conversion of glucose to glycogen 13

insulin insensitivity a condition in which target tissues (e.g. liver and muscles) do not respond to insulin 13

intensity the radiant power per unit area 34, 35, 98

intrusive rock igneous rock formed by slow cooling when magma oozes into cracks and voids in the Earth's crust 44

involuntary response an automatic response to a stimulus that you do not think about 12

ion an atom with an electrical charge (can be positive or negative) 49, 60, 61, 62, 63, 67, 68, 80, 94, 95, 107

ionic bonding chemical bonding between two ions of opposite charge 60, 62

ionic compounds compounds that contain positively charged metal ions and negatively charged non-metal ions 60, 61, 62, 63, 67

ionisation a process in which radiation transfers some or all of its energy to liberate an electron from an atom 80, 107, 111

ionosphere a region of charged particles around the Earth that reflects radio waves 75, 79

Glossary

isotopes (1) nuclei of atoms with the same number of protons but a different number of neutrons; (2) atoms with the same number of protons but different numbers of neutrons 58, 107, 109, 110, 112

J

joule the unit of work done and energy: one joule is the work done when a force of 1 N moves a distance of 1 m in the direction of the force 105, 106

K

key a chart of alternative statements that identifies organisms 5

kilowatt-hour the energy used by an appliance of power 1 kW used for 1 hour – it is a unit of energy used by electrical suppliers 88

kinetic energy the energy that moving objects have 44, 91, 92, 106, 108

kingdoms used in classification and refers to the largest grouping (except for domain) of organisms which have characteristics in common 4

L

lander a robotic space probe sent to planets and moons to carry out soil analysis 81

lattice a criss-cross structure 62, 65, 98

leguminous plants whose roots carry nodules that contain nitrogen-fixing bacteria 18

lichens a combination of fungi and algae – the relationship between the organisms is an example of mutualism 20

life cycle a term used to describe the journey of a star in time 83

ligases enzymes that catalyse reactions that insert (paste in) pieces of DNA into lengths or loops of other DNA 25

light-dependent resistor (LDR) a device in an electric circuit whose resistance falls as the light falling on it increases 98

light-year the distance travelled by light in a vacuum in one year 81

limestone sedimentary rock composed mainly of calcite or dolomite 44, 45, 47

limewater an aqueous solution of calcium hydroxide 33, 47, 52, 63

limiting factor a factor that limits a process 34

lineage known ancestry 7

locus the position of an allele on its chromosome 7

longitudinal waves waves with vibrations parallel to the direction in which they travel 77, 86

M

magma hot molten rock found in the mantle, below the Earth's surface 44, 86

magnified an image that is larger than the object 76

main sequence star an average star, just like our Sun 83

malleable capable of being shaped by hammering 51, 67

marble metamorphosed limestone produced by recrystallisation 44, 45, 71

mass the amount of matter inside an object, measured in kilograms 36, 58, 72, 102, 104, 106

mass number the total number of neutrons and protons within the nucleus of an atom (same as the nucleon number) 58, 107

meiosis a type of division of the cell nucleus that results in four daughter cells, each with half the number of chromosomes (haploid) of the parent cell (diploid) 26

melting point the temperature at which a solid becomes a liquid 46, 62, 65, 67, 68

meristem the tissue in most plants where growth occurs 36

messenger RNA (mRNA) a molecule of RNA with the code for a protein 29

metabolism all of the chemical reactions taking place in a cell 19

metallic compounds compounds composed of individual metal ions floating in a sea of electrons 67

metamorphic rock sedimentary rock that is transformed by heat and pressure 44, 45

microorganisms single-celled organisms that are only just visible in the light microscope 17, 40, 49, 111

Milky Way the name of our galaxy 74, 81

minerals solid metallic or non-metallic substances found naturally in the Earth's crust 45, 50

mitochondria structures in cells just visible in a light microscope where the breakdown of sugars begun in the cytoplasm continues, releasing energy 23

mitosis a type of division of the cell nucleus that results in two daughter cells, each with the same number of chromosomes as the parent (usually 2n) 26, 37

moderator material used to slow down the fast-moving neutrons in a nuclear reactor 108

momentum a quantity calculated by multiplying the mass of an object by its velocity 104

monohybrid refers to the inheritance of a single characteristic 8

monomer a small molecule that may become chemically bonded to other monomers to form a polymer 56

morals deciding what is right or wrong 16

MRSA methicillin-resistant *Staphylococcus aureus* 17

mucus a sticky material consisting of a mixture of substances produced by goblet cells 9, 18, 20

multicellular an organism made of many cells 4, 23

mutation a permanent change in the structure of a gene – the DNA within cells is altered (this happens in cancer) 7, 9, 13, 30

mutualism a relationship between individuals of different species where both benefit 18

N

nanotechnology engineering systems constructed at the molecular level 53

National Grid the network of pylons, high-voltage cables and transformers that carries electricity from power stations across the country 90

natural selection a process that results in the individuals of a population with characteristics suited to a particular environment surviving, reproducing and therefore passing on the genes controlling the characteristics to their offspring – natural selection is the mechanism of evolution 6, 36

nebula a cloud of dust and gas from which stars form 83

negative tropism growth movement away from a particular stimulus 14

nerve impulses electrical impulses that pass along a neurone 10, 11, 12, 37

neurone a nerve cell that carries nerve impulses 11, 12, 37

neurotransmitter a chemical released from the end of a neurone into a synapse and which stimulates the next neurone to trigger new nerve impulses 12, 15

neutralisation reaction a reaction between an acid and an alkali that produces a neutral solution 48

neutron star a very dense core of a massive star left behind after the supernova stage 83

neutrons small particles that do not have a charge, found in the nucleus of an atom 55, 58, 94, 107, 108, 109

Newton's second law a law expressed by the equation: force = mass x acceleration 102

nitrifying bacteria bacteria that convert ammonium compounds into nitrates 21

Glossary

nitrogen-fixing bacteria bacteria that convert gaseous nitrogen into nitrogen-containing compounds 18, 21

noble gases elements of group 0 in the periodic table, also called the inert gases 60, 67, 69

non-renewable resources resources which are being used up more quickly than they can be replaced, e.g. fossil fuels; they will eventually run out 19, 51, 54, 89, 112

nuclear fusion the fusing together of hydrogen nuclei to produce helium nuclei 83, 108, 109

nucleon number the total number of neutrons and protons within the nucleus of an atom (same as the mass number) 107

nucleons a term used to refer to either protons or neutrons 107

nucleotides the building block units that combine to form a strand of DNA; each nucleotide unit consists of the sugar deoxyribose, a base and a phosphate 24, 29

nucleus (1) the central core of an atom, which contains protons and neutrons and has a positive charge; (2) a distinct structure in the cytoplasm of cells that contains the genetic material 4, 23, 26, 29, 37, 55, 58, 59, 80, 94, 107, 108, 109

nutrients substances essential to maintaining living processes 21, 39

O

objective lens the lens at the front of a telescope 23, 76

order a group consisting of several families (used in classification of living things) 4

ores rocks that contain minerals, including metals, e.g. iron ore 50, 51

organ a collection of tissues joined in a structural unit to serve a common function 37

organic compounds chemicals containing carbon 55

osmoregulation the control of an organism's fluid balance 10

osmosis movement of water from a less concentrated solution to a more concentrated solution through a partially permeable membrane 35

oviparous animals that lay eggs, the embryo does not develop within the mother's body 4

oxidation reaction a reaction in which molecules gain oxygen 42, 50, 52

oxygenated blood blood containing oxyhaemoglobin 37, 38

oxyhaemoglobin the result of oxygen binding to haemoglobin in red blood cells 37

P

painkillers drugs that affect the nervous system, deadening pain 15

pancreas the organ that produces the hormones insulin and glucagons (from endocrine tissue) and digestive enzymes (from exocrine tissue) 13, 39

parallel when components are connected across each other in a circuit 87, 96

parasitism a relationship between individuals of different species where one benefits (parasite) and the other is harmed (host) 18

pathogens harmful organisms that invade the body and cause disease 17, 18

payback time the number of years it takes to get back the cost of an energy-saving method 88

peak the uppermost point of a wave 77

pedigree the known genetic line of descent from generation to generation 7, 9

peer review the process of evaluating the quality of research using anonymous review by experts in a particular field 7

pentadactyl limb the basic arrangement of five digits present in most vertebrates 36

pepsin an enzyme that catalyses reactions that break down protein to peptides (short-chain amino acids) 39

peptides short polymers of amino acids, generally 2–20 amino acid units in length 29

percentage composition the proportion of a compound's relative formula mass composed of a specific element 72

percentage yield the ratio of actual yield to theoretical yield 72

percentile the value of a variable below which a certain percentage of observations fall 36

periodic table a table of all the chemical elements based on their atomic number 58, 59, 60, 67, 68, 69

periodicity The characteristics of repeating patterns of properties in the periodic table 59

peristalsis the contraction and relaxation of muscles, which propagates in a wave down the muscular tube 39

pH a scale running from 0 to 14 that shows how acidic or alkaline a substance is 47, 48

pharmacogenomics the study of how variations in the human genome affect an individual's response to drugs; may lead to the development of drugs tailored to be the most effective according to an individual's genetic make-up 28

phenotype all of the characteristics of an organism 7, 8

phloem columns of living cells in plant stems 35

photography the process of producing permanent images 74, 78, 79

photosynthesis a process carried out by green plants where sunlight, carbon dioxide and water are used to produce glucose and oxygen 19, 21, 34, 42, 43, 53

phototropism growth movement in response to the stimulus of light 14

physical properties properties that can be observed without changing the chemical composition of a substance, e.g. colour, density, melting point and boiling point 46, 67

plant stanol esters substances found in food such as wheat and maize that reduce the absorption of harmful cholesterol 40

plantae the plant kingdom 4

plasmid DNA loops of DNA found in the cytoplasm of bacterial and yeast cells 23

Plasmodium the protozoan genus that causes malaria – *Plasmodium falciparum* is the most dangerous species, *P. vivax* the least 17

poikilotherms animals that cannot regulate their body temperature 4

pollutants substances released into the environment that are harmful to health and wildlife 20

pollution the effects of pollutants which contaminate or destroy the environment 20

polyatomic ions ions containing more than one type of atom 61

polymerisation reaction a chemical process that combines monomers to form a polymer – this is how polythene is formed 56

polymers large molecules made up of chains of monomers 55, 56

polypeptides polymers of amino acids, generally 21–50 amino acid units in length 29

population organisms of the same species that live in the same geographical area 6, 19, 35

positive tropism growth movement towards a particular stimulus 14

potential difference another term for voltage (a measure of the energy carried by an electric charge) 87, 88, 96, 97, 98, 111

power the rate of work done or the rate of energy transfer 88, 98, 105

prebiotics non-digestible food ingredients to stimulate growth of bacteria in the digestive system 40

Glossary

precipitate an insoluble solid formed in a solution during a chemical reaction 46, 62, 63

precipitation reaction a reaction that results in an insoluble product 62, 63

predators animals that feed on other animals 19

prey an animal that is eaten by a predator 19

primary coil the input coil of a transformer 90

principal focus rays parallel to the principal axis of the lens meet at this point after being refracted twice by the lens (also known as the focus or focal point) 75

principle of conservation of energy energy cannot be created or destroyed, it can simply be transferred from one form to another 91, 106

principle of conservation of momentum for a system of colliding objects, where there are no external forces, the total momentum before and after the collision remains the same 104

prioritise to rank in order 16

prism a block of glass used to split white light into a visible spectrum 78, 82

probiotics live microorganisms thought to be beneficial to health 40

producers organisms that synthesise sugars (food) by photosynthesis or chemosynthesis 19

product the substance formed in a reaction 31, 43, 46, 61, 70, 71, 72

prokaryotes single-celled organisms that lack a distinct nucleus 4

proteins molecules made up of amino acids, more than 50 amino acids in length 7, 9, 18, 21, 24, 25, 29, 30, 31, 39

protoctista single-celled organisms that have a distinct nucleus; also some simple multicellular (many-celled) organisms such as seaweed 4

proton number the number of protons inside the nucleus of an atom (same as the atomic number) 107

protons small positively charged particles found in the nucleus of an atom 55, 58, 59, 94, 107

pulse a ripple of blood as it is forced along the arteries by the beating heart 33

Punnett squares a type of genetic diagram that sets out the results of genetic crosses in the form of a table 8

pure-bred a characteristic of an organism that passes unchanged from generation to generation – the organism is homozygous for the characteristic in question 8

pyramid of biomass a diagram representing the biomass of the trophic levels of a community 19

R

radiotherapy a technique that uses gamma rays to kill cancer cells in the body 80, 111

radon a colourless, odourless and radioactive gas originating from rocks such as granite 110

rarefactions regions where particles are pulled apart and create regions of low pressure in sound waves 77

rate of reaction the speed with which a chemical reaction takes place 31, 54, 71

reactant the chemicals that react in a chemical reaction 43, 46, 61, 62, 63, 71, 72

reaction time the time taken to respond to a stimulus 15

reactivity series a list of elements in order of decreasing reactivity used in metal extraction 50

real image an image formed on the other side of the lens to the object – a real image can be formed on a screen 76

receptor part of a neurone that detects stimuli and converts them into nerve impulses 11, 12

recessive refers to an allele which controls the development of a characteristic only if its dominant partner allele is not present 7, 8, 9

recessive characteristic an allele that does not develop a particular characteristic when present with a dominant allele 8

recipient a person who receives a donated organ 16

recovery period the period during which lactic acid is removed and breathing and heart rates return to normal after exercise 32

recycling the reprocessing of materials to make new products 19, 51, 56

red blood cells blood cells which are adapted to carry oxygen 9, 15, 37, 52

red giant a huge expanded star with a cooler surface 83

reduction reaction the removal of oxygen from a compound 50

reflecting telescope a telescope with a concave mirror, a flat mirror and an eyepiece 76

reflection when a wave is bounced off a surface 75, 79

reflex arc the pathway taken by a nerve impulse from a receptor, through the nervous system, to an effector (does not go through the brain), bringing about a reflex response 12

reflex response an automatic action not controlled by the brain, made in response to a stimulus 12

refracting telescope a telescope that uses two convex lenses to collect and focus light 76

refraction the bending of a wave caused by the change in its speed – when a light ray travelling though air enters a glass block it changes direction 75, 76, 79, 82, 86

rejection destruction of a donor organ because of the activity of the recipient's immune system 16

relative atomic mass the average atomic mass of an element, taking into account the relative abundance of the isotopes of that element 58, 72

relative formula mass the sum of all the relative atomic masses of the atoms in a molecule 72

renewable resources energy resources that can be replenished at the same rate that they are used up, e.g. biofuels – they will not run out 19, 54, 89

replication when organisms or cells make copies of themselves 26, 30

residue the solid material collected in a filter paper after filtration 63

resistance refers to pathogens, e.g. bacteria, that are not affected by drugs that previously were effective treatments 17

resistance an electrical quantity determined by dividing potential difference by current 96, 97, 98

resolving power the ability to see objects close together as separate from one another 23

resources the raw materials taken from the environment and used to run industry, homes and transport, and to manufacture goods 19, 89

respiration the series of chemical reactions that oxidise (break down) glucose, releasing energy: in the presence of oxygen, glucose is oxidised to carbon dioxide and water 21, 32, 33

response the action taken as a result of a stimulus 12

restriction enzymes enzymes that catalyse reactions that cut strands of DNA into shorter pieces 25

retina the covering of light-sensitive cells at the back of the eyeball 11

retrograde motion the background motion of a planet against a background of stars 74

R_f value the ratio of how far a substance has moved up chromatography paper relative to the solvent front 66

rheostat a variable resistor 97

ribonucleic acid (RNA) a type of nucleic acid found in cells but not used to build chromosomes; some types of RNA are involved in protein synthesis 29, 30

Glossary

ribosome a component of a cell that creates proteins from all amino acids and RNA representing the protein 29

ring species a connected geographical sequence of neighbouring species that can interbreed with one another; however, the two 'end' species of the sequence cannot interbreed 5

S

sacrificial protection a method of corrosion protection using blocks of reactive metal to corrode instead of the object being protected 50

salt an ionic compound composed of positive ions (cations) and negative ions (anions) 48, 62, 63, 69

Sankey diagram a diagram showing the transfer of energy to different forms 91

saprophytic feeding from dead organic matter 4

saturated hydrocarbons hydrocarbons that contain no carbon-carbon double bonds 55

scalar a quantity that only has size or magnitude 99, 102

scavengers animals that feed on dead animals 19

scientific journals periodic publications with articles contributed by scientists reporting on their new research 7

secondary coil the output coil of a transformer 90

sedimentary rock rock formed by the sedimentation of material on riverbeds and ocean floors 36, 44, 45

seedling a germinating seed which is at a stage where shoots and roots are visibly deepening 14

seismic waves shock waves from earthquakes 77, 86

seismometer an earthquake-detecting instrument 77, 86

semiconductor a group of materials with electrical conductivity properties between metals and insulators 47, 98

sequence an arrangement in which things follow a pattern 24, 28, 30

sequencers machines that automatically determine (work out) the base sequences of DNA (or the amino acid sequence of proteins) 28

series when components are connected end-to-end in a circuit 87, 96

shells electrons are arranged in shells (or orbits) around the nucleus of an atom 55, 59, 60, 64, 68, 69

sickle cell crisis the periods of pain experienced when sickled red blood cells clump together, restricting blood flow to the organs of the body; each crisis may last for days, weeks or months 9

sickle cell disease a recessive genetic disorder in which the shape of haemoglobin molecules is altered, so they absorb less oxygen 9

silent mutation a mutation that alters the base sequence of a codon but does not result in alteration of the sequence of amino acid units of the protein in question 30

simple molecular covalent compounds small molecules consisting of non-metals covalently bonded together 67

Solar System the Sun and all the objects orbiting it (planets, asteroids, comets, etc.) 74, 81

solubility the amount of a substance that will dissolve 46, 62, 67

sonar a technique used by ships to determine the depth of water: it stands for Sound Navigation And Ranging 85

speciation an evolutionary process that results in new species 6

species a group of individuals able to mate and reproduce offspring, which themselves are able to mate and reproduce 4, 5, 6, 18, 20

specific relating to a particular thing or event, e.g. a particular hormone affects only a particular target tissue 13, 31

spectrometer a device used to analyse light from various sources 82

spectroscopy a sophisticated type of flame test: substances are heated until they produce their own unique emission spectrum 63

speed how fast an object travels, calculated using the equation: speed (metres per second) = distance / time 75, 77, 78, 85, 86, 99, 100, 101, 102, 103, 105, 106

starch large polysaccharides made by plants as a form of food storage 39, 56

state symbols the symbols that describe the state of a substance: solid, liquid, gas or aqueous (dissolved in water) 42, 62

Steady State theory a theory that proposes a Universe in which matter is created from empty space to keep its density the same 84

stem cells undifferentiated (unspecialised) cells that are able to develop (differentiate) into differentiated (specialised) cells 28

step-down transformer a device used to change the voltage of an a.c. supply to a lower voltage 90

step-up transformer a device used to change the voltage of an a.c. supply to a higher voltage 90

sterilisation a technique used to kill bacteria by exposure to radiation 111

stimulants substances that speed up responses 15

stimulus a change in the environment that causes a response by stimulating receptor nerve cells, e.g. a hot surface 10, 12

stoma (plural stomata) a pore found in plants, used in gas exchange 34, 35

stopping distance thinking distance + braking distance 103

subcutaneous fat literally means 'fat under the skin' 13

substrate molecules at the start of a chemical reaction; the substance that an enzyme helps to react 31

super red giant a huge expanded star larger than a red giant 83

supernova an exploding star 83

synapse the gap between two adjacent neurons 12

T

tangent a line drawn to a curve to determine the gradient of a curve at a point 100

target tissues tissues that respond to hormones 13

tectonic plates the several solid parts of the Earth's crust 86

telescope a device using lenses (or mirrors) to magnify distant objects 74, 75, 76, 81, 82, 84

terminal velocity the constant velocity of a falling object when the net force acting on it is zero 102

theoretical yield the predicted yield of a chemical reaction based on calculations 72

thermal decomposition the breaking down of a compound due to the action of heat 47, 56

thermistor a sensor in an electric circuit that detects temperature 98

thermograph an image produced using infrared waves 79

thermoregulation the processes than enable an animal to keep its body temperature constant 10

thinking distance the distance travelled by a car as the driver reacts to apply the brakes 103

tissue culture the growth of fragments of tissue in a liquid or on gel, which provides all the substances needed for their development; conditions are sterile and controlled 27

tissue fluid a solution that bathes and surrounds the cells of multicellular animals 38

tissues groups of cells that work together and carry out a similar task, e.g. lung tissue 13, 14, 27, 28, 32, 37, 38

titration a common laboratory method used to determine the unknown concentration of a known reactant 48

Glossary

total internal reflection a phenomenon where 100% of the light is reflected back into a material, when the ray hits the glass/air boundary at an angle that is greater than the critical angle 79

tracer a radioactive material injected into a patient for locating cancer or diagnosing a function of the body 80, 111

transfer RNA (tRNA) a molecule of RNA that transports amino acids to ribosomes 29

transformer a device that converts the voltage of an a.c. supply to another voltage 90

transition metals elements between group 2 and group 3 in the periodic table 67

transpiration the movement of water and mineral salts from roots to leaves in plants 35

transplantation tourism travelling to another country to buy organs 16

transverse waves waves with vibrations at right angles to the direction in which the wave is travelling 77, 78, 86

triple covalent bond three pairs of electrons shared in a covalent bond 64

trophic literally means relating to feeding 19

tropism growth movement by plants in response to stimuli coming mainly from one direction 14

trough the lowest point of a wave 77

turgid rigid owing to high fluid content 23, 35

U

ultrasound sound with frequencies greater than 20 kHz – too high for detection by human ears 75, 85

ultraviolet waves electromagnetic waves with a wavelength shorter than violet (blue) light 74, 76, 78, 79, 80

unicellular consisting of a single cell 4

Universe the whole of space containing all the galaxies 74, 81, 84

unsaturated hydrocarbons hydrocarbons that contain carbon-carbon double bonds 55

V

vacuole a fluid-filled space in the cytoplasm of cells – most plant cells have a permanent vacuole; if vacuoles are present in animal cells they are temporary 23

vacuum empty space that has no particles 78, 102

validated to establish the soundness of, or to corroborate, evidence 7, 109

variation the difference in characteristics between species and the range of a characteristic individuals of the same species 5, 6, 7, 30

vasoconstriction narrowing of the lumen (internal space) of blood vessels in cold conditions – this reduces the flow of blood 10

vasodilation widening of the lumen (internal space) of blood vessels in hot conditions – this increases the flow of blood 10

vector a quantity that has both magnitude and direction 99, 101, 102, 104

vegetative reproduction asexual reproduction from the vegetative parts of a plant: the roots, leaves and stem 27

veins blood vessels that carry blood back to the heart 38

velocity how fast an object is travelling in a certain direction: velocity = displacement / time 99, 100, 101, 102, 105, 106

ventricles chambers of the heart that pump blood into the arteries 38

vertebrates animals that have the characteristic of a backbone in common 4, 36

villi finger-like structures on the surface of the small intestine which give it a greater surface area for absorption 39

virtual image an image formed on the same side of the lens as the object – a virtual image can be seen by looking though the lens, it cannot be projected onto a screen 76

viruses very small infectious organisms that reproduce within the cells of living organisms and often cause disease; they consist of a protein layer surrounding a strand of nucleic acid 4, 17, 37

vitamin A a vitamin that is essential for growth and vision 25

volt the unit of voltage 87

voltage the energy transferred per unit charge – a measure of the energy carried by electric charge (also called the potential difference) 87, 90, 96, 97

voltmeter a device used to measure the voltage across a component 87, 88, 96, 98

voluntary response a response to a stimulus that you think about and can control 12

W

watt the unit for power 88, 105

wavelength the distance between neighbouring wave peaks or wave troughs 23, 77, 78, 79, 82, 84

weight the gravitational force acting on an object, measured in newtons 101, 102, 105, 106

white dwarf a hot and dense core of a star (such as our Sun) left behind after the red giant stage 83

word equations a shorthand way of representing a chemical reaction 43, 61

work done the product of force and distance moved in the direction of the force 105, 106

X

xenotransplantation transplantation of organs from another species 16

X-rays electromagnetic waves with very short wavelength of the order of 0.000 000 001 m 63, 74, 78, 80, 81, 82

xylem columns of hollow, dead reinforced cells in plant stems 35

Y

yield useful product made from a chemical reaction 72

Z

zygote a fertilised egg 26

Exam tips

The key to successful revision is finding the method that suits you best. There is no right or wrong way to do it.

Before you begin, it is important to plan your revision carefully. If you have allocated enough time in advance, you can walk into the exam with confidence, knowing that you are fully prepared.

Start well before the date of the exam, not the day before!

It is worth preparing a revision timetable and trying to stick to it. Use it during the lead up to the exams and between each exam. Make sure you plan some time off too.

Different people revise in different ways and you will soon discover what works best for you.

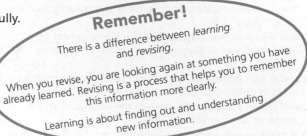

Remember!

There is a difference between *learning* and *revising*.

When you revise, you are looking again at something you have already learned. Revising is a process that helps you to remember this information more clearly.

Learning is about finding out and understanding new information.

Some general points to think about when revising

- Find a quiet and comfortable space at home where you won't be disturbed. You will find you achieve more if the room is ventilated and has plenty of light.

- Take regular breaks. Some evidence suggests that revision is most effective when tackled in 30 to 40 minute slots. If you get bogged down at any point, take a break and go back to it later when you are feeling fresh. Try not to revise when you're feeling tired. If you do feel tired, take a break.

- Use your school notes, textbook and this Revision guide.

- Spend some time working through past papers to familiarise yourself with the exam format.

- Produce your own summaries of each module and then look at the summaries in this Revision guide at the end of each module.

- Draw mind maps covering the key information on each topic or module.

- Review the Grade booster checklists on pages 256–261.

- Set up revision cards containing condensed versions of your notes.

- Prioritise your revision of topics. You may want to leave more time to revise the topics you find most difficult.

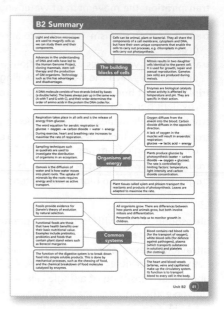

Workbook

The **Workbook** (pages 145–255) allows you to work at your own pace on some typical exam-style questions. You will find that the actual GCSE questions are more likely to test knowledge and understanding across topics. However, the aim of the Revision guide and Workbook is to guide you through each topic so that you can identify your areas of strength and weakness.

The Workbook also contains example questions that require longer answers (**Extended response questions**). You will find one question that is similar to these in each section of your written exam papers. The quality of your written communication will be assessed when you answer these questions in the exam, so practise writing longer answers, using sentences. The **Answers** to all the questions in the Workbook are detachable for flexible practice and can be found on pages 265–288.

At the end of the Workbook there is a series of **Grade booster checklists** that you can use to tick off the topics when you are confident about them and understand certain key ideas. These Grade boosters give you an idea of the grade at which you are currently working.

Classification and naming species

1 a Classify the following animals into their correct vertebrate group.

    i cane toad .... **[1 mark]**

    ii kangaroo .... **[1 mark]**

    iii basking shark .... **[1 mark]**

    iv penguin .... **[1 mark]**

    v python .... **[1 mark]**

b A tarantula does not belong to any of these groups. Explain why.

.... **[1 mark]**

2 Iguanas are reptiles that spend a large portion of the day basking on hot rocks in the Sun. Why do they need to do this?

....

.... **[2 marks]**

3 Plants are autotrophs.

a What does this mean?

.... **[1 mark]**

b What do the cells of plants contain that other cells do not?

.... **[2 marks]**

4 Lichen are compound organisms. This means that they are made up of fungi and algae cells. Explain why they are difficult to classify.

....

.... **[2 marks]**

5 The archaeopteryx was an animal that lived 150 million years ago. Archaeopteryx fossils show that it had feathers, a backbone that extended into its tail and dry scales on its face and claws. Explain why scientists classified it as a vertebrate, but found it difficult to classify it any further.

....

....

.... **[3 marks]**

6 a Tigers are a species of cat. What does the word 'species' mean?

.... **[1 mark]**

b The offspring of a male lion and a female tiger is called a liger. What do we call animals that are the offspring of two different species?

.... **[1 mark]**

7 a The brown rat has the two-part name *Rattus norvegicus*. Which part of this is its genus name?

.... **[1 mark]**

b The species name of the pacific rat is *exulans*. What is its two-part name?

.... **[1 mark]**

8 On an expedition to a rainforest, scientists discovered 50 different-looking spiders in one small area. Explain why the scientists must classify the spiders before they can make a judgement about the biodiversity of the spiders in that area.

....

.... **[2 marks]**

Identification, variation and adaptation

1 A group of students are pond-dipping. They are asked to comment on what species live in the pond. Suggest why a key would be useful to them.

....

G–E

[1 mark]

2 a Oak is a species of tree. Give two examples of variation seen in oak trees.

.... **[2 marks]**

b Animals also show variation. For each of the following examples, state whether it shows continuous or discontinuous variation.

  i weight .... **[1 mark]**

  ii blood group .... **[1 mark]**

  iii length of limbs .... **[1 mark]**

D–C

c You measured the hand spans of the students in your year group and also asked them their eye colour. Describe which type of graphs you would use to display this data and explain your choices.

....

....

....

....

[4 marks]

3 The greenish warbler is a species of bird that lives in the regions surrounding the Himalayas. There are many subspecies of the bird. Most of the subspecies can interbreed, but two cannot. Explain why scientists think this may be an example of a ring species.

B–A*

....

....

[2 marks]

4 Camels are well adapted to living in deserts. Explain how each of the features below enables a camel to survive in desert conditions.

a Fat stored in hump

.... **[1 mark]**

G–E

b Wide feet

.... **[1 mark]**

c Ability to close nostrils

.... **[1 mark]**

d Lemmings are small round animals that look like fat hamsters. They live in the Arctic. Explain how the size and shape of the lemming helps it survive the cold temperatures.

....

.... **[2 marks]**

D–C

e Another Arctic animal is the caribou. Its fur contains hollow hairs. How does this help keep it warm?

....

.... **[3 marks]**

5 Giant tube worms live around deep-sea hydrothermal vents, where temperatures can reach above 90 °C. They belong to the same phylum as earthworms. Explain why the two species would have some characteristics in common but many that are different.

B–A*

....

.... **[3 marks]**

Evolution

1 a Snakes are found in habitats all over the world, including rainforests, deserts and even the sea. All species of snake on Earth evolved from a common ancestor but they all have different characteristics. Explain why.

.... **[1 mark]**

b Rattlesnakes are a group of venomous snakes that can be found in a range of habitats in America.

i Suggest one resource that rattlesnakes compete for.

.... **[1 mark]**

ii Some species of rattlesnake have become extinct. What does this mean?

.... **[1 mark]**

iii Suggest a reason why these species became extinct.

.... **[1 mark]**

c Rattlesnakes show variation in the colour of their skin. This ranges from from pale brown to dark green. Most snakes found in the desert have pale skin. Use natural selection to explain why.

....

....

.... **[3 marks]**

d The end of a rattlesnake's tail (the rattle) is specially adapted to make a rattling noise when shaken. The snake does this to warn away predators. Rattlesnakes evolved from snakes that did not have this 'rattle'.

Suggest how this evolution occurred.

....

....

....

....

....

....

.... **[4 marks]**

2 In 1995, at least 15 iguanas survived a hurricane in their home of the Virgin Islands. They survived for a month on a raft of uprooted trees and landed on another island called Anguilla, which had no iguana population.

a Explain why the iguanas may have found it difficult to survive on Anguilla.

....

....

.... **[2 marks]**

b If they do survive, a new species of iguana may evolve. Explain how this will occur.

....

....

....

.... **[3 marks]**

G–E

D–C

B–A*

B–A*

Genes and variation

1 For each of these examples of variation, state whether they are inherited or environmental:

    a Eye colour .... [1 mark]

    b Hair length .... [1 mark]

    c Blood group .... [1 mark]

2 For each definition below, state whether it is a description of DNA, genes or chromosomes.

    a This long molecule is wound around proteins to form chromosomes. .... [1 mark]

    b Many of these are found on one chromosome. .... [1 mark]

    c These are arranged in pairs inside the nucleus of most cells. .... [1 mark]

G–E

3 Ryan and Chloe are both very tall. Their one-year-old son, Ethan, is also expected to be tall when he is older. Why is this?

....

.... [2 marks]

D–C

4 A comparison of the DNA of humans and chimpanzees show that these animals share 99% of their DNA. Explain how this is evidence for Darwin's theory of evolution.

....

.... [3 marks]

B–A*

5 Mia has alleles for both blood group O and blood group A.

    a Explain why she has two alleles.

    .... [1 mark]

    b Is Mia heterozygous or homozygous for blood group?

    .... [1 mark]

G–E

6 Mice have two alleles for fur colour, white and grey. A white mouse and a grey mouse mate. All of the babies have grey fur.

    a What is the phenotype of the babies?

    .... [1 mark]

    b Which of the alleles is the most likely to be dominant?

    .... [1 mark]

    c Explain why you think this.

    ....

    .... [2 marks]

D–C

7 This family pedigree shows the Jones family. Family members with red hair are shaded grey.

    a What is the relationship between Mark and Holly?

    .... [1 mark]

    b Explain why Holly has red hair when neither of her parents do.

    ....

    ....

    ....

    .... [3 marks]

Robert | Susan

Kirsty | Chris   Rosie | Daniel

Mark   Holly   Jacob

B–A*

Monohybrid inheritance

1 Mendel bred together pure-breeding plants with red flowers and pure-breeding plants with white flowers. He gathered the seeds and planted them. All of the first-generation offspring had red flowers.

a Which allele is dominant – the allele for red flowers or the allele for white flowers?

.... [1 mark]

b Mendel then bred together the offspring. What colour flowers would you expect this second generation to have?

.... [1 mark]

2 Scientists did not take much notice of Mendel's laws of monohybrid inheritance until 16 years after his death. Suggest a reason why.

.... [1 mark]

G–E

3 Use letters to represent the following genotypes.

a Homozygous dominant

.... [1 mark]

b Homozygous recessive

.... [1 mark]

c Heterozygous

.... [1 mark]

D–C

4 Complete the Punnett square opposite to show the likely offspring from one of Mendel's breeding experiments with the red- and white-flowered plants as discussed in Question 1.

Parental gametes	Cross: Rr × rr	
	R	r
r		
r		

[2 marks]

5 What is the probability that offspring from the cross as shown in the Punnett square in Question 4 will:

a have red flowers .... [1 mark]

b have white flowers .... [1 mark]

c be heterozygous .... [1 mark]

d be homozygous dominant .... [1 mark]

e be homozygous recessive .... [1 mark]

6 Mendel then bred the red-flowered plants from this cross to produce a second generation.

B–A*

a Draw a Punnett square to show this cross. [2 marks]

b What is the probability of offspring with white flowers?

.... [1 mark]

Genetic disorders

1 Harry is 19 years old. His sperm cells all carry a mutation.

  a What is a mutation?

  .... [1 mark]

  b Harry's future children will also have this mutation. Why?

  ....

  .... [1 mark]

G–E

2 Cystic fibrosis is an example of a genetic disorder. Those affected produce thick, sticky mucus, particularly in their lungs and digestive system.

  a Explain what we mean by the term 'genetic disorder'.

  .... [1 mark]

  b Describe the main symptoms of cystic fibrosis.

  ....

  .... [2 marks]

  c Another genetic disorder, sickle cell disease, causes red blood cells to become sickle-shaped. Explain how the disorder can lead to organ damage.

  ....

  ....

  ....

  .... [3 marks]

D–C

3 The diagram below is a pedigree analysis for sickle cell disease. A coloured shape indicates an affected individual.

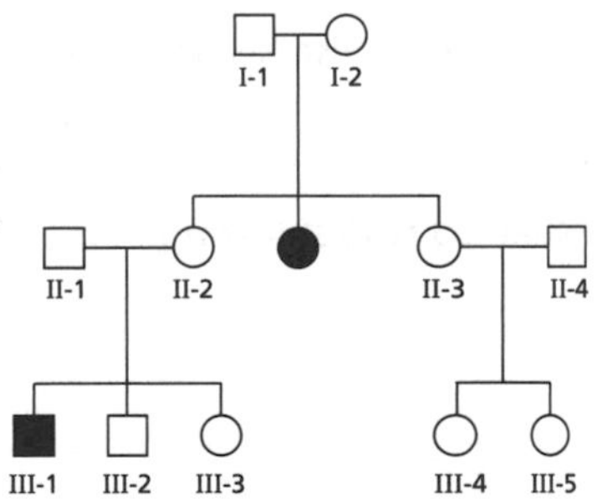

  a What is the function of a pedigree analysis?

  .... [1 mark]

  b Which individuals *must* be a carrier of the sickle cell allele?

  .... [1 mark]

  c Draw a genetic diagram such as a Punnett square to show the cross between II-1 and II-2.

  [2 marks]

  d What is the probability that II-4 is a carrier of the sickle cell allele?

  .... [1 mark]

B–A*

Homeostasis and body temperature

G–E

1 Name two things that are kept constant inside the body.

.... [2 marks]

D–C

2 Jade is at the cinema. She drinks a litre of cola, has to go to the toilet and misses the end of the film.

   a What change will happen to Jade's blood as she drinks the cola?

.... [1 mark]

   b What is her kidneys' response to this stimulus?

.... [1 mark]

   c Describe how this response is brought about by the brain.

....

.... [2 marks]

   d Why is this response known as 'self-regulating'?

.... [1 mark]

B–A*

   e This response in Jade's body is an example of negative feedback. What would the response have been if her brain gave positive feedback?

....

.... [2 marks]

G–E

3 James goes outside into the snow without a coat on. His body temperature starts to decrease.

   a What is normal body temperature for a human?

.... [1 mark]

   b James starts to shiver. How does this help to raise his body temperature?

.... [1 mark]

D–C

   c Explain how James's body knows when to stop shivering.

....

....

....

.... [3 marks]

B–A*

   d Vasoconstriction also takes place in James's body. Explain how this response is brought about and how it helps James to maintain a normal body temperature outside in the snow.

....

....

....

....

....

.... [4 marks]

Senses and the nervous system

1 a Which **two** parts of the body make up the central nervous system?

.... **[2 marks]**

b Name **two** sense organs.

.... **[2 marks]**

c How does information travel from sense organs to the brain?

.... **[2 marks]**

G–E

2 a In which sense organ would you find receptors for:

  i touch .... **[1 mark]**

  ii light .... **[1 mark]**

  iii vibrations .... **[1 mark]**

  iv chemicals .... **[1 mark]**

b Explain why the lips are the most sensitive part of the face.

.... **[1 mark]**

3 Raj is playing as a goalkeeper in a football game. The ball is heading towards him. Study this diagram of his eye.

optic nerve

a In which labelled part of his eye will receptors be stimulated?

retina

blind spot

.... **[1 mark]**

b Describe how this image reaches the brain.

lens

....

....

....

.... **[2 marks]**

D–C

4 The diagrams below show two different types of neurone.

cell body

direction of nerve impulse

A

receptor

dendrites

muscle

direction of electrical impulse

B

a Which is a sensory neurone?

.... **[1 mark]**

b What is the name of the other neurone?

.... **[1 mark]**

B–A*

Responses and coordination

G–E

1 For each of the following responses, say whether they are involuntary or voluntary.

a Blinking

.... [1 mark]

b Pulling your hand away from a hot saucepan

.... [1 mark]

c Chewing a sandwich

.... [1 mark]

2 For one of the involuntary responses in Question 1, explain how it protects the body.

.... [1 mark]

D–C

3 A nerve impulse reaches a gap (synapse). Describe what happens in order for the impulse to start again in the next neurone in the chain.

→ direction of nerve impulses

....

.... [2 marks]

4 Reflex actions do not involve the brain.

a How does this allow them to be fast?

.... [1 mark]

b Why do they need to be fast?

....

.... [2 marks]

B–A*

5 You touch a hot pan. Almost immediately, your arm moves away from the pan. Describe the sequence of events that occurs to bring this about.

....

....

....

....

....

.... [5 marks]

Hormones and diabetes

1 Testosterone is produced by the testes. It controls the development of male characteristics during puberty.

　a Which word in the sentence above is a hormone? .... **[1 mark]**

　b Which word is a gland?... **[1 mark]**

　c Testosterone affects muscles situated all over the body. How does it reach them?

　.. **[1 mark]**　G–E

2 Padma has just eaten a chocolate bar, which causes her blood glucose level to rise. Her pancreas releases a hormone.

　a What is the hormone called?.. **[1 mark]**

　b What is its function?

　.. **[1 mark]**

3 The graph opposite shows changes in the blood glucose level of a person throughout a day.

　Use the graph to state a time the following events occurred:

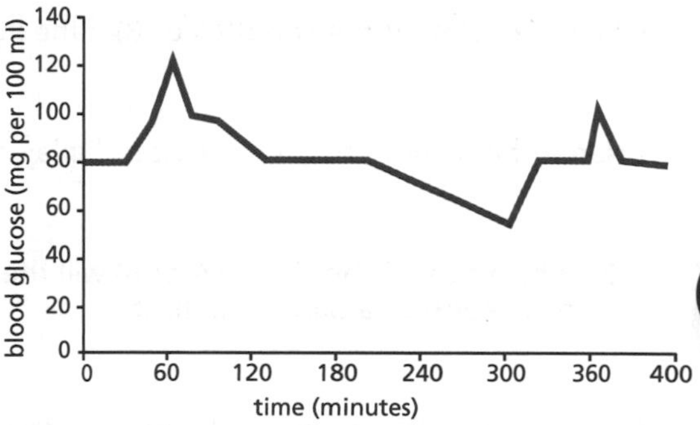

　a Food was eaten

　.. **[1 mark]**

　b The person exercised

　.. **[2 marks]**

　c Explain why this change happened in blood glucose levels when the person exercised.

　..

　.. **[2 marks]**　D–C

　d Explain what happened in order for the blood glucose level to go back to 80 mg per 100 ml between 60 and 120 minutes.

　..

　.. **[3 marks]**　B–A*

　e What hormone was released at 300 minutes?

　.. **[1 mark]**

4 Emma has diabetes. Her body does not produce a certain hormone.

　a Name this hormone.

　.. **[1 mark]**　G–E

　b Which type of diabetes does Emma have?

　.. **[1 mark]**

　c In order to stay healthy Emma's body needs this hormone. How does she supply her body with it?

　.. **[1 mark]**

　d What would happen if she did not do this?

　..

　.. **[2 marks]**　D–C

　e Emma inherited diabetes. Explain how.

　..

　.. **[2 marks]**　B–A*

Plant hormones

1 Tropisms can be positive or negative. Complete each of these statements with the correct word.

a A shoot growing towards the light is _El_ .... phototropism. **[1 mark]**

b A shoot growing away from gravity is ___Heather___ gravitropism. **[1 mark]**

2 For each of the following examples of tropisms, explain how it helps the plant to survive.

a Shoots growing towards the light.

Trees **[1 mark]**

b Roots growing downwards into the soil.

Pineappl **[1 mark]**

3 The picture opposite shows the shoot of a plant being exposed to light coming from one direction.

light

a On which side of the shoot (A or B) is the light brightest?

.... **[1 mark]**

b Name the tropism that the plant is displaying.

.... **[1 mark]**

A   B

c On which side of the shoot (A or B) will the concentration of auxin be highest?

.... **[1 mark]**

d Explain how this causes the bending of the shoot.

....

....

....

....

[3 marks]

4 Plant growers may use rooting powder on their plant cuttings.

a What is the function of rooting powder?

.... **[1 mark]**

b Why would a plant grower want to take cuttings?

....

....

....

[2 marks]

5 In 1926, scientists found that a substance produced by a type of fungus increased the distance along a shoot between leaves. They called the substance gibberellin.

a What evidence is there that gibberellin is a plant hormone?

.... **[1 mark]**

b Scientists treated seedlings with different concentrations of gibberellin. Predict the results.

....

.... **[1 mark]**

Drugs, smoking and alcohol abuse

1 State what type of drug each of the following is (painkiller, hallucinogen, stimulant or depressant).

  a Sleeping pills .... [1 mark]

  b Cocaine .... [1 mark]

  c Paracetamol .... [1 mark]   **G–E**

  d Magic mushrooms .... [1 mark]

  e Nicotine .... [1 mark]

2 There are around 250,000 heroin addicts in the UK.

  a Explain why people become addicted to heroin.

    ....

    .... [2 marks]   **D–C**

  b Why do many heroin addicts turn to crime?

    ....

    .... [2 marks]

3 A group of people were asked to carry out a reaction-time test before and after drinking a cup of coffee.

  a Predict what happened to their reaction times after drinking the coffee.

    .... [1 mark]   **B–A***

  b Explain what effect caffeine has on the nervous system.

    ....

    .... [2 marks]

4 State the substance found in tobacco smoke that contributes to the following illnesses.

  a Lung cancer .... **[1 mark]**   **b** Heart disease .... [1 mark]

5 Why do many smokers constantly have a bad cough?   **G–E**

    ....

    .... [2 marks]

6 Paul drinks four pints of beer in one night.

  a State one short-term effect of drinking this much alcohol.

    .... [1 mark]

  b Half a pint of beer contains one unit of alcohol. How many units has Paul drunk?

    .... [1 mark]   **D–C**

  c Explain why frequently drinking this much alcohol could be dangerous to Paul's health.

    ....

    .... [2 marks]

7 Research data was collected about the causes of people dying from lung conditions.

  a The number of people dying from tuberculosis fell. Suggest why.

    .... [1 mark]

  b However, the number of people dying from lung cancer rose. Suggest why.   **B–A***

    .... [1 mark]

  c This data could not, by itself, prove that smoking caused lung cancer. Why not?

    .... [1 mark]

Ethics of transplants

1 Maria has kidney failure. This means that both of her kidneys have stopped working.

a Maria needs a kidney transplant. What does this mean?

.... [1 mark]

b Maria has a sister who has two healthy kidneys. Explain why she would be a first choice of donor.

....

.... [1 mark]

2 Bernard needs a lung transplant. His brother comes to visit him and suggests he becomes a donor. Why is this not possible?

.... [1 mark]

3 The search for alternative sources of human organs has raised many ethical issues.

Suggest ethical arguments against buying organs from people in developing countries (transplantation tourism).

....

....

.... [2 marks]

4 Another possible source of human organs are animal organs (xenotransplantation).

a Why are pig organs considered the best for human transplants?

.... [1 mark]

b Explain why xenotransplantation is not currently used as an option for transplantation.

.... [1 mark]

c What technology may make xenotransplantation possible in the future?

.... [1 mark]

d Give **one** ethical argument against xenotransplantation.

.... [1 mark]

5 At any one time there will be several people waiting for transplants.

Explain why people have to wait.

.... [1 mark]

6 Doctors have to decide which patient will be next to receive an organ. Discuss the factors that doctors will consider when making this difficult decision.

....

....

....

....

.... [3 marks]

Infectious diseases

1 Tuberculosis is an infection of the lungs caused by bacteria.

  a What word describes a microbe that causes disease?

  .... [1 mark]

  b Tuberculosis is an infectious disease. What does this mean?

  .... [1 mark]

  c Why do bacteria grow and multiply so well inside human lungs?

  .... [1 mark]

G–E

2 Infectious diseases can be spread in many different ways. For each method below, state a disease that is spread in this way.

  a Through moisture droplets in the air .... [1 mark]

  b Eating or touching undercooked food .... [1 mark]

  c Sewage in drinking water .... [1 mark]

D–C

3 Some types of bacteria can cause diarrhoea, which can be fatal. Explain why.

  ....

  ....

  ....

  .... [4 marks]

B–A*

4 Some insects can be vectors.

  a What does the term 'vector' mean?

  .... [1 mark]

  b A person in an African village becomes ill with malaria. Describe how the disease gets passed to other people in the village.

  ....

  ....

  ....

  .... [3 marks]

G–E

5 Britney falls over and cuts her knee. Her Mum puts antiseptic cream on the cut. Why?

  ....

  .... [2 marks]

G–E

6 Ortis goes to his doctor with a bad cold. The doctor does not prescribe him any antibiotics. Why not?

  ....

  .... [2 marks]

D–C

7 There is an outbreak of MRSA in a hospital ward. MRSA is an antibiotic-resistant bacteria.

  a Why is MRSA dangerous to the patients in the ward?

  ....

  .... [2 marks]

  b How did resistance in MRSA arise?

  ....

  ....

  .... [3 marks]

B–A*

Defences and interdependency

1 Why is it vital that skin heals itself after a cut?

....

....

[2 marks]

G–E

2 The human body produces many different chemicals that are used to defend against pathogens. For each of the following chemicals, state where in the body it is made.

a Lysozyme .... **[1 mark]**

b Sebum .... **[1 mark]**

c Hydrochloric acid .... **[1 mark]**

D–C

3 HIV is a virus that destroys white blood cells. Explain why people infected with HIV are at high risk of dying from infectious diseases.

....

....

....

[2 marks]

4 Many areas of tropical rainforest are destroyed each year. Explain why it is important that they are conserved to help in the fight against infectious diseases.

....

....

....

[2 marks]

B–A*

5 a Aphids are tiny insects that feed on the sap of plants. Sap contains sugars that the plants make through photosynthesis and use as food. Is this an example of mutualism or parasitism? Explain the reason for your choice.

....

....

[2 marks]

D–C

b Aphids give the sap they collect to ant colonies. The ants defend the aphids against attacks from their predators, such as ladybirds. Is this an example of mutualism or parasitism? Explain the reason for your choice.

....

....

....

[2 marks]

6 A gardening technique called companion planting makes use of mutualism between plants. The image opposite shows an example called 'the three sisters' technique.

a The bean plant is leguminous. Describe the mutualism between it and nitrogen-fixing bacteria.

....

....

....

....

....

.... **[3 marks]**

corn has tall, sturdy structure

beans are a climbing leguminous plant. Leaves are covered with prickly hairs.

squash leaves cover the ground

B–A*

b Choose one other example of mutualism between two of the plants, and explain the relationship.

....

....

[2 marks]

Energy, biomass and population pressures

1 A food chain that exists in the ocean is: seaweed ⟶ limpet ⟶ octopus ⟶ seal

  a Choose the organism in the food chain that is an example of a:

    i producer ....   **ii** consumer .... **[2 marks]**

  b A limpet is a herbivore. What does this mean?

  .... **[1 mark]**

  c Fill in the gaps in the following sentence.

    A seal is a ...., its prey is the .... **[2 marks]**

  d Crabs feed on bits of dead plants and animals that fall to the ocean floor.

    What type of consumer are they? .... **[1 mark]**

G–E

2 a Why are food webs a more accurate representation of feeding relationships than food chains?

  ....

  .... **[2 marks]**

  b Explain why food chains are rarely longer than five organisms.

  ....

  .... **[2 marks]**

D–C

3 Sketch a pyramid of biomass for the food chain in Question 1. Label each level with the name of the organism only.

B–A*

[2 marks]

4 A population pyramid for Germany is shown opposite.

  a Compare the numbers of children and adults living in Germany.

  .... **[1 mark]**

  b Is the population of Germany set to increase or decrease in the future? Use the pyramid to explain your prediction.

  ....

  .... **[2 marks]**

G–E

5 Plastic is a resource made from the raw material crude oil.

  a Predict how the demand for plastics will change in the near future. Give a reason for your prediction.

  .... **[2 marks]**

  b Scientists are researching ways of making plastics from raw materials other than oil. Give a reason why this is a good idea.

  .... **[1 mark]**

D–C

6 Adil has to sort out his rubbish into two bags. Paper, metal and some plastic goes into a recycling bag. The rest goes into a normal black refuse bag.

  a State the two places that the rubbish in the black bag could end up.

  .... **[2 marks]**

  b Adil thinks that recycling wastes energy as so much is used up collecting and sorting the contents of the recycling bag. Counteract his argument.

  ....

  .... **[2 marks]**

B–A*

Water and air pollution

1 Farmers use fertilisers to increase crop yield. These fertilisers may contain pollutants.

a Define the word 'pollutant'.

....

[1 mark]

G–E

b Name one pollutant that may be found in fertilisers.

....

[1 mark]

c These pollutants could enter drinking water. How?

....

....

[2 marks]

2 Ponds and rivers that have increased amounts of pollutants in their water will also contain higher than normal numbers of bacteria.

a Explain why.

....

....

[3 marks]

D–C

b Why is this a problem for animals living in the polluted water?

....

....

[2 marks]

3 A sample of water is taken from a river and is found to contain high numbers of bloodworms. Comment on the oxygen concentration in the water and link this to how polluted it is.

B–A*

....

....

[2 marks]

4 Acid rain or snow is a major environmental concern.

a Describe how burning fossil fuels can produce acid rain or snow.

....

....

[2 marks]

b State one way that acid rain damages the environment

....

[2 marks]

5 A scientist surveyed the distribution of lichens. She counted the number of different species that grew on trees at various distances from the centre of a polluted city. Her results are shown opposite.

D–C

Distance to the town centre (km)	0	1	2	3	4	5	6	7	8
Number of different species of lichen	0	1	2	4	4	5	8	10	13

a Describe the relationship between the number of species and the distance from the centre of the city.

....

[1 mark]

b What do these results show about the levels of sulfur dioxide in the air as you get further from the city centre? Explain how you know.

....

....

....

[3 marks]

6 Explain how cars in the UK can cause acid rain in Sweden.

B–A*

....

....

....

[3 marks]

Recycling carbon and nitrogen

1 Food for decomposers is dead plants and animals.

a Name an organism that is a decomposer.

.... **[1 mark]**

b Without decomposers plants couldn't grow. Explain why.

....

.... **[2 marks]**

G–E

2 This diagram of the carbon cycle shows how carbon is recycled around the Earth.

Name the process missing in:

A .... **[1 mark]**

B .... **[1 mark]**

D–C

3 Animals need carbon in order to build tissue.

a How do they get the carbon they need?

.... **[1 mark]**

b A carbon atom that makes up the muscle of an animal will one day return to the atmosphere. Explain how this will happen.

....

....

.... **[3 marks]**

4 Both nitrifying and denitrifying bacteria live in the soil. Nitrifying bacteria require high levels of oxygen to survive. Denitrifying bacteria are anaerobic. This means they do not require oxygen and high levels will kill them.

a Describe the role of nitrifying bacteria in the nitrogen cycle.

....

.... **[2 marks]**

b What do denitrifying bacteria do?

.... **[1 mark]**

B–A*

c The soil beneath Sandra's lawn is very waterlogged. She notices that her grass goes yellow and does not grow well. Explain why.

....

....

....

....

....

.... **[5 marks]**

B1 Extended response question

In the 1920s, Dutch biologist Frits Went investigated the development of cereal seedlings. He suggested that a substance – which he called auxin – regulated growth of the shoots. He carried out experiments to investigate the effects of auxin on shoot growth in response to light. One of the hypotheses that Went wanted to test was: **The tip of the shoot detects the direction of light**.

Write a plan that will enable you to test this hypothesis.

The quality of written communication will be assessed in your answer to this question.

[6 marks]

Seeing cells and cell components

1 Gareth wanted to see what one of his cheek cells looked like.

  a Why must he use a microscope to do this?

  .... **[1 mark]**

  b He used an objective lens with magnification ×20 and an eyepiece lens with magnification ×10. What is the total magnification he used to view the cells?

  .... **[1 mark]**

2 Give **two** advantages of using an electron microscope over a light microscope.

....

.... **[2 marks]**

3 Look at the diagram of a cell opposite.

  a Name the missing components (1 and 2).

    1..

    2.. **[2 marks]**

  b Is this an animal or a plant cell? Explain your decision.

  ....

  .... **[2 marks]**

cell wall
chloroplast
1
2

4 The type and number of components of cells differ according to their function. Explain why:

  a Sperm cells have many mitochondria in their cytoplasm.

  ....

  .... **[2 marks]**

  b The cell walls of the cells in a plant stem are thicker than the walls of other plant cells.

  ....

  .... **[2 marks]**

  c Plant root cells have no chloroplasts but leaf cells have many.

  ....

  ....

  .... **[3 marks]**

5 A scientist studied a cell under a microscope and drew a diagram of what he saw. He labelled some of the components.

flagella

nucleus

cell wall

Which component(s) is found:

  a In plant and animal cells?

  .... **[1 mark]**

  b In both plant and bacterial cells?

  .... **[1 mark]**

  c Is this a bacterial cell? Explain your decision.

  ....

  .... **[2 marks]**

DNA

1 a DNA has a double helix structure. What does this mean?

....

....

....　**[2 marks]**

b The bases of a single strand of DNA are AATGCTTA.
Write out the order of the bases in the other
complementary strand.

....

....　**[1 mark]**

c In a gene, a nucleotide with the base C changes into an A. Explain why this may affect the
protein made from this gene.

....　**[2 marks]**

2 Explain the purpose of each of the following stages in the DNA extraction process.

a Adding a detergent/salt solution to the cells.

....

....　**[2 marks]**

b Using a protein-denaturing enzyme.

....

....　**[2 marks]**

c Adding cold methanol.

....

....　**[2 marks]**

3 The hydrogen bonds holding the complementary base pairs together are weak.

Explain why this is important.

....

....　**[2 marks]**

4 Explain how the work of Franklin and Wilkins helped Watson and Crick to discover the
structure of DNA.

....

....　**[2 marks]**

5 Explain how genetic engineering relies on the fact that the genetic code is universal.

....

....

....

....　**[3 marks]**

Genetic engineering and GM organisms

1 Human insulin can be produced by bacteria grown in a fermenter.

a Why are these bacteria known as a GM organisms?

.... **[1 mark]**

b The culture inside the fermenter has to be maintained at a temperature of around 37 °C. Why?

....

.... **[2 marks]**

G–E

2 Bacteria are genetically engineered by following the procedure shown in the diagram below.

strand of DNA carrying a gene which enables
cells to produce useful proteins

bacterial cell

bacteria have pieces of circular DNA
called plasmid DNA

A   B

C

plasmid is put back
into the bacterial cell

bacteria multiply and produce millions of
identical clones, all with the DNA coding
for the required protein

bacteria grow in special tanks called fermenters.
The end product is removed from the fermenter

For each of the stages A, B and C, state what type of enzyme is used.

....

.... **[3 marks]**

D–C

3 State **one** benefit and **one** drawback of using genetically engineered insulin in the treatment of diabetes.

....

.... **[2 marks]**

B–A*

4 A type of GM cotton plant called Bt cotton produces a toxin in its leaves, which kills any insect pests that feed on it.

a Explain why farmers may want to grow Bt cotton instead of normal cotton.

....

.... **[2 marks]**

b State **one** reason why people may be worried about farmers growing Bt cotton.

.... **[1 mark]**

G–E

5 Vitamin A deficiency is believed to cause the death of around two million people a year in developing countries.

a How can the GM crop golden rice help people in these countries?

....

.... **[2 marks]**

b Some people are against golden rice being available for consumption. State **one** of their arguments.

....

.... **[2 marks]**

D–C

6 Scientists have developed GM wheat that is resistant to herbicides. Explain how growing herbicide-resistant wheat helps the farmers to make more profit.

....

....

.... **[3 marks]**

B–A*

Mitosis and meiosis

1 Mitosis is a type of cell division. Describe why:

a Mitosis results in two daughter cells.

.... [1 mark]

G–E

b The daughter cells are genetically identical to the parent cell.

.... [1 mark]

c Skin cells frequently undergo mitosis.

.... [1 mark]

2 Hydras are organisms that can carry out both asexual and sexual reproduction.

Which type of cell division:

a Is used to form offspring in asexual reproduction?

.... [1 mark]

b Forms gametes for sexual reproduction?

.... [1 mark]

D–C

3 An adult hydra cell has 30 chromosomes. The diagram opposite shows the stages that happen during mitosis.

a For each cell (1–3) state how many chromosomes are present.

.... [2 marks]

b The daughter cells are clones. What does this mean?

.... [1 mark]

c The daughter cells are diploid. What does this mean?

.... [1 mark]

4 When the hydra reproduces sexually it carries out meiosis.

a Why is meiosis needed in order for sexual reproduction to take place?

....

.... [2 marks]

The diagram opposite shows the stages that happen during this process.

b For each stage (1–7), state the number of chromosomes present in each cell.

....

B–A*

....

.... [3 marks]

c Which cells are haploid? Why are haploid cells essential for successful sexual reproduction?

....

....

.... [3 marks]

d Each daughter cell is genetically different. Explain how this occurs.

....

.... [2 marks]

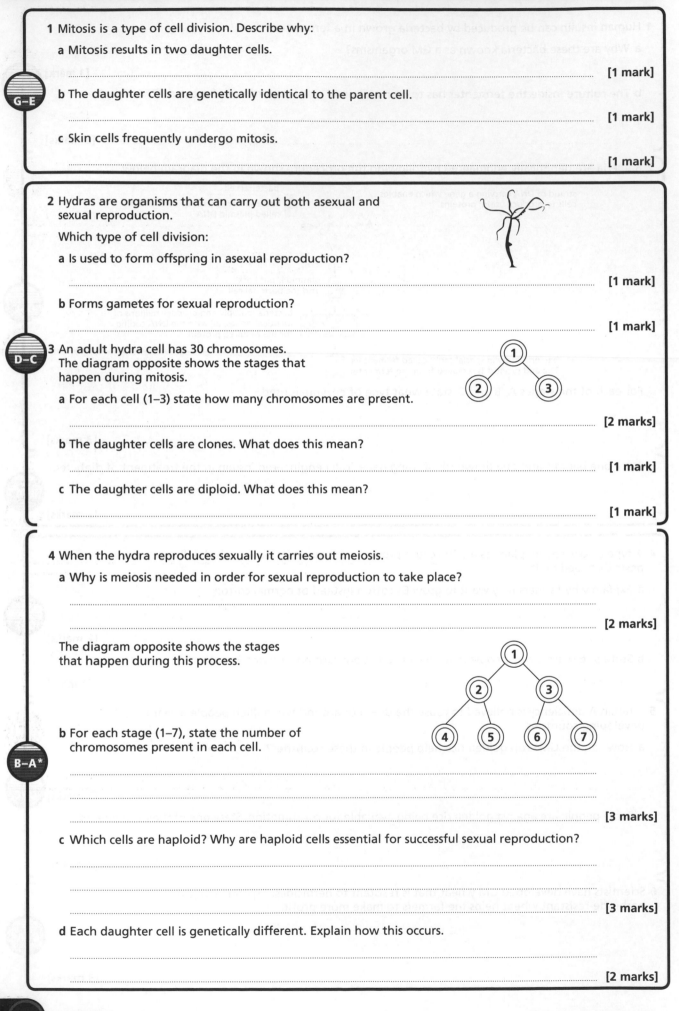

Cloning plants and animals

1 Susan has successfully bred an award-winning plant. She wants to clone it.

a What are clones?

.... [1 mark] G–E

b Why does Susan want to clone her plant?

.... [1 mark]

c State **one** method she can use to clone her plant, and explain how to carry it out.

....

....

.... [3 marks] D–C

2 An egg was taken from a cow and fertilised in the laboratory. The resulting embryo was split into individual cells and each cell put into the womb of a host mother.

a Explain why the host cow will give birth to more than one baby.

.... [1 mark]

b The babies will be clones. Why?

.... G–E

.... [2 marks]

c Why might a cattle farmer want to carry out this procedure?

....

.... [2 marks]

d Why might the farmer want to consider cloning rather than embryo transplants?

....

.... D–C

.... [2 marks]

3 The Pyrenean ibex, a form of wild mountain goat, was officially declared extinct in 2000, but scientists have preserved tissue samples and can use these to create clones. D–C

a Give **one** risk of making clones of the ibex.

.... [1 mark]

b Outline the process that could be used to clone the ibex.

....

.... B–A*

....

.... [4 marks]

Stem cells and the human genome

1 Choose the correct words to complete the following statements about stem cells:

a Stem cells are cells that have the potential to do any job.

They are .... **[1 mark]**

b Stem cells then change into the different types of cell that have a specific job.

This is called .... **[1 mark]**

G–E

2 We can use stem cells to treat people whose tissues are damaged.

a Why are embryonic stem cells more useful than adult cells in stem cell therapy?

....

.... **[2 marks]**

b Some people feel that using embryonic stem cells is ethically wrong. Why do they think this?

....

.... **[2 marks]**

D–C

3 You see a website advertising a cure for blindness using stem cell therapy in a clinic in China. Outline reasons why people should be cautious about going there for treatment.

....

....

....

....

.... **[3 marks]**

B–A*

4 Scientists are working on stem cell therapy using adult stem cells that have been programmed for reverse differentiation.

Explain the reasons why these stem cells would be suitable for stem cell therapy.

....

....

....

....

.... **[3 marks]**

5 a What was the primary aim of the Human Genome Project?

.... **[1 mark]**

b Many scientists from different countries worked together on the project. Why was this collaboration helpful?

....

.... **[2 marks]**

B–A*

c Scientists hope to use the results to work out the protein that each gene in the genome codes for. State **one** potential application for this information.

.... **[1 mark]**

Protein synthesis

1 The image below represents a protein molecule.

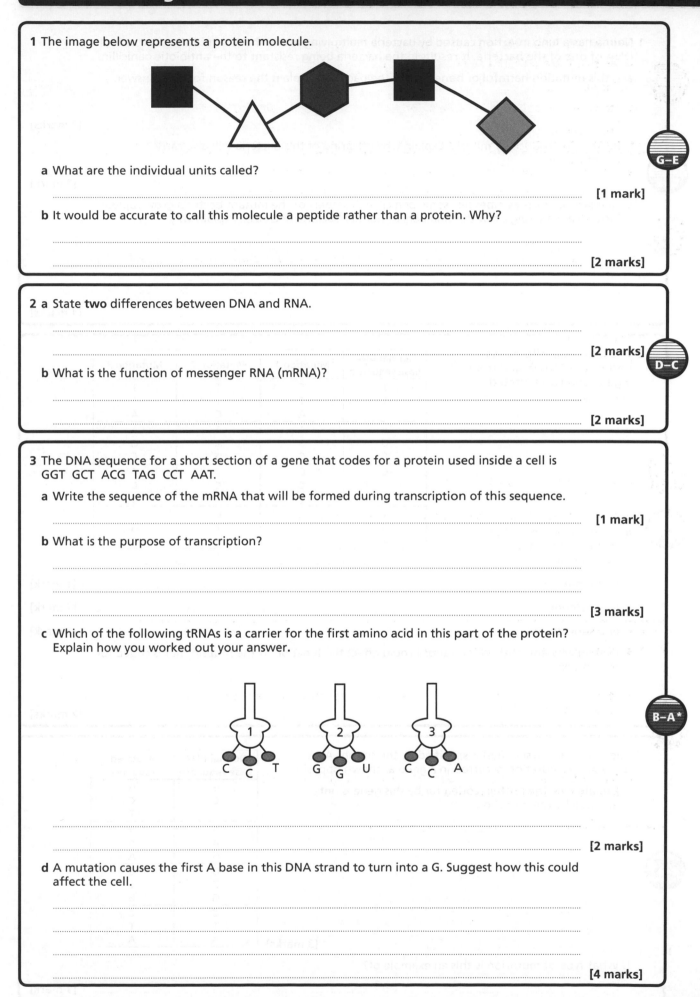

a What are the individual units called?

.... **[1 mark]**

b It would be accurate to call this molecule a peptide rather than a protein. Why?

....

.... **[2 marks]**

2 a State **two** differences between DNA and RNA.

....

.... **[2 marks]**

b What is the function of messenger RNA (mRNA)?

....

.... **[2 marks]**

3 The DNA sequence for a short section of a gene that codes for a protein used inside a cell is
GGT GCT ACG TAG CCT AAT.

a Write the sequence of the mRNA that will be formed during transcription of this sequence.

.... **[1 mark]**

b What is the purpose of transcription?

....

....

.... **[3 marks]**

c Which of the following tRNAs is a carrier for the first amino acid in this part of the protein?
Explain how you worked out your answer.

....

.... **[2 marks]**

d A mutation causes the first A base in this DNA strand to turn into a G. Suggest how this could
affect the cell.

....

....

....

.... **[4 marks]**

Mutations

1 Naima has a lung infection caused by bacteria multiplying inside them. A mutation occurs in the DNA of one of the bacteria. It results in the bacteria being resistant to the antibiotic penicillin.

a Is this mutation harmful or beneficial to the bacteria? Explain the reason for your answer.

....

.... **[2 marks]**

b Bacteria reproduce by mitosis. Explain why offspring of this bacteria will also carry the mutation.

.... **[1 mark]**

c Naima's doctor prescribes her some penicillin. Numbers of the mutant bacteria grow rapidly inside Naima's lungs, but the numbers of normal bacteria decrease. Explain why.

....

....

....

.... **[4 marks]**

G–E

2 The table opposite shows part of the DNA strand from a gene and two mutated versions.

Normal DNA base sequence	Mutation 1	Mutation 2	Mutation 3
T	T	T	T
T	T	T	T
A	A	G	A
A	A	A	A
G	G	A	G
C	C	G	C
C	C	C	T
C	C	C	C
C	T	C	C
T	G	C	T
G	A	T	G
A		G	A
		A	

D–C

a State the number of the mutation that is an example of:

i an insertion .... **[1 mark]**

ii a deletion .... **[1 mark]**

iii a substitution .... **[1 mark]**

b Explain why any of these mutations could affect the function of the protein that is coded for by this gene.

....

.... **[2 marks]**

3 The amino acid lysine (lys) is specified by the codons AAA and AAG. A mutation occurred in a gene as shown opposite.

a Explain why the protein coded for by this gene is not affected by the mutation.

Normal DNA base sequence	Mutated sequence
G	G
C	C
T	T
A	A
A	A
G	A
G	G
T	T
G	G
C	C
T	T
A	A

B–A*

....

....

....

....

....

.... **[3 marks]**

b What type of mutation is this an example of?

.... **[1 mark]**

Enzymes

1 Amylase and pepsin are examples of enzymes that are found in the human digestive system.

a Enzymes are biological catalysts. What does this mean?

.... [2 marks]

b Why are enzymes needed for human digestion?

.... [2 marks] **G–E**

c Pepsin is found in the stomach, which contains acid. Describe what the activity of pepsin will be like at pH 10. Give a reason for your answer.

....

.... [3 marks]

2 Pepsin catalyses the breakdown of proteins into amino acids. In this reaction what is the:

a substrate? .... [1 mark]

b product? .... [1 mark] **D–C**

c Explain why the way enzymes work is sometimes referred to as the 'lock and key' hypothesis.

....

.... [2 marks]

3 Describe the role of **one** enzyme involved in protein synthesis and explain why an enzyme is needed.

.... **B–A***

.... [2 marks]

4 The graph opposite shows the effect of temperature on an enzyme-controlled reaction.

a Describe what is happening to the activity of the enzyme between –5 °C and 0 °C.

.... [1 mark]

b What is the optimum temperature for this enzyme?

.... [1 mark] **G–E**

c The enzyme was taken from a fish. Would you say the fish lived in warm or cold water? Give a reason for your answer.

....

.... [2 marks]

5 Kuba was investigating the action of the enzymes. He kept the concentration of enzyme the same but slowly increased the concentration of substrate, measuring the rate of reaction each time. All other factors, such as temperature, were kept constant.

a At first he found that increasing the amount of substrate speeded up the rate of the reaction. Why?

.... **D–C**

.... [3 marks]

b Kuba then discovered that the rate of reaction stopped increasing, even though he was still increasing the concentration of substrate. Why?

.... [1 mark]

6 Amylase is an enzyme that breaks starch down into glucose. It has an optimum pH of around 8.

Explain why starch is not broken down by amylase at pH 1.

.... **B–A***

....

.... [4 marks]

Respiring cells and diffusion

1 The table opposite compares the percentage by volume of inhaled and exhaled air in humans.

Gas	Inhaled air (%)	Exhaled air (%)
Nitrogen	78	78
Oxygen	21	16
Carbon dioxide	0.035	4
Other gases	about 1–2	about 1–2

a State the difference in the percentage of carbon dioxide between inhaled and exhaled air.

.... **[1 mark]**

b Why is the percentage of oxygen in exhaled air lower than in inhaled air?

.... **[2 marks]**

c Why is there no difference in the percentage of nitrogen?

.... **[1 mark]**

2 Lydia is out doing her weekly run.

a During the first 10 minutes she runs at a fast pace. What type of respiration are her leg muscle cells carrying out?

.... **[1 mark]**

b After 10 minutes she notices she has less energy and her legs start to ache. What type of respiration is being carried out now in her leg muscles?

.... **[1 mark]**

c Why did Lydia's leg muscles start to ache?

.... **[1 mark]**

d Why did she have less energy?

.... **[1 mark]**

e Lydia decides to rest after 20 minutes. She continues to breathe fast and her legs start to feel better. Explain why.

....

.... **[2 marks]**

3 A pot of fresh coffee is brewing in the kitchen. Use ideas about diffusion to explain why the smell travels throughout the house.

....

.... **[3 marks]**

4 Glucose molecules pass through the wall of the small intestine and enter blood capillaries via diffusion.

a Explain why this process is vital to human survival.

....

.... **[3 marks]**

b Where is the concentration of glucose highest: in the small intestine or the bloodstream? Explain why this is important.

....

.... **[3 marks]**

5 a Explain how a concentration gradient is maintained across the small intestine.

....

.... **[3 marks]**

b Why is it important that this concentration gradient is maintained?

.... **[1 mark]**

Effects of exercise

1 During exercise breathing rate increases, enabling more oxygen to enter the blood through the lungs.

a Why is more oxygen needed in the blood during exercise?

....

.... [3 marks]

b Give another reason why it is important that breathing rate increases during exercise.

.... [1 mark]

c Apart from the lungs, which other organ is important in increasing the amount of oxygen getting to exercising muscles?

.... [1 mark]

G–E

2 You are asked to carry out investigations into how exercise affects the body. For each of these questions, suggest a simple method to collect results and predict what they will show.

a How does exercise affect the amount of carbon dioxide exhaled in one breath?

....

....

.... [3 marks]

b How does exercise affect heart rate?

....

.... [2 marks]

D–C

3 Three people had breathing measurements taken.

a Use the following equation to fill in the missing values in the table.

$$\frac{\text{number of breaths in a}}{\text{minute (breathing rate)}} \times \frac{\text{volume of air}}{\text{per breath}} = \text{volume of air exchanged}$$

Person	Volume of air per breath (dm³)	Breathing rate (breaths per minute)	Volume of air exchanged per minute (dm³)
Tom	4.4	50	i
Vishram	0.5	ii	9
Chelsea	iii	20	6

[3 marks]

b Which person is:

i at rest?

.... [1 mark]

ii exercising?

.... [1 mark]

iii a young child?

.... [1 mark]

B–A*

4 Peter has a stroke volume of 0.07 l. His resting heart rate is 72 beats per minute.

Use the following equation to calculate his cardiac output.

cardiac output = stroke volume × heart rate

....

.... [3 marks]

Photosynthesis

1 An apple tree is a green plant that carries out photosynthesis.

a Why is photosynthesis an essential process for the tree?

.... **[2 marks]**

b The flowers of the tree are white. Can the petal cells carry out photosynthesis? Give a reason for your answer.

.... **[2 marks]**

G–E

2 Explain **one** way that a leaf is adapted for photosynthesis.

.... **[2 marks]**

D–C

3 Greenhouses are used to increase the rate of photosynthesis of the plants growing inside.

a Suggest how tomatoes grown inside a greenhouse would compare to those grown outside.

.... **[1 mark]**

b For each labelled part of the greenhouse, explain how it maximises the rate of photosynthesis of the plants growing inside.

i .... **[1 mark]**

ii .... **[1 mark]**

iii .... **[1 mark]**

i shades removed from ceiling

ii heater

iii automatic watering system

c Suggest one more addition that would maximise photosynthesis even more.

.... **[1 mark]**

G–E

4 An experiment was set up as shown in the diagram below. The lamp was moved towards the plant and the average number of bubbles of oxygen released by the plant per minute was counted and recorded at each distance. The results are shown in the table below.

Distance between the lamp and plant (cm)	Number of bubbles of oxygen released per minute
50	1
45	2
40	5
35	10
30	16
25	32
20	54
15	56
10	56
5	56

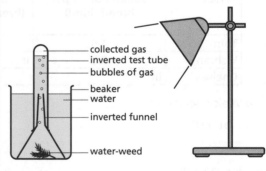

collected gas
inverted test tube
bubbles of gas
beaker
water
inverted funnel
water-weed

a Describe how you could prove that the collected gas was oxygen.

.... **[2 marks]**

b Use the results to describe how light intensity affects the rate of photosynthesis.

.... **[1 mark]**

c Explain why the number of bubbles does not increase any further once the lamp gets closer than 15 cm away.

....

.... **[2 marks]**

D–C

d The switched-on lamp was left 5 cm away from the water-weed for a further 30 minutes. After this time, the number of bubbles released in one minute was counted again. The result was only 5 bubbles. Explain why.

....

.... **[3 marks]**

B–A*

Transport in plants, osmosis and fieldwork

1 This diagram shows the tubes that exist within a plant's stem.

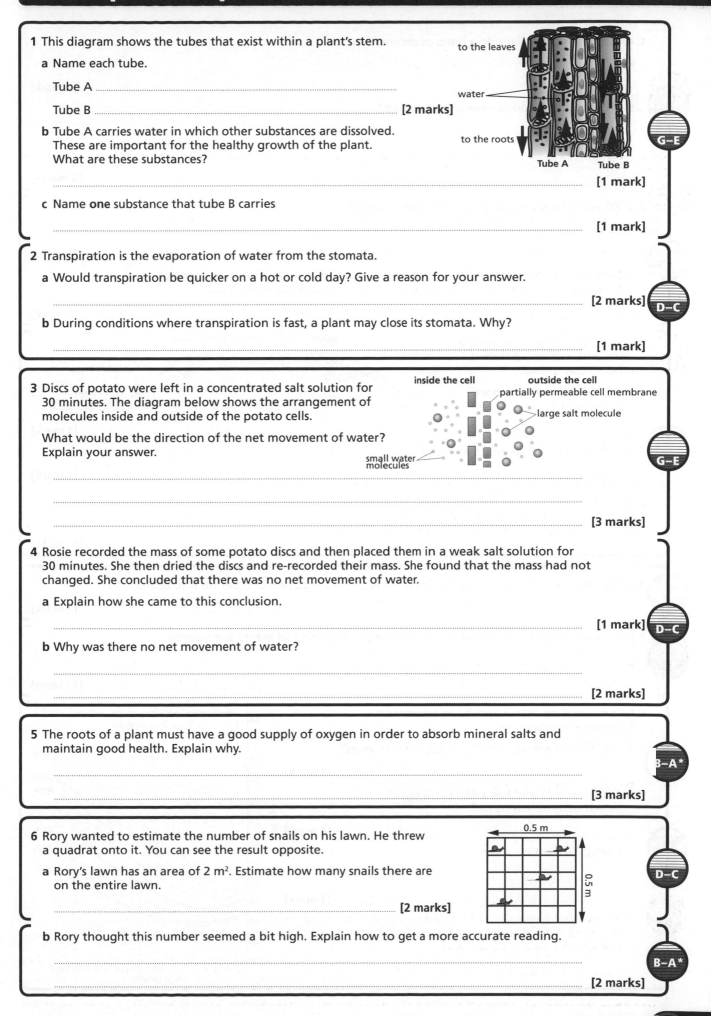

a Name each tube.

Tube A ....

Tube B .... **[2 marks]**

b Tube A carries water in which other substances are dissolved. These are important for the healthy growth of the plant. What are these substances?

....
[1 mark]

c Name **one** substance that tube B carries

....
[1 mark]

to the leaves

water

to the roots

Tube A   Tube B

G–E

2 Transpiration is the evaporation of water from the stomata.

a Would transpiration be quicker on a hot or cold day? Give a reason for your answer.

.... **[2 marks]**

b During conditions where transpiration is fast, a plant may close its stomata. Why?

....
[1 mark]

D–C

3 Discs of potato were left in a concentrated salt solution for 30 minutes. The diagram below shows the arrangement of molecules inside and outside of the potato cells.

What would be the direction of the net movement of water? Explain your answer.

....

....

....
[3 marks]

inside the cell   outside the cell
partially permeable cell membrane

large salt molecule

small water
molecules

G–E

4 Rosie recorded the mass of some potato discs and then placed them in a weak salt solution for 30 minutes. She then dried the discs and re-recorded their mass. She found that the mass had not changed. She concluded that there was no net movement of water.

a Explain how she came to this conclusion.

.... **[1 mark]**

b Why was there no net movement of water?

....

....
[2 marks]

D–C

5 The roots of a plant must have a good supply of oxygen in order to absorb mineral salts and maintain good health. Explain why.

....

....
[3 marks]

B–A*

6 Rory wanted to estimate the number of snails on his lawn. He threw a quadrat onto it. You can see the result opposite.

a Rory's lawn has an area of 2 m². Estimate how many snails there are on the entire lawn.

.... **[2 marks]**

0.5 m

0.5 m

D–C

b Rory thought this number seemed a bit high. Explain how to get a more accurate reading.

....

....
[2 marks]

B–A*

Fossil record and growth

1 Amira found a fossil inside a lump of sedimentary rock.

  a What is a fossil?

  ....

                                                           [1 mark]

G–E

  b Why are fossils found in sedimentary rock?

  ....

  ....

  ....

                                                           [3 marks]

2 a What do we mean by 'gaps in the fossil record'?

  ....

                                                           [2 marks]

D–C

  b Some people believe that the gaps in the fossil record are evidence that **disproves** Darwin's theory of natural selection. Give **one** way that scientists explain the gaps.

  ....

                                                           [1 mark]

3 The images opposite show the 'arm' limb of three different modern-day animals.

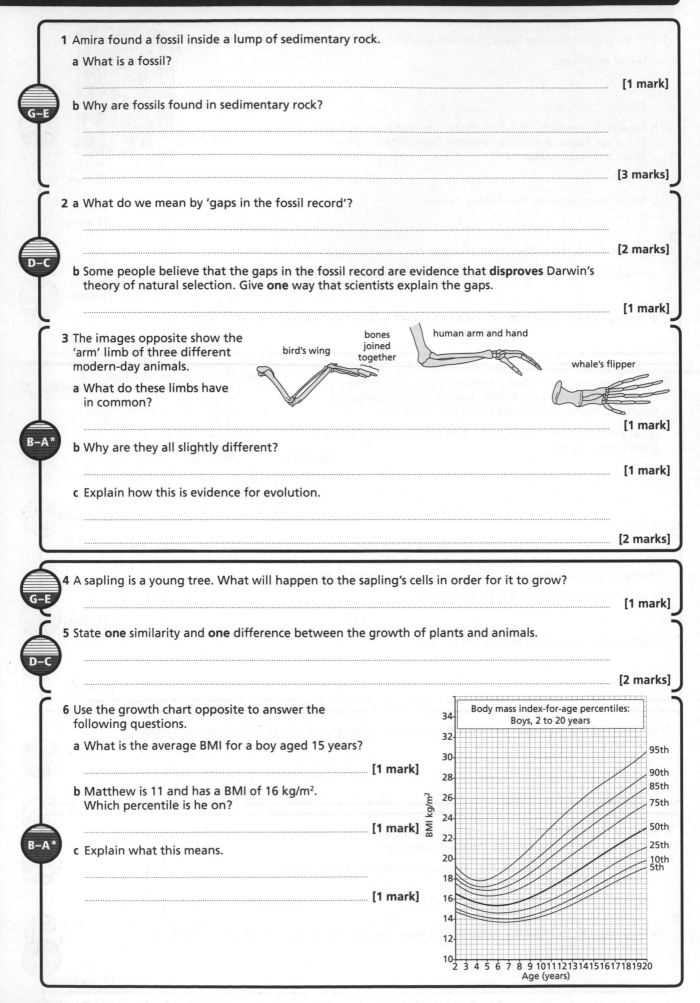

bird's wing   bones joined together   human arm and hand   whale's flipper

  a What do these limbs have in common?

  ....

                                                           [1 mark]

B–A*

  b Why are they all slightly different?

  ....

                                                           [1 mark]

  c Explain how this is evidence for evolution.

  ....

  ....

                                                           [2 marks]

4 A sapling is a young tree. What will happen to the sapling's cells in order for it to grow?

G–E

  ....   **[1 mark]**

5 State **one** similarity and **one** difference between the growth of plants and animals.

D–C

  ....

                                                           [2 marks]

6 Use the growth chart opposite to answer the following questions.

  a What is the average BMI for a boy aged 15 years?

  .... **[1 mark]**

  b Matthew is 11 and has a BMI of 16 kg/m². Which percentile is he on?

  .... **[1 mark]**

B–A*

  c Explain what this means.

  ....

  .... **[1 mark]**

Body mass index-for-age percentiles: Boys, 2 to 20 years

BMI kg/m²

95th, 90th, 85th, 75th, 50th, 25th, 10th, 5th

Age (years)

Cells, tissues, organs and blood

1 Classify each of these as a cell, tissue or organ.

  a Muscle .... [1 mark]

  b Neurone .... [1 mark]

  c Sperm .... [1 mark]

  d Liver .... [1 mark]

  e Skin .... [1 mark]

2 Match each of the organs below to the organ system it is part of.

a	ovary	1	digestive system
b	stomach	2	nervous system
c	heart	3	circulatory system
d	brain	4	reproductive system

[4 marks]

3 Most cells in the adult human body have undergone differentiation. What does this mean?

....

.... [2 marks]

4 For each of the following cells, give **one** way that it is adapted to carry out its function.

  a Red blood cell.

.... [2 marks]

  b Sperm cell.

.... [2 marks]

5 Why are patterns in gene activity vital in the process of differentiation?

....

.... [3 marks]

6 Name these **two** blood cells.

  a .... [1 mark]

  b .... [1 mark]

  c Name one other component of blood.

.... [1 mark]

7 Describe how red blood cells transport oxygen from the lungs to respiring tissues.

....

....

.... [3 marks]

8 Michael has cut his leg. In a few minutes a clot forms over the wound.

  a Which component of the blood is responsible for the clot?

.... [1 mark]

  b Explain how the clot protects Michael's body.

....

.... [2 marks]

The heart and circulatory system

1 For each of the following parts of the heart, state its function.

a Pulmonary artery.

.... [1 mark]

b Valves.

.... [1 mark]

c Vena cava.

.... [1 mark]

2 Fill in the missing words to complete the order of the flow of blood through the heart.

Vena cava → right .... → .... ventricle → .... artery. [3 marks]

3 Why is the wall of the left ventricle thicker than that of the right ventricle?

....

....

....

.... [3 marks]

4 Explain why the human heart is known as a 'double pump'.

....

.... [2 marks]

5 During one pumping cycle blood flows through each side of the heart.

a Describe how blood enters the ventricles from the atria.

....

.... [3 marks]

b How is blood prevented from returning into the atria?

.... [1 mark]

6 Name **three** types of blood vessel.

.... [3 marks]

7 Name the types of blood vessel being described below.

a Carries blood from tissues to the heart.

.... [1 mark]

b Has the largest diameter.

.... [1 mark]

c Carries blood at a high pressure.

.... [1 mark]

d Walls are only one cell thick.

.... [1 mark]

8 Why do veins need valves?

....

.... [2 marks]

9 What is the function of a 'capillary bed'?

....

....

.... [3 marks]

The digestive system

1 Name the parts of the digestive system on the diagram.

a .... **[1 mark]**

b .... **[1 mark]**

c .... **[1 mark]**

2 Give a reason why:

  a Food is chewed in the mouth.

.... **[1 mark]**

  b Food is mixed with saliva in the mouth.

.... **[1 mark]**

a

b

c

G–E

3 What is the function of peristalsis?

.... **[1 mark]**   D–C

4 Janet has a gall stone blocking her bile duct.

  a What is the function of the bile duct?

.... **[2 marks]**

  b Predict how a blockage will affect Janet's health.

....
....
.... **[3 marks]**

B–A*

5 What is the function of enzymes in digestion?

....
.... **[3 marks]**   G–E

6 A science class set up an experiment using the equipment opposite and left it for 30 minutes.

For each of these substances, predict if the class will detect it in the warm water surrounding the Visking tubing, giving a reason for your answer.

thread to attach to support
warm water
mixture of protein, starch and protease
Visking tubing

  a Protein.

....
.... **[2 marks]**

  b Amino acids.

....
....
.... **[3 marks]**

D–C

  c Glucose.

....
.... **[2 marks]**

7 Explain how the structure of the villi of the small intestine is related to its function.

....
....
....
.... **[4 marks]**

B–A*

Functional foods

1 Yakult is a drink that contains the bacteria *Lactobacillus*. The makers of Yakult claim that drinking it improves the health of the digestive system.

G–E

a Why is Yakult classed as a functional food?

.... [1 mark]

b What type of functional food is Yakult?

.... [1 mark]

2 Another type of functional food is prebiotics.

a Explain how these increase the amount of 'good' bacteria in the alimentary canal.

....

....

.... [3 marks]

b Why does an increase in 'good' bacteria improve the health of the alimentary canal?

D–C

....

.... [2 marks]

3 The margarine Benecol has added plant stanols.

a What are the claimed health benefits of this functional food?

.... [1 mark]

b How believable are these claims? Explain your answer.

....

....

.... [2 marks]

4 Brainactive cereal bars are a functional food containing plant omega-3, which is a type of fat that may enhance concentration and learning in children.

A study was carried out where a class of Year 5 children ate a Brainactive cereal bar before school every day for a month. At the end of the month their teacher was asked if she saw an improvement in the learning of each child. The teacher said that 75% of the children had an improvement.

Brainactive cereal bars used this as evidence to make the claim that eating their cereal bars enhances learning in children. Evaluate this claim.

B–A*

....

....

....

....

....

....

.... [4 marks]

B2 Extended response question

Homeostasis is used to keep human temperature at around 37 °C. Explain why it is dangerous for our body temperature to go much higher. Use what you know about enzymes in your answer.

The quality of written communication will be assessed in your answer to this question.

[6 marks]

The early atmosphere and oceans

1 a Which of the following gases was not in the atmosphere 4.5 billion years ago, but is in our modern atmosphere?

i Water vapour   **ii** Carbon dioxide   **iii** Oxygen   **iv** Nitrogen

.... **[1 mark]**

2 Scientists analyse the gases that come from volcanoes. How does this help them to understand what the early atmosphere was like?

....

.... **[2 marks]**

3 Scientists analysed some rocks that contain iron compounds. Those formed more than 4 billion years ago contained iron sulfide (FeS), while rocks that formed only 1.8 billion years ago contain iron oxide (Fe_2O_3).

a Name the type of reaction that converts iron to iron oxide.

.... **[1 mark]**

b Explain how these rocks have helped scientists understand how the atmosphere of the Earth has changed.

....

....

....

.... **[3 marks]**

4 a In the modern atmosphere there is about 0.03% carbon dioxide. How does this compare to the amount of carbon dioxide that was in the atmosphere 4.5 billion years ago?

.... **[1 mark]**

b Suggest what happened to cause the change.

....

.... **[2 marks]**

5 The diagram opposite shows the gases in the atmosphere 3 billion years ago.

a Calculate the percentage of oxygen in the atmosphere at that time. Show your working.

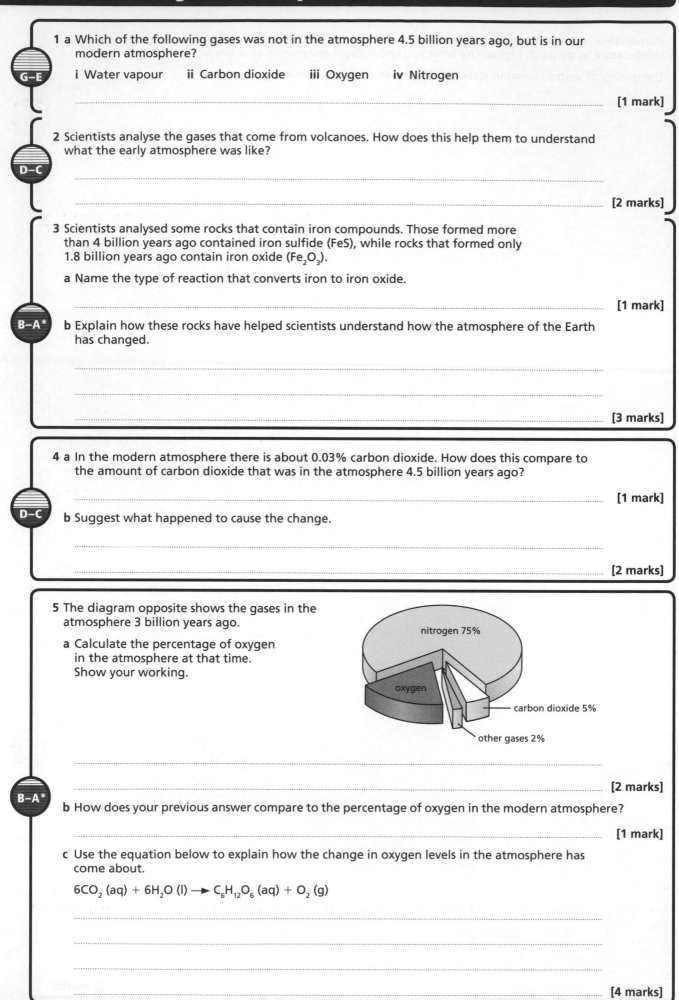

nitrogen 75%

oxygen

carbon dioxide 5%

other gases 2%

....

.... **[2 marks]**

b How does your previous answer compare to the percentage of oxygen in the modern atmosphere?

.... **[1 mark]**

c Use the equation below to explain how the change in oxygen levels in the atmosphere has come about.

$6CO_2$ (aq) + $6H_2O$ (l) ⟶ $C_6H_{12}O_6$ (aq) + O_2 (g)

....

....

....

.... **[4 marks]**

Today's atmosphere

1 a Which of the following would not be used to measure changes in atmospheric gases?

   i Satellites   **ii** Mass spectrometer   **iii** Digital thermometer   **iv** Fractional distillation

   .... [1 mark]

   b Name **two** human activities that can cause changes in the gases in Earth's atmosphere.

   ....

   .... [2 marks]

   c Explain why scientists want to measure the amount of each gas in the atmosphere.

   .... [1 mark]

G–E

2 a Over the past 100 years there has been a noticeable increase in the percentage of carbon dioxide in the atmosphere. Select the answer that gives the most likely reason for this change.

   i Carbon dioxide has dissolved into the oceans.

   ii Oxygen has been converted into carbon dioxide by respiration.

   iii Combustion of hydrocarbons has released carbon dioxide.

   iv Volcanoes have erupted, releasing carbon dioxide.

   .... [1 mark]

   b It is estimated that 40% of the rainforests of Central America have been cut down over the last 40 years.

   i What effect does cutting down forests have on carbon dioxide and oxygen levels in the atmosphere?

   ....

   .... [2 marks]

   ii Explain why removing trees results in these changes.

   ....

   .... [2 marks]

D–C

3 a Balance the following equation.

   ____ CH_4 + ____ O_2 ⟶ ____ CO_2 + ____ H_2O [4 marks]

   b Methane, oxygen and carbon dioxide are all gases. Water is a liquid. Explain how you could show this on the equation.

   ....

   .... [3 marks]

   c Many scientists consider that the problem of increasing carbon dioxide levels in the atmosphere would be improved by using hydrogen (H_2) as a fuel instead of hydrocarbons (C_xH_y).

   Explain why burning hydrogen would be better. Use an equation to illustrate your answer.

   ....

   ....

   .... [3 marks]

B–A*

Types of rock

1 a Which sentence below describes metamorphic rock?

   i It forms when lava solidifies.   **ii** It forms when magma solidifies.

   iii It forms when rock comes under heat and pressure. **iv** It forms when sediment hardens.

                                                           [1 mark]

b John has two samples of igneous rock. They formed in different places. One has large crystals and the other has tiny crystals.

How does the size of the crystals help John to understand where the rocks were formed?

                                                           [2 marks]

c Asha reads that chalk is made from the skeletons of plankton that lived in the sea millions of years ago. Describe to Asha what processes occurred to change the skeletons into chalk.

                                                           [3 marks]

2 Chen tried an experiment to see what happens when copper sulfate crystals form. First he dissolved as much copper sulfate as he could in 50 cm³ of hot water. Then he poured 25 cm³ of the solution into boiling tube 1 and put the tube in a beaker of ice and water. He poured the remaining 25 cm³ into boiling tube 2 and put the tube into a beaker packed with cotton wool.

The following day he looked at the contents of the tubes and found that both had formed crystals. In one tube the crystals were large and in the other they were small.

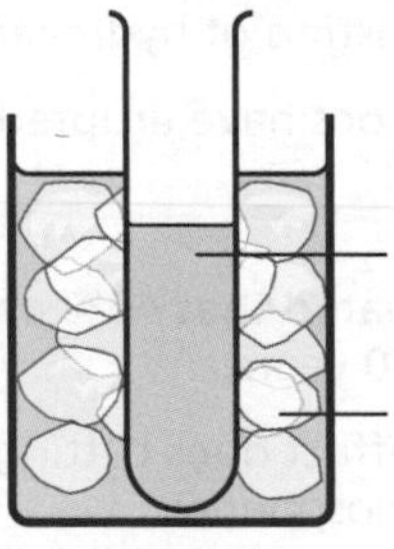

a Which tube would have cooled the fastest and why?

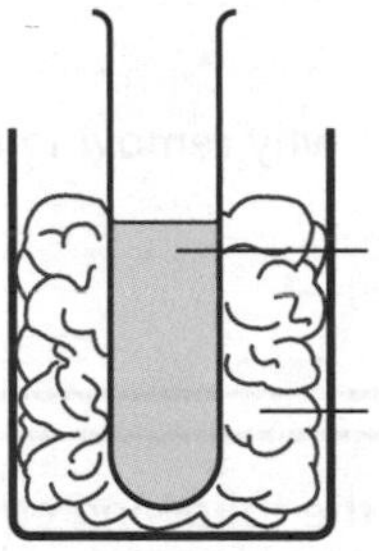

                                      [2 marks]

b Which tube would contain the large crystals?

                                      [1 mark]

c Chen said that tube 2 is a good model for the formation of granite rock. Do you agree or disagree? Explain why.

                                      [4 marks]

Tube 1 — Copper sulfate solution, Ice and water

Tube 2 — Copper sulfate solution, Cotton wool

3 Yellowstone National Park in the USA is the site of an ancient super-volcano.

Geologists studied the rock from different sites around the park and obtained the results shown opposite.

Location	Crystal size (mm)
Obsidian Cliffs	No crystals, smooth glass-like texture
Tower Falls	0.78, 0.92, 1.20, 0.93, 1.1, 0.86, 0.76, 0.61
Hell Roaring Mountain	2.15, 1.76, 0.89, 1.24, 0.98, 1.35, 1.29, 1.17

a What is the difference in average crystal size between the rocks at Tower Falls and those at Hell Roaring Mountain? Show your working.

                                      [3 marks]

b Suggest a difference in the way these three rocks may have formed and explain your suggestion.

                                      [4 marks]

Sedimentary rock and quarrying

1 a Link the correct statements.

Cementation	Rock fragments are removed from the surface of the rock.
Weathering	Concentrated solutions of salts stick rock fragments together.
Sedimentation	Rocks break into small fragments.
Erosion	Small fragments of rock are deposited on the sea bed.

[2 marks] G–E

b Put the four words above in the correct order to describe the formation of sedimentary rock.

.... [2 marks]

c Explain how chalk is converted into limestone.

....

.... [3 marks]

2 The ancient Greeks believed that animals grew from the Earth. They thought that fossils were animals that had failed to break out on to the surface of the Earth.

Explain what we believe today about the way fossils form and give any evidence that supports this theory.

....

....

....

[4 marks] D–C

3 The picture shows Idol Rock at Brimham Rocks in Yorkshire.

a Use evidence from the picture to support the view that Idol Rock is made from sedimentary rock.

.... [1 mark]

b Use your knowledge of the properties of sedimentary rock and of the way that it forms to suggest how Idol Rock came to be balanced on such a tiny point.

....

.... [2 marks]

B–A*

4 A mining company has asked the local council for permission to quarry limestone just outside a small town. The local residents have prepared a petition against the quarry. What arguments could they use to prevent the quarrying?

....

.... [3 marks]

G–E

5 Tick the correct boxes for each product.

Ingredient	Glass	Cement	Concrete	Ingredient	Glass	Cement	Concrete
Limestone				Gravel			
Sand				Water			
Clay				Sodium carbonate			

[3 marks] D–C

6 Some forms of industry are unavoidably disruptive to local communities. Contractors are required to provide plans showing how they intend to minimise disruption and long-term damage to the countryside. Write the outline of a suitable plan for a company that wants to build a limestone quarry near a town.

....

....

....

....

[4 marks] B–A*

Atoms and reactions

1 Which of the following is **not** true?

  i Atoms are the smallest particle of an element.

  ii Atoms from one element change to another element during a chemical reaction.

  iii Atoms are rearranged during chemical reactions.

  iv Different elements have different types of atoms.

.... **[1 mark]**

2 Below are the instructions that Kelly followed during a chemistry lesson.

  • Hold a piece of grey magnesium metal in the tongs.

  • Put it into the Bunsen flame until it lights.

  • **DO NOT LOOK DIRECTLY AT THE FLAME**

  • When it has finished burning, place the white powdery solid down on the heat-resistant mat.

  a What evidence tells you that Kelly had carried out a chemical reaction?

....

.... **[2 marks]**

  b The reactants for this reaction were magnesium and oxygen and the product was magnesium oxide. The mass of the magnesium and oxygen added together was 2.0 g. What was the mass of the magnesium oxide?

.... **[1 mark]**

3 Read the information. Copper carbonate is a green solid. If it is heated in a boiling tube it begins to melt when the temperature reaches 200 °C. If heating is continued to 290 °C it reacts to form copper oxide and carbon dioxide. Copper carbonate reacts with acid to form a salt, water and carbon dioxide.

Identify **one** physical and **one** chemical property of copper carbonate.

....

.... **[2 marks]**

4

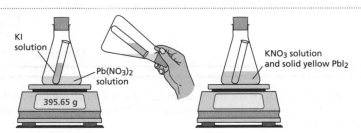

KI solution

$Pb(NO_3)_2$ solution

KNO_3 solution and solid yellow PbI_2

395.65 g

  a Fill in the missing reading on the second balance. **[1 mark]**

  b Name **one** difference in the physical properties of KNO_3 and PbI_2.

.... **[1 mark]**

5 Sarah carried out an investigation to discover what mass of carbon dioxide is made when copper carbonate reacts with acid. The equation for the reaction is below:

$$2HCl\ (aq) \quad + \quad CuCO_3\ (s) \quad \longrightarrow \quad CuCl_2\ (aq) \quad + \quad H_2O\ (l) \quad + \quad CO_2\ (g)$$

hydrochloric acid   copper carbonate   copper chloride

She measured the mass of the acid and the copper carbonate in a beaker at the start of the reaction.

Describe the other measurements Sarah must make and the calculation that she must do. Explain why this method enables her to find out how much carbon dioxide is made.

....

....

....

.... **[5 marks]**

Thermal decomposition and calcium

1 Calcium oxide is a very useful chemical that is used to make cement. The starting material for making calcium oxide is limestone.

a What is the chemical compound in limestone?

.... [1 mark]

b How can limestone be converted into calcium oxide?

.... [1 mark] G–E

c What is the other product of this reaction?

.... [1 mark]

d What name is given to this type of reaction?

.... [1 mark]

2 Magnesium oxide can also be made by a chemical reaction. Suggest a reactant for this reaction and explain how you made your choice.

.... D–C

.... [3 marks]

3 a Write the chemical formula for calcium hydroxide.

.... [2 marks]

b Calcium hydroxide solution is also known as limewater. Describe what limewater is used for and explain the chemistry that allows it to be used in this way.

.... D–C

....

....

.... [4 marks]

4 A student decided to investigate which substance decomposed more easily – zinc carbonate ($ZnCO_3$) or copper carbonate ($CuCO_3$). She set up the apparatus as in the diagram opposite.

retort stand — clamp
metal carbonate   test-tube   delivery tube

a Write a word equation for the thermal decomposition of zinc carbonate.

.... [2 marks]

test-tube

b Write the instructions for experiments that will allow the student to find out whether zinc or copper carbonate decomposes more easily.

....

....   — Bunsen burner

.... B–A*

.... [3 marks]

limewater

c Predict the results of the experiment and explain why you made this prediction.

....

.... [2 marks]

5 a Some farmers add calcium compounds to soil. Explain why.

.... [1 mark]

b Calcium oxide and calcium carbonate are both used to improve some soils. Farmers often choose to use calcium oxide. Explain why.

G–E

.... [2 marks]

c Describe **one** advantage and **one** disadvantage of using calcium oxide rather than limestone to neutralise soil.

D–C

....

.... [2 marks]

Acids, neutralisation and their salts

1 a Which of the following would not neutralise an acid?

    i Sodium hydroxide

    ii Sodium chloride

    iii Calcium oxide

    iv Calcium carbonate

.... **[1 mark]**

b Joe and Folu are discussing the best way to decide how much acid is in two brands of vinegar. They decide to see how much base it takes to neutralise the vinegar. Write a word equation to show the type of products that are formed when an acid and a base react together.

.... **[2 marks]**

c Folu says they must use a base that will dissolve in water. What name is given to this kind of base?

.... **[1 mark]**

d Joe and Folu decide to add drops of base to 5 cm³ of vinegar. Explain how they could tell when the acid had been neutralised and which vinegar contained the most acid.

....

....

.... **[2 marks]**

(G–E)

2 A class is planning an experiment to measure the effectiveness of different antacid tablets.

Elsa's plan is as follows:

- Put an antacid tablet in a beaker.
- Add some hydrochloric acid to the beaker.
- When the fizzing has stopped, measure the pH.
- Repeat these steps with a different type of antacid.

a Improve Elsa's method so that she will get a valid result.

....

....

.... **[3 marks]**

b How will Elsa tell which antacid was the most effective?

.... **[1 mark]**

c Miriam suggests that it would be better to use 25 cm³ of acid and add spatulas of crushed powder to it until the acid is neutralised, then repeat with fresh acid and the second tablet. Do you agree? Explain your answer.

....

....

....

.... **[2 marks]**

(D–C)

3 Calcium chloride ($CaCl_2$) is a salt used as a drying agent to remove water from liquids. It can be manufactured by adding acid to limestone ($CaCO_3$).

a Which acid would be added to limestone to form calcium chloride?

.... **[1 mark]**

b The other products of the reaction are carbon dioxide and water. Write a symbol equation for this reaction.

.... **[3 marks]**

(B–A*)

Electrolysis and chemical tests

1 a Which of the following describes a method of obtaining chlorine from sea water?

    i Electroplating   **ii** Filtering

    iii Precipitation   **iv** Electrolysis

.... **[1 mark]**

b Explain why industries that use chlorine gas must be very careful about safety. Suggest **two** safety precautions an industry should use.

....

.... **[3 marks]**

c Name a polymer that requires chlorine for its manufacture.

.... **[1 mark]**

2 a Chlorine can be produced by passing an electrical current through sea water. Which of the following statements is **not** true.

    i Chlorine can be collected from the cathode.

    ii A direct current must be used.

    iii Hydrogen can be collected from the other electrode.

    iv Electrical energy is used to decompose the compound.

.... **[1 mark]**

b Explain what causes substances in the sea water to separate out when an electric current passes through it.

....

.... **[3 marks]**

3 Class 10A carried out an investigation into what happens when electricity passes through water. They recorded the results shown opposite.

a Calculate the average volume of gas collected at the anode and the average at the cathode.

....

.... **[2 marks]**

Group	Volume of gas at the anode (cm³)	Volume of gas at the cathode (cm³)
1	2.2	4.5
2	4.9	10.0
3	3.2	6.5
4	2.3	4.8
5	2.5	10.3
6	4.6	9.2
7	2.3	4.7
8	4.4	9.1
9	3.3	6.7

b What does the relationship between these two values tell you about the compound that is being decomposed?

....

.... **[1 mark]**

c Suggest the identity of the two gases and a test that you could carry out to confirm your suggestions.

....

....

....

.... **[4 marks]**

d Suggest a reason why different groups obtained very different volumes.

.... **[1 mark]**

e Identify the group that obtained an anomalous result and suggest a possible explanation for it.

.... **[1 mark]**

Metals – sources, oxidation and reduction

1 a Which of the following describes a mineral?

   i A rock that contains a metal.   **ii** A naturally occurring compound.

   iii An unreactive metal.   **iv** A mixture of crystals.

G–E

.... **[1 mark]**

b Why can gold sometimes be found in the ground as an uncombined element?

.... **[1 mark]**

2 a What factors need to be considered when estimating the cost of producing metals from their ores?

....

.... **[3 marks]**

D–C

b What is meant by electrolysis and why is it useful in metal extraction?

....

.... **[2 marks]**

3 Daniel and Joyce carried out an experiment to simulate the extraction of copper from its ore. The instructions were as follows.

- Weigh out 2 g of charcoal and 2 g of copper oxide.
- Mix the two black powders together in a crucible and heat strongly above a roaring flame for 15 minutes.
- Allow the contents to cool and then pour out onto a heat-resistant mat.

a What is the purpose of adding charcoal to the mixture?

B–A*

.... **[2 marks]**

b What would Daniel and Joyce have seen at the end of the experiment?

.... **[2 marks]**

c Joyce suggested repeating the experiment with a different metal oxide. Suggest **one** metal oxide that might give a good result and **one** that would not work in this experiment.

.... **[2 marks]**

4 a Name the substance oxidised and the substance reduced in the following equation.

   $Al (s) + Fe_2O_3 (s) \longrightarrow 2Fe (s) + Al2O_3 (s)$

.... **[2 marks]**

b This method could be used to extract iron from iron ore. Suggest why it is not the usual method.

.... **[1 mark]**

D–C

c Describe the method usually used. Include an equation in your answer.

....

....

....

.... **[4 marks]**

5 Steel dustbins are often coated with a layer of zinc to help prevent corrosion.

   a Suggest how coating steel helps to prevent corrosion.

....

B–A*

.... **[2 marks]**

   b Explain why zinc is chosen as the coating material.

....

.... **[2 marks]**

Metals – uses and recycling

1 Link the metal with its properties and uses.

Gold		Low density		Electronic connectors
Aluminium		Malleable		Plumbing
Copper		Does not corrode		Aircraft

G–E

[3 marks]

2 a Iron is extracted from its ore by heating with carbon. This leaves a product with about 4% carbon atoms mixed in with the iron. What name is given to metals that contain a mixture of atoms?

.... [1 mark]

b The iron is treated to reduce the level of carbon to below 0.3%, which results in a softer metal that is easier to work with. Explain why having carbon atoms mixed in with the iron makes a harder metal. Include a diagram in your answer.

....

....

.... [3 marks]

D–C

c Suggest what metal might be added to make the iron suitable for use in cutlery, and say what property this would give the iron.

....

.... [2 marks]

3 White gold has become a popular choice for jewellery. Explain how gold can be white, and describe the advantages that white gold may have over pure gold.

....

....

.... [3 marks]

B–A*

4 A local business manufactures aluminium drinks cans. They are eager to win a Green Award for environmentally friendly practice. Suggest **three** ways in which they could make their business as environmentally friendly as possible, and explain how your suggestions will help the environment.

....

....

.... [3 marks]

D–C

5 The table below shows the UK percentages for recycling of metals, and the energy savings.

Metal	New metals made using recycled metals	Energy saving
Aluminium	39%	95%
Copper	32%	85%
Lead	74%	60%
Steel	42%	62–74%
Zinc	20%	60%

Explain briefly why these totals should be increased in the coming years. Include a justified opinion on which metal should be the first to target for improved recycling figures.

B–A*

....

....

....

....

.... [5 marks]

Hydrocarbons and combustion

1 a Why is crude oil described as a *mixture of hydrocarbons*?

....

.... **[3 marks]**

b What property of the mixture allows it to be separated into fractions?

.... **[1 mark]**

2 Explain why crude oil is processed before use. Illustrate your answer with the names and uses of at least two substances that are made from crude oil.

....

.... **[3 marks]**

3 Use ideas of kinetic energy and intermolecular forces to explain how the molecules in crude oil can be separated into different fractions.

....

.... **[4 marks]**

4 a Which of the following does **not** describe the combustion of a hydrocarbon?

   i It is a reduction reaction.   **ii** It releases energy as heat and light.

   iii It produces water.   **iv** It requires oxygen.

.... **[1 mark]**

b Explain why we cannot see the products of combustion of methane.

.... **[2 marks]**

5 Jonas carried out an investigation to find out what happens when a tea light candle burns in a closed system. He measured the height of the wax in the candle. Then he lit the candle on a heatproof mat, put a 250 cm³ beaker over the top and sealed the edges with plasticine so that no gases could get in or out. After 9 minutes the candle went out. The candle wax was 3 mm lower than at the start.

a Where did the wax go to? .... **[1 mark]**

b Why did the candle go out? .... **[1 mark]**

c What was produced when the candle first began to burn? .... **[2 marks]**

d What different gas was made just before the candle went out? .... **[1 mark]**

e Suggest a test to detect **one** of the gases produced.

....

.... **[2 marks]**

6 a How is carbon monoxide produced and why is it dangerous?

....

.... **[2 marks]**

b Linda suspects that the gas water heater in the holiday flat she is thinking about renting is producing carbon monoxide. What might indicate that Linda is correct, why does this happen and what is it called?

....

.... **[3 marks]**

7 A recent report suggested that Earth may be suffering from global dimming. The underlying cause is thought to be tiny particles (particulates) in the atmosphere. Describe a possible source of these particulates and explain why they might be causing global dimming.

....

.... **[2 marks]**

Acid rain and climate change

1 a Which of the following correctly describes damage caused by acid rain?

 i It destroys the ozone layer.　　　　　ii It raises the pH of ponds.

 iii It causes breathing problems for asthmatics.　　iv It reduces the yield of some crops.

.... [1 mark] G–E

b Explain why northern Norway has problems with acid rain, even though there is no industry in that area.

.... [2 marks]

2 a Explain why acid rain is linked with burning hydrocarbon fuels.

....

.... [3 marks]

b Give **two** reasons why acid rain might affect the growth of trees.

.... [2 marks]

3 The table opposite shows the pH values at which various animals are able to live.

a Which species would be the first to be affected by acid rain?

.... [1 mark]

b Suggest a reason why the population of perch decreases very fast in water at pH 5.

....

.... [1 mark]

D–C

	pH6.5	pH5.0	pH5.5	pH5.0	pH4.5	pH4.0
Trout						
Bass						
Perch						
Frogs						
Salamanders						
Clams						
Crayfish						
Snails						
Mayfly						

4 What advice about use of fuels would you give to an industrialist who wanted to set up a factory with low sulfur emissions?

....

....

.... [3 marks]

B–A*

5 Put the statements about the greenhouse effect into the correct order.

 i The warm Earth radiates infrared radiation.　　ii Some heat radiates back to the surface.

 iii UV radiation passes though Earth's atmosphere and warms the surface of the planet.　　iv Greenhouse gases absorb some of the radiation.

.... [3 marks]

G–E

6 a Describe the general pattern of global temperature change over the past 50 years.

.... [1 mark]

b Why do most scientists believe that this temperature change is a result of human activity?

....

.... [2 marks]

D–C

7 Discuss how seeding the oceans with iron might help to solve the problem of global warming.

....

....

....

.... [4 marks]

B–A*

Biofuels and fuel cells

G–E

1 a Why are scientists trying to find alternatives to fossil fuels?

....

....
[3 marks]

b Suggest the name of a biofuel that can be made from sugar beet.

....
[1 mark]

c Why do some people think that using biofuels will put up the cost of food?

....
[1 mark]

D–C

2 A company has produced a new fuel from waste products. Suggest **three** factors that should be investigated to see if the fuel is suitable to develop into a petrol substitute.

....

....

....
[3 marks]

B–A*

3 Use the information in the table below to compare the suitability of different crops as a raw material for biofuel production and to suggest how biofuel should be made.

Crop	Fuel made	Energy needed to grow and harvest crops and produce the fuel	% of farmland needed to make 50% of current fuel needs
Sugar cane	Ethanol	Medium	46–57
Wood residue	Ethanol or biodiesel	Low	150–250
Rapeseed	Biodiesel	Medium-low	30
Algae	Biodiesel	High	1–2

....

....

....
[4 marks]

G–E

4 Give **two** advantages of using hydrogen as a fuel in place of petrol.

....

....
[2 marks]

D–C

5 A class tested two fuels to see which one gave out the most energy. They put the fuel in a burner under a beaker holding 100 cm³ of water and measured the temperature of the water. Then they lit the burner and allowed it to burn for two minutes. They measured the highest temperature that the water reached. The results are shown below.

a Group C got a much higher final temperature for fuel 2 than the other groups. Suggest a possible reason for this result.

Group	Fuel 1 starting temperature (°C)	Fuel 1 finishing temperature (°C)	Fuel 2 starting temperature (°C)	Fuel 2 finishing temperature (°C)
A	20	39	21	35
B	21	41	21	34
C	20	38	*	45
D	20	39	21	34
E	20	38	22	34

*forgot to measure

....

....
[1 mark]

b Which fuel gave out the most energy? Explain your answer.

....

....
[2 marks]

c What improvement to the experiment would you suggest to get a more valid result?

....
[1 mark]

B–A*

6 Explain some problems in achieving carbon-neutral, hydrogen-fuelled cars.

....

....
[3 marks]

Topic 5: 5.17, 5.18, 5.19, 5.20, 5.21, 5.22, 5.23, 5.24

Alkanes and alkenes

1 a Which of the three structures shown is ethane?

.... **[1 mark]**

b Which of the three structures are found in natural gas?

.... **[1 mark]**

A   B   C

c Draw the structure of propane.

[1 mark]

d How can you tell that propane is a saturated molecule?

.... **[2 marks]**

2 a An oil refinery analysed a sample of natural gas. Complete the table of results opposite. **[5 marks]**

b Describe the difference between a molecule of methane and a molecule of propane.

Substance	Mass (g)	% of total
Natural gas	80	100
Methane		80
Ethane	8	
Propane		3
Butane		

....

....

.... **[2 marks]**

c A liquid mix of propane and butane can be used as a fuel for vehicles. It is called LPG and is stored under pressure. Why is LPG stored under pressure?

....

.... **[2 marks]**

3 Give the structural formula of butane and describe the way that the atoms in a molecule of butane are bonded together.

....

.... **[3 marks]**

4 Complete the table opposite.

Propene	True	False
Has four carbon atoms		
Is an unsaturated molecule		
Contains a double bond		
Does not react with bromine water		

[4 marks]

5 Some green activists have suggested that burning crude oil fractions as fuel is madness. Say whether you agree and suggest other uses for crude oil.

....

....

.... **[3 marks]**

6 a Select an unsaturated hydrocarbon with four carbon atoms from the following:

   i Propane   **ii** $C_4H_8Cl_2$

  iii Butene   **iv** C_4H_{10}

.... **[1 mark]**

b Write the name and formula of the unsaturated hydrocarbon with two carbon atoms.

.... **[2 marks]**

Cracking and polymers

1 a The diagram shows the laboratory apparatus to crack hydrocarbons. Label the hydrocarbon and the catalyst. [2 marks]

product gas

b What would you have to do to make the cracking reaction happen in this apparatus?

....

.... [2 marks]

c What two types of hydrocarbons would be in the product gases?

.... [2 marks]

G–E

2 a Give a reason to use catalysts in industry.

.... [2 marks]

b One process in converting crude oil into petrol uses platinum as a catalyst. Platinum is one of the most expensive metals. How can the industry afford to use such an expensive metal?

.... [2 marks]

D–C

3 Use the data in the charts to explain why oil companies carry out cracking.

....

....

....

....

....

.... **[4 marks]**

Proportion of each fraction from a sample of crude oil.

bitumen LPG fuel oil petrol diesel kerosene

Percentage demand for crude oil fractions.

bitumen fuel oil LPG diesel kerosene petrol

B–A*

4 Use the information in the table to choose a suitable polymer for the following uses:

Polymer name	Polymer properties
Poly(ethene)	Waterproof, easily moulded
Poly(propene)	Can be spun into fibres, strong
Poly(tetrafluoroethene)	Water repellent, shiny
Poly(chloroethene)	Flexible, good insulator

a Rope .... **[1 mark]**

b Plastic bowls .... **[1 mark]**

c Waterproof fabrics .... **[1 mark]**

d Coverings for electrical wires .... **[1 mark]**

G–E

5 Put the statements in order so that they describe addition polymerisation.

i A very long chain of carbon atoms forms, called a polymer.

ii One bond of the double bond breaks.

iii The monomers have a double carbon-carbon bond.

iv A new bond forms between one monomer and the next.

.... **[2 marks]**

D–C

6 Most people think that recycling polymers is a good idea. Give **two** reasons to support this view.

....

.... **[2 marks]**

G–E

7 Humans only discovered how to manufacture polymers in the 20th century but plants have been making them for millions of years.

a Give the names of two plant polymers.

.... **[2 marks]**

b Explain why plant polymers do not cause the same disposal problems as man-made polymers.

.... **[1 mark]**

c What other advantage would there be to developing polymers from plant materials?

.... **[1 mark]**

B–A*

Topic 5: 5.30, 5.31, 5.32, 5.33, 5.34, 5.35, 5.36, 5.37

C1 Extended response question

Power stations that burn fossil fuels releases gases into the atmosphere. Some of these gases may have damaging effects. Explain how these gases form and why they are damaging to the environment. Suggest ways that this damage could be reduced.

The quality of written communication will be assessed in your answer to this question.

[6 marks]

Atomic structure and the periodic table

1 a Which of the following statements correctly describes the nucleus of a helium atom?

  i It contains only positively charged particles.   **ii** It has no overall charge.

  iii It is where the neutrons are found.   **iv** It is the control centre of the atom.

.... **[1 mark]**

b Hydrogen and helium are both elements. Explain **one** way in which an atom of hydrogen and an atom of helium are the same and **one** way in which they are different.

....

.... **[2 marks]**

c In 1808, the chemist John Dalton first suggested that elements were made from atoms. He thought that atoms were like tiny snooker balls.

Describe **two** ways in which modern ideas about atoms differ from Dalton's view.

....

.... **[2 marks]**

G–E

2 a Describe the difference between protons and neutrons.

....

.... **[2 marks]**

b What is the relative mass of an atom of lithium that contains three protons, three electrons and four neutrons?

.... **[1 mark]**

D–C

3 a Write the symbol of the element that is in group 2 and period 3 of the periodic table.

.... **[1 mark]**

b Use the periodic table to decide the following:

  i The number of protons in an atom of fluorine .... **[1 mark]**

  ii The number of neutrons in an atom of sodium .... **[1 mark]**

  iii The number of electrons in an atom of beryllium .... **[1 mark]**

G–E

4 There are two types of chlorine atom. One has a relative mass of 35 and the other a relative mass of 37. Although these atoms are different, they are still atoms of the same element.

a Explain why the two atoms have a different relative mass.

.... **[1 mark]**

b Explain why they are still considered to be the same element.

....

.... **[2 marks]**

D–C

5 Explain why the relative atomic mass of magnesium is not a whole number.

....

....

....

.... **[4 marks]**

B–A*

Electrons

1 a Complete the table opposite.

Element symbol	Electronic configuration
F	2.7
Al	
	2.8.6

[2 marks]

b The diagram opposite shows an atom.

i What is the name of the particle labelled a?

.... [1 mark]

ii How many protons would you find in b?

.... [1 mark]

iii What do the circles labelled c represent?

.... [1 mark]

2 a Write the name and electron configuration of the element that is in group 2 and period 3 of the periodic table.

.... [2 marks]

b In which group of the periodic table would you find the element that has 17 protons in the nucleus? Explain how you can tell.

....

....

....

.... [4 marks]

c Complete the table below.

Element	Atomic number	Electronic configuration
Ne	10	
Ar	18	

[2 marks]

d Use the electronic configuration of Ne and Ar to explain why they have similar properties.

....

.... [3 marks]

e What is the name of the element with electronic structure 2.8.2?

.... [1 mark]

f In which period would you find the element with electronic structure 2.8.8.1?

.... [1 mark]

3 a Discuss how Mendeleev was able to create a periodic table similar to the modern one, even though nothing was known about the structure of the atom at the time.

....

....

....

.... [4 marks]

b Explain why Mendeleev's periodic table was eventually accepted by scientists of the day, when other proposed arrangements of elements had been rejected.

....

....

....

.... [4 marks]

Ionic bonds and naming ionic compounds

1 Draw lines to match the descriptions with the statements below.

Electrostatic attraction	Atom that has gained electrons
Negative ion	Atom that has lost an electron
Positive ion	Force that holds ions together
Ionic compound	Made from positive and negative ions

[4 marks]

2 a Use the diagram opposite to explain how magnesium reacts with oxygen to form a positive and a negative ion.

....

....

....

.... **[5 marks]**

b Complete the table to show the formula of the ion formed by each element.

Atom	Ion
F	
Na	
S	
Ca	

[4 marks]

c Explain why potassium forms an ion with a 1+ charge but chlorine forms an ion with a 1– charge.

....

.... **[2 marks]**

3 a Calcium carbonate is an ionic compound with the formula $CaCO_3$. Give the formula of the ions that it is made from.

Cation .... **[1 mark]**

Anion .... **[1 mark]**

b Draw a diagram that shows the arrangement of electrons in a fluoride ion. Include the charge on the ion.

[2 marks]

c Suggest why a sodium ion is more stable than a sodium atom.

....

.... **[2 marks]**

4 a Complete the table below.

Name	Ions	Elements
Sodium chloride	Sodium and chloride	Sodium and chlorine
Potassium fluoride		Potassium and fluorine
		Calcium and oxygen
	Magnesium and nitrate	

[5 marks]

b Name the compounds with the following formulae: $Ca(OH)_2$, NH_4Cl, $MgBr_2$, Na_2CO_3

....

.... **[4 marks]**

c Explain the difference between potassium sulfide and potassium sulfate.

....

.... **[2 marks]**

Writing chemical formulae

1 a Complete the table below.

Name of ion	Charge	Formula	Name of ion	Charge	Formula
Sulfide	2–	S^{2-}			OH^-
Fluoride					NO_3^-
Hydrogen					Sn^{2+}
Ammonium			Copper	2+	
Potassium			Silver	1+	
Sulfate			Iron	3+	
Carbonate			Iron	2+	

[14 marks]

b Underline **three** errors in the sentence below.

Barium sulfate is an ionic compound that contains barium and sulfur ions. The barium ion is a polyatomic cation and the sulfate ion is a polyatomic cation. **[3 marks]**

---

2 a James and Alex cannot agree on the formula for calcium chloride. James thinks it is $CaCl_2$, but Alex thinks it is CaCl. State who is correct and explain why.

....

.... **[4 marks]**

b Complete the table below to show the formula of each compound named.

Name	Formula
Potassium fluoride	
Calcium nitrate	
Aluminium oxide	
Ammonium sulfate	

[4 marks]

---

3 The word equation and formula equation for the reaction between ammonia and sulfuric acid are given below.

Ammonia + sulfuric acid ⟶ ammonium sulfate

$$2NH_3 + H_2SO_4 \longrightarrow (NH_4)_2SO_4$$

a What does the large number in front of the formula for ammonia mean?

.... **[1 mark]**

b What does the small number after the H in the formula for sulfuric acid mean?

.... **[1 mark]**

c Explain why there are brackets around the (NH_3) in the formula for ammonium sulfate.

.... **[1 mark]**

4 a Write word equations for the following formula equations.

i $NaCl + AgNO_3 \longrightarrow AgCl + NaNO_3$

.... **[2 marks]**

ii $KOH + HCl \longrightarrow KCl + H_2O$

.... **[2 marks]**

b Write balanced formula equations for the following word equations.

i Sodium sulfate + barium chloride ⟶ barium sulfate + sodium chloride

.... **[4 marks]**

ii Magnesium nitrate + sodium hydroxide ⟶ magnesium hydroxide + sodium nitrate

.... **[4 marks]**

Ionic properties and solubility

1 a Which one of the following best describes an ionic compound?

    i Soft, high melting point.   **ii** Soft, low melting point.

    iii Forms crystals, high melting point.   **iv** Forms crystals, low melting point.

.... **[1 mark]**

b Draw a diagram of a sodium chloride crystal and label the two ions.

[2 marks]

2 Keshma wrote the following notes in her science notebook:

We took a sample of the substance in a test tube and heated it. It melted easily in the Bunsen flame. Then we put two electrodes into the liquid. It did not conduct electricity.

Was Keshma testing an ionic compound? Explain your answer.

....

.... **[2 marks]**

3 a Bauxite is a mineral that contains mostly aluminium oxide, an ionic compound. Aluminium can be extracted from it by electrolysis. Extraction is very expensive because large amounts of energy are needed to melt the compound.

Use ideas about the structure of aluminium oxide to explain why large amounts of energy are needed to melt aluminium oxide and why it must be melted before electrolysis.

B–A*

....

.... **[3 marks]**

b From the list below, choose the compound with the strongest ionic bonding. Explain your choice.

magnesium chloride   sodium fluoride   calcium oxide   potassium sulfide

.... **[3 marks]**

4 Complete the table below.

Name	Soluble? ✓ or ✗
Sodium chloride	
Magnesium nitrate	
Calcium carbonate	
Ammonium carbonate	
Silver chloride	

[5 marks]

5 Which of the following equations describes a precipitation reaction?

    i $CaCO_3$ (s) ⟶ CaO (s) + CO_2 (g)

    ii HCl (aq) + NaOH (aq) ⟶ NaCl (aq) + H_2O (l)

    iii $AgNO_3$ (aq) + NaCl (aq) ⟶ AgCl (s) + $NaNO_3$ (aq)

    iv CH_4 (g) + $2O_2$ (g) ⟶ CO_2 (g) + H_2O (l)

.... **[1 mark]**

6 Which of the following reactants would you expect to result in a precipitation reaction?

    i Potassium nitrate and sodium chloride   **ii** Silver nitrate and sodium chloride

    iii Potassium sulfate and ammonium carbonate   **iv** Ammonium chloride and sodium carbonate

.... **[1 mark]**

Preparation of ionic compounds

1 a Liz wrote this description of how to make an insoluble salt. It has **two** mistakes.

Choose two insoluble compounds. Then dissolve them and mix them together. The insoluble salt that you want will be dissolved in the solution.

Write the description again, but correct Liz's errors.

....

.... **[2 marks]**

b Name the type of reaction used to make an insoluble salt.

.... **[1 mark]**

G–E

2 The diagram opposite shows one stage in the preparation of an insoluble salt. Label the diagram.

D–C

[3 marks]

3 Mrs Brown is worried because she has been sent to the hospital to have a barium meal. Explain to her how this will help doctors to spot any abnormalities in her bowels and reassure her about possible toxic effects.

....

....

.... **[3 marks]**

B–A*

4 Draw lines to match the metal ions to the correct flame colour.

Calcium		Lilac
Sodium		Brick red
Potassium		Green/blue
Copper		Yellow/orange

G–E

[4 marks]

5 a Leroy wants to test a sample of tap water for sodium chloride. Describe the tests that he could carry out and state what the results would be if sodium chloride is present.

....

....

.... **[5 marks]**

b Sarah added some hydrochloric acid to a sample of white powder. The powder started to fizz and bubble. What do these results tell you about the white powder? What further tests could Sarah do to confirm the result?

....

....

.... **[3 marks]**

D–C

6 The element helium was first discovered using spectroscopy on the Sun during an eclipse. Explain how it was possible to tell that the Sun contained an unknown element.

....

....

.... **[3 marks]**

B–A*

Covalent bonds

G–E

1 a Which of the following does **not** have covalent bonds?

   i carbon dioxide   **ii** oxygen   **iii** methane   **iv** iron oxide

   .... [1 mark]

b Water is a covalent molecule.

   i Write the formula of water. .... [1 mark]

   ii How many covalent bonds does water have? .... [1 mark]

D–C

2 a A water molecule contains three atoms. Describe what holds the atoms together in water.

   .... [2 marks]

b Why do the atoms in water form covalent bonds?

   .... [1 mark]

B–A*

3 Hydrogen always forms H_2 molecules.

   a Explain why hydrogen forms molecules rather than remaining as atoms.

   .... [1 mark]

   b Explain why hydrogen forms H_2 molecules and not H_3 or H_4.

   ....

   ....

   .... [3 marks]

G–E

4 Draw a dot and cross diagram
of a hydrogen molecule.

[2 marks]

D–C

5 a Look at the dot and cross diagram
opposite. Circle two errors.

[2 marks]

b Draw a dot and cross diagram for a
molecule of methane (CH_4).

[2 marks]

c i How many electrons surround the carbon atom in methane? .... [1 mark]

   ii How many electrons surround each hydrogen atom in methane? .... [1 mark]

   iii Explain why the two atoms have different numbers of electrons surrounding them.

   ....

   .... [2 marks]

B–A*

6 a Explain what is meant by a double bond. Use a diagram of oxygen to illustrate your answer.

   ....

   ....

   ....

   .... [4 marks]

b Ethene (C_2H_4) is a molecule with a double bond
between carbon atoms. Draw a dot and cross
diagram of ethene.

[2 marks]

Properties of elements and compounds

1 a Which of the following properties **do not** help us to classify sodium chloride as an ionic substance?

i It dissolves in water

ii It is white

iii It has a high melting point

iv It conducts electricity when molten

.... **[1 mark]**

b Draw a diagram to show how to test whether a solution conducts electricity.

[4 marks] G–E

c Which of the substances opposite has the highest melting point?

.... **[1 mark]**

Name	State at room temperature
Methane	Gas
Hexane	Liquid
Sucrose	Solid

d Name **one** simple molecular substance and **one** giant molecular substance.

.... **[2 marks]**

2 a James investigated the properties of three different solids. His results are shown in the table below. Use the results to name the structure and bonding in each of the substances.

Solid	Melted in the Bunsen flame?	Dissolved easily?	Conducted electricity when dissolved?	Structure and bonding
1	Yes	No	No	
2	No	Yes	Yes	
3	No	No	No	

[3 marks]

b Suzanne and Laurence are investigating the properties of sucrose. They discover that it dissolves easily in water. Suzanne says that this proves it is ionic. What further tests could they do to get more evidence? What results would you expect if Suzanne is right?

....

....

[4 marks] D–C

c Explain why sodium chloride conducts electricity when it is dissolved but carbon dioxide does not.

....

....

.... **[3 marks]**

3 a Use ideas about structure and bonding to explain why graphite can be used as a lubricant.

....

....

.... **[3 marks]**

b Carbon dioxide and silicon(IV)oxide are both covalent substances. Carbon dioxide is a gas and silicon dioxide is a solid. What do the different physical states of these compounds tell us about their structures?

B–A*

....

....

....

.... **[5 marks]**

Separating solutions

1 Perfume manufacturers need to extract the essential oils from lavender flowers. The first step in the extraction gives a layer of lavender oil floating on water. Describe how the lavender oil could be separated from water.

....

....

[3 marks]

2 Put the following statements describing fractional distillation into the correct order.

i The vapour condenses and is collected as a liquid.

ii The lowest boiling point fraction vaporises.

iii The flask is heated and the temperature rises.

iv The vapour is cooled to below its boiling point in the condenser.

v The temperature of the flask rises again until the boiling point of the next fraction is reached.

....

[3 marks]

3 Use the data in the table below to explain how a sample of oxygen could be prepared from air.

	Oxygen	Nitrogen	Argon	Carbon dioxide
Melting point (°C)	−219	−210	−189	Sublimes at −78
Boiling point (°C)	−183	−196	−186	

....

....

....

[5 marks]

4 Draw a diagram of a paper chromatogram of some food colouring that contains two different colours.

[2 marks]

5 Food scientists can identify the sugars present in foods using paper chromatography. Opposite is a diagram of a paper chromatogram of the sugars in grapes. Which sugars are present?

....

.... [2 marks]

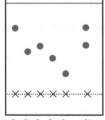

6 A groups of students was investigating how quickly the sweetener aspartame breaks down. They boiled aspartame in hydrochloric acid and removed a small sample every five minutes. A chromatogram was made from each sample. The results are shown in the table below.

Time of sample (mins)	0	5	10	15	20	25
Substance A R$_f$ value 0.85	no	small	medium	large	large	large
Substance B R$_f$ value 0.50	no	small	medium	large	large	large
Substance C R$_f$ value 0.15	large	large	medium	small	no	no

a Which substance travelled the furthest up the chromatography paper?

.... [1 mark]

b Which substance is aspartame?... [1 mark]

c Research suggests that one of the spots is phenylalanine. How could the students confirm which of the substances is phenylalanine?

....

....

[3 marks]

Classifying elements

1 a Which of the following structures would you expect to have the highest melting point?

   i Simple molecular covalent or metallic?

.... **[1 mark]**

   ii Ionic lattice or simple molecular covalent?

.... **[1 mark]**

b Maltose melts at 160 °C and does not conduct electricity under any circumstances. What is the most likely structure for maltose?

.... **[1 mark]**

2 a Describe the structure of sodium chloride.

....

.... **[2 marks]**

b Explain why sodium chloride has a high melting point.

.... **[2 marks]**

3 Complete the table below.

	Melting point	Solubility	Electrical conductivity	Structure
Silicon carbide	2730 °C		Non-conductor	
Boron trifluoride	−127 °C	Very soluble	Non-conductor	
Copper(II)oxide	1201 °C	Insoluble		Ionic
Cerrosafe	74 °C		Conducts as solid	

[6 marks]

4 a What name is given to the elements in group 1 of the periodic table

.... **[1 mark]**

b Which group in the periodic table is known as the noble gases?

.... **[1 mark]**

c Which of the statements below does not describe a transition metal?
   i It has a low melting point.
   ii It forms coloured compounds.
   iii It lies between group 2 and group 3 of the periodic table.
   iv It conducts electricity.

.... **[1 mark]**

5 a Draw a diagram which shows how metal atoms are arranged in a transition metal.

[3 marks]

b Use your diagram to explain why metals are malleable and can conduct electricity.

....

....

.... **[4 marks]**

6 Copper is a transition metal. Explain why the chemical structure of copper makes it particularly good for use in household wiring.

....

....

.... **[4 marks]**

Alkali metals

1 a Which line correctly describes sodium?

A	Less reactive than potassium	Reacts with water to produce oxygen
B	More reactive than potassium	Reacts with water to produce hydrogen
C	Less reactive than lithium	Reacts with water to produce oxygen
D	More reactive than lithium	Reacts with water to produce hydrogen

[1 mark]

b Describe **two** properties of alkali metals that are different from properties of typical metals.

[2 marks]

2 a Kumar watched a demonstration of the reaction of lithium with water. His teacher told him that hydrogen and lithium hydroxide were produced.

i What would Kumar have seen that suggests that hydrogen was produced?

[1 mark]

ii What test could he do to confirm it was hydrogen?

[2 marks]

iii What test could he do to support the claim that lithium hydroxide was produced?

[2 marks]

iv Write a formula equation for this reaction. Include state symbols.

[5 marks]

b Draw a diagram of a lithium atom and use it to explain why alkali metals always form 1+ ions.

[4 marks]

3 a Describe the trend in reactivity seen in alkali metals.

[1 mark]

b Explain the difference in the reactivity of lithium and potassium.

[4 marks]

c Predict the reaction of rubidium with water. Include an equation in your answer.

[3 marks]

4 In 1817, Johann Döbereiner put forward a 'law of triads'. He had observed that lithium, sodium and potassium had similar properties. He noted that the trend in reactivity in these elements matched the trend in their atomic mass. This early observation was an important contribution to the organisation of the periodic table.

a Describe how the trend in atomic mass of lithium, sodium and potassium matches other trends in these elements.

[2 marks]

b Suggest **two** ways in which Döbereiner might have shared his ideas with other scientists.

[2 marks]

Halogens and noble gases

1 Draw lines to link the halogen with the correct properties.

Fluorine		grey		solid
Chlorine		brown		liquid
Bromine		yellow		gas
Iodine		green		gas

G–E

[6 marks]

2 a Which of the following is the correct balanced equation for the formation of hydrogen bromide?

   i $H (g) + Br (g) \longrightarrow HBr (g)$ **ii** $2H (g) + 2Br (l) \longrightarrow 2HBr (g)$

   iii $H_2 (g) + Br_2 (l) \longrightarrow 2HBr (g)$ **iv** $H (g) + Br_2 (g) \longrightarrow 2HBr (s)$

D–C

.... [1 mark]

  b Suggest the pH of a solution made by dissolving hydrogen bromide in water.

.... [1 mark]

3 Pravin carried out an experiment. She mixed chlorine water and then iodine water with potassium bromide solution.

  a Describe the outcome of the experiments. Include any relevant equations.

....

.... [3 marks]

B–A*

  b Explain how the results of this experiment show the relative reactivity of the halogens.

....

.... [3 marks]

4 List the following noble gases in order of increasing atomic mass:

                Ar   **He**   **Ne**   **Xe**

G–E

.... [2 marks]

5 a Write the electronic structure of neon and argon.

.... [2 marks]

  b Use the electronic structure you have written above to explain why neon and argon are very chemically inert.

D–C

....

.... [3 marks]

6 a Use the data in the table below to predict the density of argon.

Element	Density g/dm³	Boiling point °C
Helium	0.1786	−268.93
Neon	0.9002	−246.08
Argon		
Xenon	5.894	−108.12

[1 mark]

  b Use the periodic table and the data in the table above to explain how you could predict the maximum temperature at which radon exists as a liquid.

B–A*

....

....

.... [4 marks]

Endothermic and exothermic reactions

1 a Which of the statements below correctly describes an endothermic reaction?

    i The temperature at the end of the reaction is higher than at the start.

    ii The reaction only involves breaking bonds.

    iii Chemical energy is transferred to heat energy.

    iv The reaction takes in energy overall.

    .... **[1 mark]**

b For each of the reactions below, state whether it is endothermic or exothermic.

    i Photosynthesis .... **[1 mark]**

    ii Burning wood .... **[1 mark]**

    iii Dissolving ammonium chloride .... **[1 mark]**

    iv Lighting a firework .... **[1 mark]**

2 Alice and Sanjay want to find out whether the reaction between sodium hydroxide and hydrochloric acid is exothermic or endothermic.

    a Draw a diagram of the apparatus they could use.

                                                      [2 marks]

    b What measurements would they need to take?

    .... **[1 mark]**

    c What results would they see if the reaction was exothermic?

    .... **[1 mark]**

3 Eli's teacher has posters on the wall of the classroom. One says 'Breaking bonds requires energy, making bonds releases energy.' Use this information to explain why some reactions are exothermic.

    ....

    ....

    ....

    .... **[4 marks]**

4 The results of an experiment are shown below. 25 cm³ of silver nitrate solution was mixed with 25 cm³ of sodium chloride solution.

	Silver nitrate	Sodium chloride
Temperature at start of reaction (°C)	22	24
Temperature at end of the experiment (°C)	43	

    a Is the reaction endothermic or exothermic? .... **[1 mark]**

    b Calculate the energy change in the reaction. Show your working.

    .... **[4 marks]**

    c Write a word equation for the reaction.

    .... **[2 marks]**

    d Draw a labelled energy level diagram for this reaction.

                                                      [3 marks]

5 Use ideas about bond breaking and bond making to explain why burning methane is exothermic.

    ....

    ....

    ....

    .... **[4 marks]**

Reaction rates and catalysts

1 Sam is cooking carrots. He wants them to cook quickly. Which of the following methods could Sam use to speed up the cooking?

  i   Reduce the temperature of the water.   ii Cut the carrots into small pieces.

  iii Use large pieces of carrot.   iv Use a large volume of water.

.... **[1 mark]**

G–E

2 Ella added acid to chalk and the chalk slowly disappeared. When she added water and acid to chalk it took longer to disappear. Why did this happen?

.... **[2 marks]**

3 Magnesium reacts with hydrochloric acid to form hydrogen. Design an experiment that would show the rate of reaction for two different concentrations of acid.

  a Draw a diagram of the apparatus you will use.

[3 marks]

  b Say what you will measure.

D–C

.... **[2 marks]**

  c Say how you will present the results.

.... **[1 mark]**

  d Say what the results will look like.

.... **[2 marks]**

4 Hydrogen and iodine react together to make hydrogen iodide. Explain why the number of reactions that take place at 25 °C is only a small percentage of the total number of collisions.

B–A*

.... **[2 marks]**

5 a Modern cars are fitted with catalytic converters. Describe what a catalyst does.

.... **[2 marks]**

G–E

  b Why are modern cars fitted with catalytic converters?

.... **[2 marks]**

  c Catalytic converters reduce the level of unburned hydrocarbons that are emitted from cars. Write a word equation for the reaction they catalyse.

D–C

.... **[1 mark]**

6 a Explain how the honeycomb design of car catalytic converters enables them to give the best performance.

.... **[2 marks]** **B–A***

  b Write a balanced equation to show how catalytic converters reduce the quantity of carbon monoxide emitted from cars.

.... **[2 marks]**

Mass and formulae

1 Calculate the relative formula mass of the following compounds. Use a separate piece of paper for your working if necessary.

  a HCl .... [1 mark]

  b Br_2 .... [1 mark]

  c $NaNO_3$ .... [1 mark]

  d $Ca(OH)_2$ .... [1 mark]

2 Calculate the percentage by mass of the following. Use a separate piece of paper for your working if necessary.

  a Carbon in CO_2 .... [2 marks]

  b Oxygen in $KBrO_3$ .... [2 marks]

  c Nitrogen in $Mg(NO_3)_2$ .... [2 marks]

3 A group of students carried out an experiment to calculate the formula for copper oxide. They heated some copper powder in an open crucible to turn it into copper oxide. The results are recorded in the table below.

Mass of empty crucible (g)	10.10g
Mass of crucible + copper before heating (g)	15.05
Mass of copper (g)	
Mass of copper + crucible after heating (g)	16.30
Mass of copper oxide (g)	

  a Complete the table of results. [2 marks]

  b Calculate the mass of oxygen that was added to the copper during heating.

  .... [1 mark]

  c Calculate the empirical formula of the copper oxide.

  .... [3 marks]

4 The theoretical yield for a reaction is 65 g. The actual yield was 43 g. Calculate the percentage yield for the reaction.

  .... [2 marks]

5 George is making oxygen from hydrogen peroxide. The theoretical yield from 10 cm^3 of his hydrogen peroxide solution is 100 cm^3 of oxygen. The percentage yield for the reaction is 50%. What volume of hydrogen peroxide solution must George use to make 100 cm^3 of oxygen?

  .... [2 marks]

6 What mass of copper sulfate can be made from 5 g of copper oxide in the following reaction?
$CuO(s) + H_2SO_4 (aq) \rightarrow CuSO_4 (aq) + H_2O (l)$

  .... [3 marks]

7 Nylon can be made in two different ways. Method 1 is slow and produces water as a waste product. Method 2 is fast and produces hydrochloric acid as a waste product.

  Suggest **one** reason why manufacturers might choose method 1 and **one** why they might chose method 2.

  ....

  .... [2 marks]

Topic 6: 6.1, 6.2, 6.3, 6.4, 6.5, 6.6, 6.7, 6.8, 6.9, 6.10, 6.11

C2 Extended response question

Zain is given three white powders. All have different structures. One is sucrose (sugar), one is barium chloride (toxic) and one is powdered sand.

Plan an experiment that Xain could carry out to decide which is which. You should describe any apparatus that is needed and explain why the substances will give different results.

The quality of written communication will be assessed in your answer to this question.

[6 marks]

The Solar System

1 The ancient Greeks used a geocentric model of the Solar System.

G–E

a What is meant by the term 'geocentric'?

....

.... **[2 marks]**

b Name the five planets in the geocentric model.

....

.... **[3 marks]**

c Why are there only five planets in the geocentric model when there are more in the modern model of the Solar System?

....

.... **[2 marks]**

D–C

d Describe an observation that the ancient Greeks found difficult to explain using the geocentric model, but which was easier to explain using the heliocentric model.

....

....

....

.... **[3 marks]**

2 Why was Galileo ostracised by the Catholic Church after he claimed that there were mountains on the Moon and that Jupiter had four moons of its own?

B–A*

....

....

....

.... **[3 marks]**

3 Until the telescope was invented in the 17th century, astronomers observed the night sky with the naked eye.

Describe **two** advantages of using a telescope over the naked eye.

G–E

....

.... **[2 marks]**

4 Describe the nature and position of the asteroid belt.

D–C

....

.... **[2 marks]**

5 Why can the Hubble space telescope take much clearer photographs than any telescopes on Earth?

B–A*

....

....

.... **[2 marks]**

Reflection, refraction and lenses

1 Use words from the list to complete the law of reflection.

ray   normal   incidence   reflected   angle   mirror

The angle of .... is equal to the ....

of reflection. The incident ray, the normal and the .... ray are all

in the same plane. The .... is an imaginary line at right angles to

the surface of the mirror where the reflection happens.   **[4 marks]**

G–E

2 The diagram opposite shows a periscope.
Complete the diagram by drawing
in the path of the rays of light
from the tree towards the eye.

[3 marks]

3 Which of the following diagrams correctly shows how a ray of light will travel through a
rectangular glass block? Put a tick in the box below the correct diagram.

D–C

[1 mark]

4 The speed of light in different materials is shown in the table below. Which material refracts
light by the largest angle? Explain why.

Material	Speed of light (m/s)
Air	3.0×10^8
Crown glass	2.0×10^8
Water	2.3×10^8
Quartz	2.1×10^8

B–A*

....

....   **[2 marks]**

5 Converging lenses are used in many optical instruments, including microscopes and telescopes.

a Explain what is meant by the term 'focal point' of a converging lens.

....

....   **[2 marks]**

b Which of these two lenses will have the longer focal length?

G–E

A   B

....   **[1 mark]**

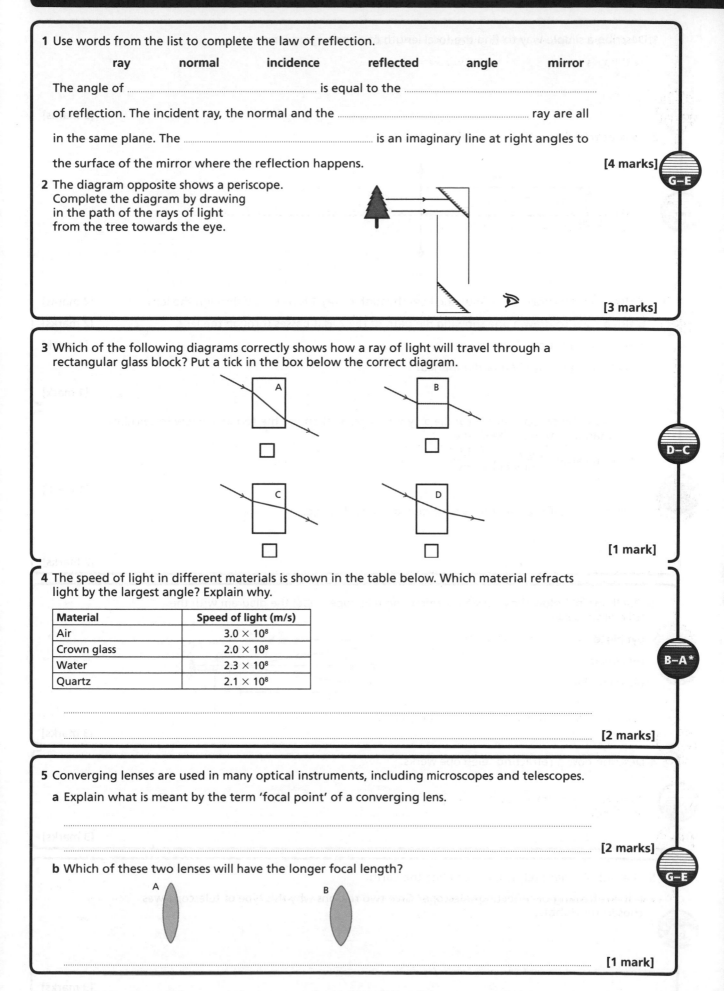

Lenses in telescopes

1 Describe a simple way to find the focal length of a converging lens.

....

....

.... **[3 marks]**

2 Look at the incomplete ray diagram below.

a Use a ruler to draw a line showing how the path of ray 1 is refracted through the lens. **[2 marks]**

b Use a ruler to draw a line showing the path of ray 2 as it passes through the lens. **[2 marks]**

c Mark the position of the image. **[1 mark]**

d Is the image magnified or diminished?

.... **[1 mark]**

e Measure the object and the candle on your diagram, then use the formula below to calculate the magnification of the image.

$$\text{Magnification} = \frac{\text{image height}}{\text{object height}}$$

.... **[1 mark]**

f Explain the difference between a virtual and a real image.

....

....

.... **[2 marks]**

3 The diagram below shows a simple refracting telescope. Label the diagram with the following words:

eyepiece

real image

objective lens

[3 marks]

4 Describe how a refracting telescope works.

....

....

.... **[3 marks]**

5 The Hubble space telescope is orbiting the Earth.

Is it a refracting or reflecting telescope? Give **two** reasons why this type of telescope was chosen for Hubble.

....

....

....

.... **[3 marks]**

Waves

1 Complete the sentences describing waves, using words from the list below.

vibrate	**oscillations**	**energy**	**speed**	**matter**

Waves transfer ...., without transferring

.... Waves are caused by .... **[3 marks]** G–E

2 The diagram opposite shows a transverse wave.

Write the letter of the line that represents:

a The amplitude of the wave .... **[1 mark]**

b The wavelength of the wave .... **[1 mark]**

3 Tom is watching waves on a beach. He counts 15 waves in a minute. The distance between the crests is 8 m. Calculate the speed of the wave.

....

.... **[3 marks]**

4 Class 10C are trying to measure the speed of sound. Lizzie stands at the far side of the playing field, 200 m away, with a starting pistol. The rest of the class all record the time between seeing the smoke and hearing the shot.

a Explain why they all get slightly different results.

.... D–C

.... **[2 marks]**

b Here are five pupils' results: 0.55 s; 0.62 s; 0.58 s; 0.56 s; 0.58 s. Work out the average time, then calculate the speed of sound.

....

.... **[3 marks]**

5 Sound waves travel at 1500 m/s in water. If the frequency of a sound is 500 Hz, calculate the wavelength in water. B–A*

.... **[2 marks]**

6 There are two main types of wave: longitudinal and transverse. Draw lines joining the types of wave to the correct descriptions. **[6 marks]**

Longitudinal	**Transverse**

Sound wave	**Vibrations at right angles to the direction of travel**	**Transfer energy**

G–E

Radio wave	**Light waves**

7 There are two main types of seismic wave.

a What is a seismic wave?

.... D–C

.... **[2 marks]**

b Describe how a seismometer works.

.... B–A*

.... **[2 marks]**

The electromagnetic spectrum

1 The electromagnetic spectrum is a family of waves.

a Complete the following statements about electromagnetic waves.

Electromagnetic waves are t.. waves. They all travel

through a v.. at the same speed. **[2 marks]**

b Put the following waves into the correct spaces in the electromagnetic spectrum:

infrared   **gamma rays**   **ultraviolet**   **radio waves**

	X-rays		Visible light		Microwaves	

[4 marks]

2 a Which type of wave has the longest wavelength?

.... **[1 mark]**

b Which type of wave has the most energy?

.... **[1 mark]**

c How do we protect ourselves from the harmful effects of ultraviolet waves?

....

.... **[2 marks]**

3 Explain what is meant by a spectrum of white light and describe how you can produce one in a school laboratory.

....

....

.... **[2 marks]**

4 a The physicist William Herschel carried out experiments to find out which colour of the spectrum was the warmest.

Explain how this led to the discovery of infrared waves.

....

....

.... **[3 marks]**

b Silver chloride turns black when it is exposed to sunlight. Johann Ritter carried out experiments to find out which colour of light reacted fastest with silver chloride.

Explain how this led to the discovery of ultraviolet waves.

....

....

.... **[3 marks]**

5 Electromagnetic waves travel at a speed of 3×10^8 m/s in a vacuum. If the frequency of an EM wave is 60 MHz:

a Calculate the wavelength of the wave.

....

.... **[2 marks]**

b Which type of electromagnetic wave is it?

.... **[1 mark]**

Uses of EM waves

1 Draw lines to join the type of electromagnetic radiation with its use.

| Microwaves | | Mobile-phone communication |

| X-rays | | TV remote control |

| Infrared | | Airport security scanner |

| Ultraviolet |

| Water sterilisation |

[4 marks]

2 You phone your friend in the USA on your mobile phone. The sound is coded and sent using microwaves.

Explain how the microwave signal travels to the USA.

....

....

.... [3 marks]

3 All objects emit some infrared waves.

Explain how infrared imaging can be used to compare home insulation.

Explain how you can tell that one house has much better wall insulation than the others.

....

....

.... [3 marks]

4 a On each diagram below, draw the path of the ray of light as it leaves the semicircular block.

[4 marks]

b The diagram shows light travelling down an optical fibre.

Continue the path of the light ray until it emerges from the optical fibre.

[3 marks]

Gamma rays, X-rays, ionising radiation

1 X-rays are dangerous, but they are used in hospitals.

Explain why they are used, despite being harmful.

G–E

....

.... **[2 marks]**

2 Gamma rays and X-rays have similar wavelengths and frequencies.

Name **one** similarity and **one** difference between gamma rays and X-rays.

D–C

....

.... **[2 marks]**

3 Describe how X-rays are produced.

B–A*

....

....

.... **[3 marks]**

4 To a physicist, ionising radiation can mean electromagnetic waves or radiation from the nucleus of atoms.

Name the **three** different types of ionising radiation that come from the nucleus of atoms.

G–E

.... **[3 marks]**

5 Joanna and Michael are investigating the ionising radiation emitted from some rocks. Joanna suggests that they lower the temperature of the rocks because this will reduce the amount of radiation emitted. Michael disagrees with her, and says that lowering the temperature will make no difference.

a Who is correct? Explain your answer.

D–C

....

....

.... **[3 marks]**

b Joanna and Michael use a Geiger–Müller tube to measure the radioactivity emitted from the rocks.

Describe how a Geiger–Müller tube detects ionising radiation.

B–A*

....

....

....

.... **[4 marks]**

The Universe

1 List the objects below in order of size, from smallest to largest:

Jupiter   Mercury   Moon   Sun   comet   Milky Way

G–E

....

.... **[3 marks]**

2 The table below contains information about some of the bodies in the Solar System.

Use the data to answer the questions below.

Body	Average distance from Sun (AU)	Diameter relative to Earth
Mercury	0.39	0.38
Earth	1.00	1.00
Mars	1.52	0.53
Jupiter	5.20	11.20
Moon	1.00	0.27

D–C

a How many times larger than Mercury is Jupiter?

.... **[1 mark]**

b How many times further away from the Sun than Mars is Jupiter?

.... **[1 mark]**

c How many times larger than the Moon is the Earth?

.... **[1 mark]**

3 Light from the Sun takes approximately 8 minutes to get to Earth. The speed of light is 3×10^8 m/s.

a Calculate the average distance of the Earth from the Sun. Show your working.

....

....

B–A*

.... **[3 marks]**

b Explain why looking at distant stars is like looking back in time.

....

.... **[2 marks]**

4 Scientists have studied space for thousands of years.

Early telescopes only used light to obtain images of space. Why do we now use other parts of the electromagnetic spectrum, such as radio waves, to study space?

....

G–E

....

.... **[2 marks]**

5 a Scientists have sent a robotic lander to Mars.

What sort of information can a lander find out about Mars?

....

.... **[2 marks]**

b What is the Search for Extraterrestrial Intelligence (SETI) project?

D–C

....

....

.... **[3 marks]**

Analysing light

1 a What is a spectrometer?

.... **[1 mark]**

b Describe how to make a simple spectrometer.

....

....

....

.... **[3 marks]**

G–E

2 Astronomers use spectrometers to analyse the light from stars.

a Why do different stars show different spectra?

....

.... **[2 marks]**

b What information can astronomers obtain by studying the spectra from stars?

....

.... **[2 marks]**

D–C

3 Earth's atmosphere protects us from harmful radiation from space.

Use the graph opposite to explain why holes in the ozone layer have been linked to a rise in the number of cases of skin cancer.

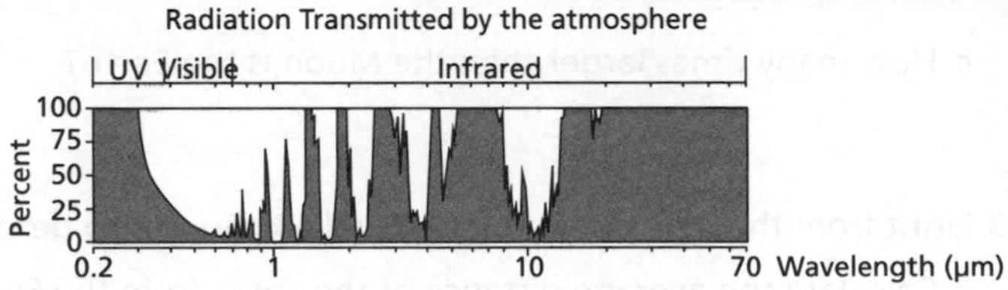

Radiation Transmitted by the atmosphere

UV | Visible | Infrared

Percent: 100, 75, 50, 25, 0

0.2   1   10   70   Wavelength (μm)

....

....

....

.... **[3 marks]**

B–A*

4 The left-hand diagram opposite shows some circular wave fronts from a stationary source.

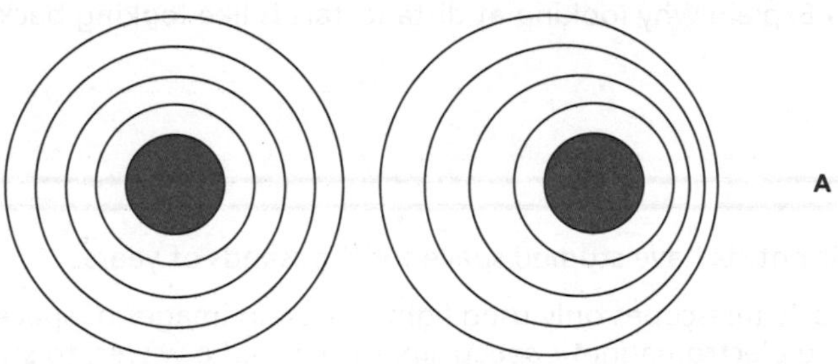

A

a What is the wavelength of the waves in mm?

.... **[1 mark]**

b The right-hand diagram shows the same wave fronts, but now the source is moving to the right. If you were observing the waves from position A, the wave source would be moving towards you.

What has happened to the wavelength?

....

.... **[1 mark]**

c If the waves were sound waves, what would happen to the pitch of the sound?

.... **[1 mark]**

D–C

5 Explain what is meant by red-shift.

....

....

.... **[3 marks]**

B–A*

The life of stars

1 Put the following stages in the birth of a star into the correct order.

The first one has been done for you.

A A nebula contains clouds of dust, ice and gas.

B All the clouds disappear and a glowing ball of gas remains.

C The particles start to collide with each other and the temperature starts to rise.

D Nuclear fusion reactions occur, giving out energy in the form of electromagnetic radiation.

E The force of gravity attracts all the matter together

A .... **[3 marks]**

G–E

2 Our Sun is a main sequence star in a steady state.

a Explain why the Sun does not collapse in on itself due to the force of gravity.

....

.... **[2 marks]**

D–C

b Towards the end of its life, the Sun will cool and expand to become a red giant.

What will happen once the fuel for the nuclear reactions runs out?

....

.... **[2 marks]**

3 Massive stars like Rigel in the constellation Orion look blue in the night sky.

a Why do they look bluer than average-sized stars like the Sun?

.... **[1 mark]**

b Describe what will happen to Rigel when it comes to the end of its life.

....

....

....

....

.... **[4 marks]**

B–A*

4 Put these objects in order of size, from smallest to largest:

supernova   **white dwarf**   **red giant**   **main sequence star**

....

.... **[3 marks]**

D–C

5 Scientists have predicted that there is a black hole in the centre of the Milky Way.

Explain why it is difficult to observe a black hole.

....

....

.... **[2 marks]**

B–A*

Theories of the Universe

1 In the 1940s there were two opposing theories of the Universe: the Steady State theory and the Big Bang theory.

Draw lines to join the main theories to the correct ideas.

G–E

Big Bang theory

Steady State theory

| 1 The Universe spontaneously creates matter in empty space. |

| 2 The Universe is expanding. |

| 3 The Universe is constantly changing. |

| 4 The Universe had a beginning about 14 billion years ago, and will eventually end. |

| 5 The Universe has no beginning and no end. |

[6 marks]

2 a Describe the evidence that proves that the Universe is expanding.

....

.... **[2 marks]**

b What is meant by 'cosmic background radiation'?

D–C

....

.... **[2 marks]**

c Why does the observation of cosmic background radiation support the Big Bang theory?

....

.... **[2 marks]**

3 Read the paragraph below about the discovery of cosmic background radiation.

In 1964, Penzias and Wilson accidentally discovered the existence of cosmic background radiation. They were actually measuring radio waves reflected from orbiting satellites with a very large horn antenna. They found some 'interference', which they could not account for. They eliminated all known sources of the radiation, but still recorded microwave interference. They believed it might be caused by bird droppings inside the antenna, so they spent several hours cleaning it off. They still recorded the 'interference', and found that it was equal in strength in all directions. They eventually linked this to Gamov's predicted cosmic background radiation.

a What is meant by 'interference'?

....

.... **[2 marks]**

B–A*

b Why did Penzias and Wilson need to eliminate all known sources of microwave radiation?

....

.... **[2 marks]**

c Why is cosmic background radiation equal in strength in all directions?

....

.... **[2 marks]**

d Why should cosmic background radiation appear in the microwave section of the electromagnetic spectrum?

....

....

.... **[3 marks]**

Ultrasound and infrasound

1 Sounds can have a range of frequencies.

a Which of these frequencies can most humans hear: 6 Hz, 50 Hz, 2000 Hz, 50000 Hz?

.... [2 marks]

b As you get older, your hearing deteriorates. Which frequencies do you lose the ability to hear – higher or lower?

.... [1 mark]

c Explain what is meant by the terms infrasound and ultrasound.

....

.... [2 marks]

G–E

2 Ultrasound is used to scan unborn babies in the womb.

a Describe how ultrasound can be used to create an image.

....

.... [3 marks]

b Why is ultrasound used to scan unborn babies?

....

.... [2 marks]

3 Name **one** man-made and **one** natural source of infrasound.

....

.... [2 marks]

D–C

4 A boat sends out pulses of ultrasonic waves and receives echoes. Both the outgoing pulses and reflected pulses are displayed on an oscilloscope screen as shown below.

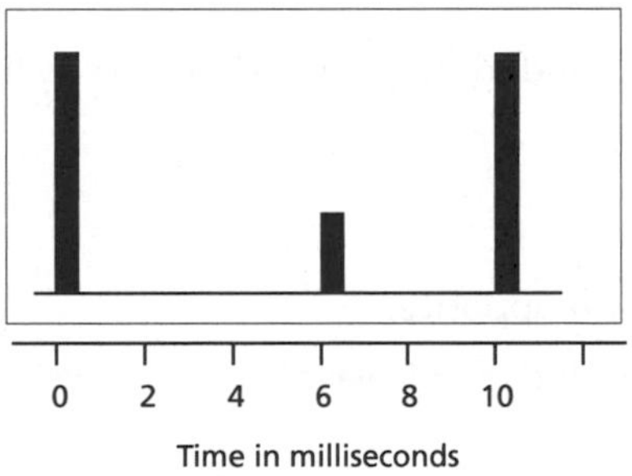

Time in milliseconds

a How long did it take the sound to travel to the sea bed and back again?

.... [1 mark]

b If the speed of sound in water is 1500 m/s, calculate the depth of water below the bottom of the boat.

....

.... [3 marks]

c How will the pattern on the screen change as the tide comes in?

.... [2 marks]

d Suggest how you could adjust the equipment so that greater depths could be measured.

.... [1 mark]

B–A*

Earthquakes and seismic waves

1 The Earth is made up of several layers.

   a Label the diagram showing four layers of the Earth.

G–E

                                                          [4 marks]

   b What is magma?

   .... **[1 mark]**

2 The Earth's crust is made up of several solid plates that float on top of the magma.

   a What is the name of these plates?

   .... **[1 mark]**

D–C

   b How do the plates cause earthquakes?

   ....

   ....

   .... **[3 marks]**

   c Alfred Wegener first suggested this theory about Earth's plates in 1915.

       Describe the evidence that supports his theory.

B–A*

   ....

   ....

   ....

   .... **[3 marks]**

3 There are two main types of seismic wave: S waves and P waves.

G–E

   a Which seismic waves are longitudinal? .... **[1 mark]**

   b Which seismic waves can travel through Earth's liquid outer core? .... **[1 mark]**

   c Which seismic waves can travel through Earth's semi-solid mantle?........... **[1 mark]**

4 It is difficult to predict the exact time and location an earthquake will occur. Explain why.

   ....

   .... **[2 marks]**

D–C

5 The trace from a seismograph is shown opposite.

   a Which vibration is the P wave and which the S wave?

   .... **[1 mark]**

6 The diagram opposite shows how P waves travel through the centre of the Earth.

   a Explain why there is a shadow zone where no P waves are detected.

   ....

   .... **[2 marks]**

   b Complete the diagram opposite to show how S waves travel through the centre of the Earth.

B–A*

                                                          [3 marks]

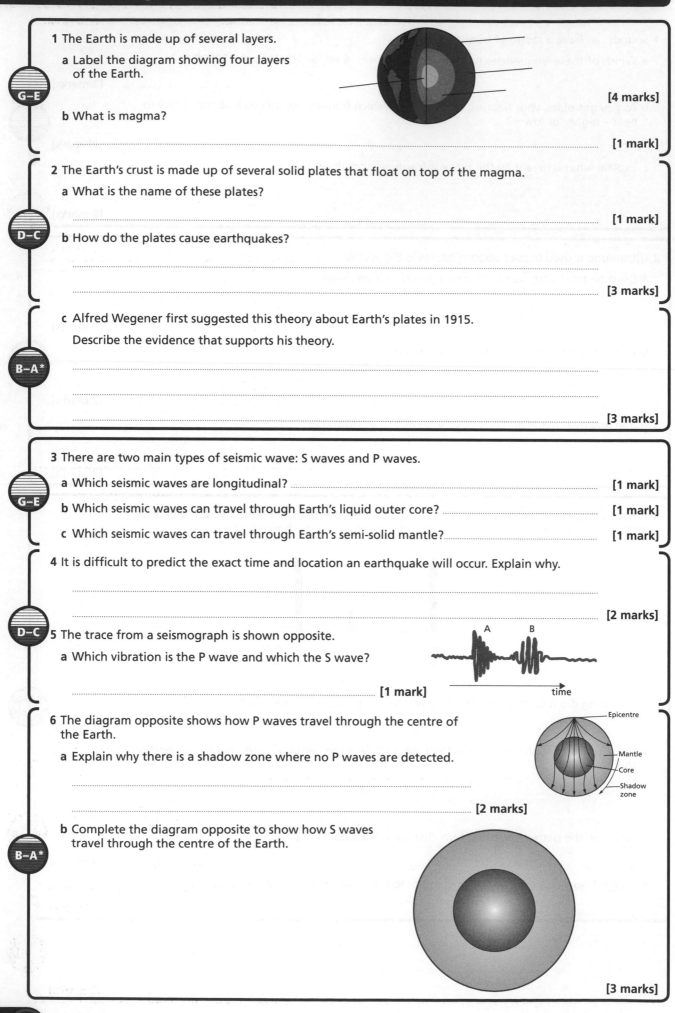

Electrical circuits

1 a Draw an electric circuit for a bulb with a battery and a switch.

G–E

[3 marks]

b Describe what happens to the charged particles in the above circuit.

....

....

....

D–C

[3 marks]

2 In the circuit opposite, both the bulbs are identical.

What is the current at positions X, Y and Z?

0.6A .... Z

X

B–A*

X = ...

Y = ...

Z = ...

Y

[3 marks]

3 In the circuit below, bulb A has a voltage of 3 V across it, and bulb B has a voltage of 4 V across it.

A   B

G–E

Which bulb, A or B, is converting the most energy?

....

[1 mark]

4 Complete the following sentences using the words **current** or **voltage**.

When lamps are connected in series, the .... through them is

the same. The .... across each lamp will add up to the total

supply ....

When lamps are connected in parallel, the .... across each

bulb will be the same as the .... across the power supply.

D–C

[3 marks]

5 What is meant by the term 'potential difference'?

....

....

....

B–A*

[3 marks]

Electrical power

1 The power ratings of some domestic appliances are listed in the table below.

Appliance	Power rating in watts	Average daily use
Kettle	2000	10 minutes
Iron	1200	30 minutes
Vacuum cleaner	1000	20 minutes

a Which appliance uses the most energy per second?

.... [1 mark]

b Which appliance uses the most energy in an average day?

.... [1 mark]

c Which appliance will cost the most to run?

.... [1 mark]

d Complete the table below showing the amount of energy used by each appliance in both joules and kilowatt-hours.

Appliance	Energy used in joules	Energy used in kW h
Kettle		
Iron		
Vacuum cleaner		

[6 marks]

e Explain why electricity supply companies use kW h instead of joules as units of energy.

.... [1 mark]

f The mains electricity supply is 230 V. Calculate the current that each appliance will use, and decide which size fuse should be used for each appliance. Fuses available are 1 A, 3 A, 5 A and 13 A.

Kettle .... [2 marks]

Iron .... [2 marks]

Vacuum cleaner .... [2 marks]

2 The local council is encouraging people to save energy in the home. One suggestion is to use energy-saving light bulbs instead of filament bulbs.

a Give **two** advantages of using energy-saving light bulbs.

....

.... [2 marks]

b Mr Hicks says that he does not want to use energy-saving light bulbs as they cost about twice as much as filament bulbs.

Explain the idea of 'payback time', to encourage him to use them.

....

.... [3 marks]

3 The table below shows the initial cost of installing some other energy-saving devices in the home.

a Complete the table below to show the annual savings and payback time for each device.

Device	Initial cost (£)	Annual saving (£)	Payback time (years)
Double glazing	7000	350	
Loft insulation	450		6
Draught excluders	40	5	
Cavity-wall insulation	550		5

[4 marks]

b Which device has the longest payback time? Give **one** other reason why householders should choose to install this device.

....

.... [2 marks]

Topic 5: 5.2, 5.3, 5.4, 5.16, 5.17, 5.18, 5.19, 5.20, 5.21

Energy resources

1 Fossil fuels are used to produce electricity in power stations. Put the following sentences in the correct order to describe how they are used. The first one has been done for you.

A The fuel is burnt to produce heat.   **D** The steam turns turbines.
B The turbines are connected to generators.   **E** The heat is used to boil water to produce steam.
C The generators produce electricity.

A, ... , ... , ... , ...   **[3 marks]**

G–E

2 a Circle the two gases that are produced when fossil fuels are burnt.

oxygen   carbon dioxide   ammonia   sulfur dioxide   **[2 marks]**

b What problems are associated with these two gases?

....

....   **[2 marks]**

c Nuclear power stations do not produce these gases. Name **two** disadvantages of using nuclear power stations.

....

....   **[2 marks]**

D–C

3 Explain what is meant by the term 'greenhouse effect'.

....

....

....   **[3 marks]**

B–A*

4 Next to each of the following energy resources, write whether it is renewable or non-renewable.

wind ....   biomass ....

coal ....   nuclear ....

hydroelectric ....   oil ....

solar ....   tidal ....   **[4 marks]**

G–E

5 a Explain briefly how wind turbines are used to create electricity.

....

....   **[2 marks]**

b Name **two** disadvantages of wind turbines.

....

....   **[2 marks]**

D–C

6 The graph opposite shows how the power output from a wind turbine varies with wind speed.

a What is the maximum power output of the wind turbine?

.... **[1 mark]**

b At what wind speed is the maximum power obtained?

.... **[1 mark]**

c The graph only shows data from wind speeds of about 4 m/s up to 25 m/s. Suggest why only this range of data has been given.

....

....   **[2 marks]**

Wind speed in m/s

B–A*

Generating and transmitting electricity

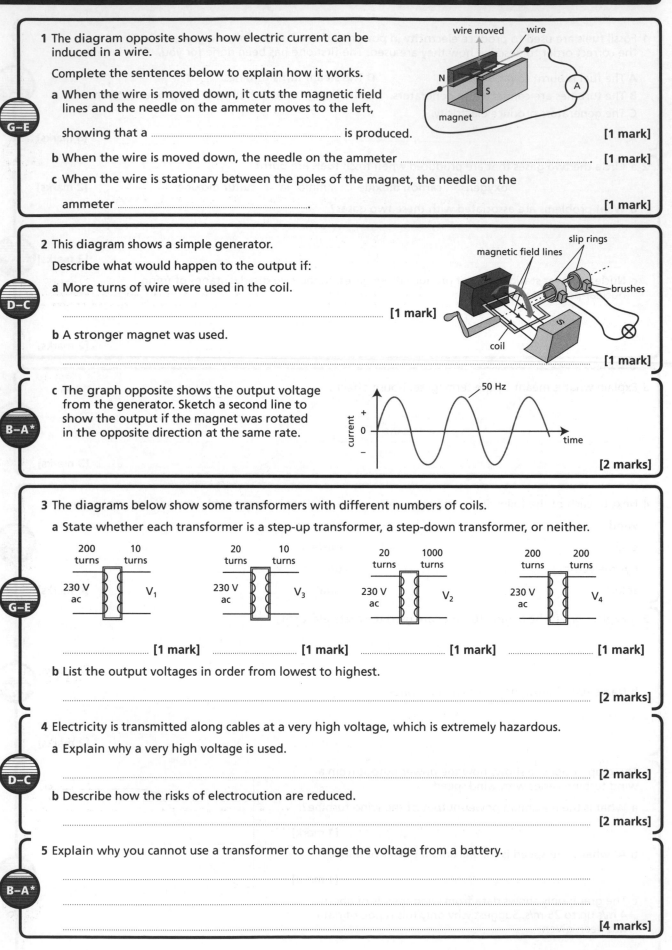

1 The diagram opposite shows how electric current can be induced in a wire.

Complete the sentences below to explain how it works.

a When the wire is moved down, it cuts the magnetic field lines and the needle on the ammeter moves to the left,

showing that a .... is produced. **[1 mark]**

b When the wire is moved down, the needle on the ammeter .... **[1 mark]**

c When the wire is stationary between the poles of the magnet, the needle on the

ammeter .... **[1 mark]**

2 This diagram shows a simple generator.

Describe what would happen to the output if:

a More turns of wire were used in the coil.

.... **[1 mark]**

b A stronger magnet was used.

.... **[1 mark]**

c The graph opposite shows the output voltage from the generator. Sketch a second line to show the output if the magnet was rotated in the opposite direction at the same rate.

[2 marks]

3 The diagrams below show some transformers with different numbers of coils.

a State whether each transformer is a step-up transformer, a step-down transformer, or neither.

.... **[1 mark]**   .... **[1 mark]**   .... **[1 mark]**   .... **[1 mark]**

b List the output voltages in order from lowest to highest.

.... **[2 marks]**

4 Electricity is transmitted along cables at a very high voltage, which is extremely hazardous.

a Explain why a very high voltage is used.

.... **[2 marks]**

b Describe how the risks of electrocution are reduced.

.... **[2 marks]**

5 Explain why you cannot use a transformer to change the voltage from a battery.

....

....

....

....

[4 marks]

Energy and efficiency

1 Draw a line to match the object to the type of energy.

A roller coaster at the top of the ride	Elastic potential energy
A stretched rubber band	Gravitational potential energy
A battery	Heat energy
A cup of tea	Chemical energy

[4 marks]

2 Complete the table below, identifying the energy input, useful energy output and the wasted energy output for a variety of devices. Some have been done for you.

Device	Energy input	Useful energy output	Wasted energy output
Electric fan	Electricity		
Television		Light and sound	
Catapult		Kinetic	Heat
Gas ring on a cooker			Light and sound

[7 marks]

3 Explain what is meant by the principle of conservation of energy.

....

.... **[3 marks]**

4 The diagram below shows how a coal-fired power station generates electricity.

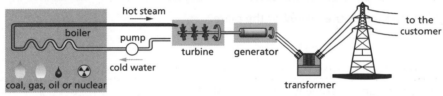

a Complete the energy transfer chain for this process.

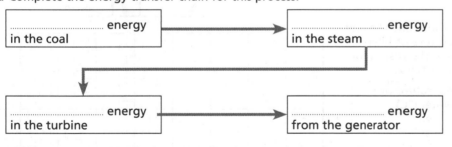

[4 marks]

b For every 360 MJ of energy stored in the coal, only 144 MJ of electricity is generated. Calculate the efficiency of the power station.

.... **[2 marks]**

c What happens to the rest of the energy from the coal?

.... **[2 marks]**

5 Sketch a labelled Sankey diagram for the power station in Question 4.

[4 marks]

Radiated and absorbed energy

1 a Which of the following contains the most heat energy?

   i 100 ml of water at 20 °C   **ii** 200 ml of water at 40 °C

   iii 100 ml of water at 40 °C

.... **[1 mark]**

b Which of the following would emit the most heat energy?

   i A cup of tea in a black cup   **ii** A cup of tea in a white cup

   iii A cup of milky tea in a white cup

.... **[1 mark]**

c Which of the following would absorb the most heat energy on a sunny day?

   i A black water bottle   **ii** A transparent water bottle

   iii A white water bottle

.... **[1 mark]**

G–E

2 Ellie got an ice cube out of the freezer (at about –5 °C) and put it on a plate (at about 50 °C), which she had just got out of the dishwasher. The room temperature in her kitchen was 20 °C. She left the ice cube there for several hours and the ice melted to a pool of water.

a What was the final temperature of the plate? .... **[1 mark]**

b What was the final temperature of the water? .... **[1 mark]**

c Describe what happened to the energy of the plate?

....

.... **[3 marks]**

3 Sam and Joe carried out an experiment to find out what type of surface lost the most heat energy. They coated one beaker with silver paint and a second beaker with matt black paint.

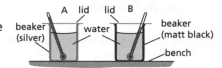

They put some hot water into each beaker and recorded the temperature at intervals as the water cooled down. The results for the silvered beaker were as follows:

Time (minutes)	Temperature of water (°C)
0	80
5	71
10	65
15	61
20	58
25	56
30	55

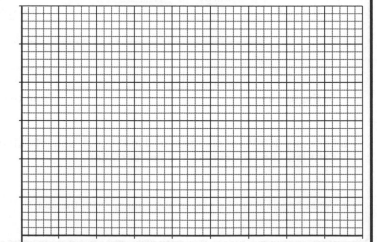

D–C

a Plot a line of time on the *x*-axis against temperature on the *y*-axis on the grid provided. **[4 marks]**

b Name **two** ways in which Sam and Joe could make sure the test was fair.

....

.... **[2 marks]**

c Sketch a line on the graph showing the results for the blackened beaker. **[2 marks]**

d Explain your answer to part C.

....

.... **[2 marks]**

B–A*

P1 Extended response question

Our Sun was born in a nebula almost five billion years ago. Our Sun is an average-sized star and it will eventually die.

Describe the main stages in an average-sized star's life.

The quality of written communication will be assessed in your answer to this question.

[6 marks]

Electrostatics

1 Scientists believe that all matter is made of atoms.

a Which two particles are present in the nucleus of an atom?

.... [2 marks]

b Explain why the nucleus has a positive charge.

....

.... [2 marks]

2 When Brendan takes off his jumper it crackles. Electrostatic charge has built up on his jumper.

a Explain how this could have happened.

.... [1 mark]

b Which particles cause this crackling sound?

.... [1 mark]

c After Brendan has removed his jumper, he finds that his hair has become charged. Explain why the strands of hair repel each other.

....

.... [2 marks]

3 If you rub a balloon against your clothes and then hold it against a wall, it sticks.

a Explain how the transfer of charge causes this to happen.

....

....

.... [3 marks]

b What name is given to the separation of charges by another charged object?

.... [1 mark]

4 An atom can be charged or uncharged.

a What can you say about the number of protons and electrons in an uncharged atom?

.... [1 mark]

b What name is given to a charged atom?

.... [1 mark]

c Explain how an atom becomes positively charged.

.... [1 mark]

5 When a charged glass rod is brought near to the cap of a gold-leaf electroscope, the leaf moves away from the metal rod. Explain why this occurs.

....

....

....

.... [3 marks]

G–E

D–C

B–A*

Topic 1: 1.1, 1.2, 1.3, 1.4, 1.5c

Uses and dangers of electrostatics

1 A farmer uses an electrostatic plant sprayer to spray his crops with insecticide.

Give **three** advantages to using an electrostatic sprayer.

....

....

.... **[3 marks]**

2 The diagram below shows a negatively charged sphere with an electric field around it.

a What name is given to the lines?

.... **[1 mark]**

b If the lines were closer together, what change would this represent?

.... **[1 mark]**

3 Adrian brings his finger close to a charged insulator.

Describe what Adrian will experience and why.

....

....

.... **[2 marks]**

4 A lorry is fuelled using a plastic fuel pipe.

a Explain how static electricity could cause an explosion during refuelling.

....

.... **[2 marks]**

b How could a metal cable be used to reduce the risk of such an explosion?

....

.... **[2 marks]**

5 A lightning conductor can help protect a tall building from lightning strikes.

a The top of the conductor points upwards from the top of the building. Where is the other end of the conductor?

.... **[1 mark]**

b A negatively charged cloud passes overhead. What charge is induced at the top of the spike?

.... **[1 mark]**

Current, voltage and resistance

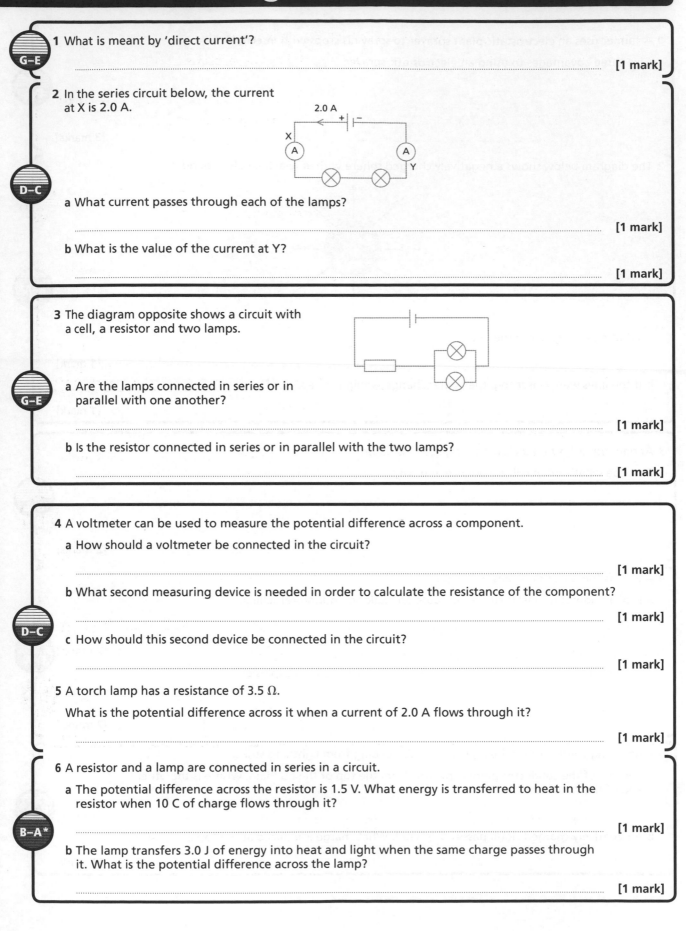

G–E

1 What is meant by 'direct current'?

.... **[1 mark]**

D–C

2 In the series circuit below, the current at X is 2.0 A.

2.0 A

X

A

A

Y

a What current passes through each of the lamps?

.... **[1 mark]**

b What is the value of the current at Y?

.... **[1 mark]**

G–E

3 The diagram opposite shows a circuit with a cell, a resistor and two lamps.

a Are the lamps connected in series or in parallel with one another?

.... **[1 mark]**

b Is the resistor connected in series or in parallel with the two lamps?

.... **[1 mark]**

D–C

4 A voltmeter can be used to measure the potential difference across a component.

a How should a voltmeter be connected in the circuit?

.... **[1 mark]**

b What second measuring device is needed in order to calculate the resistance of the component?

.... **[1 mark]**

c How should this second device be connected in the circuit?

.... **[1 mark]**

5 A torch lamp has a resistance of 3.5 Ω.

What is the potential difference across it when a current of 2.0 A flows through it?

.... **[1 mark]**

B–A*

6 A resistor and a lamp are connected in series in a circuit.

a The potential difference across the resistor is 1.5 V. What energy is transferred to heat in the resistor when 10 C of charge flows through it?

.... **[1 mark]**

b The lamp transfers 3.0 J of energy into heat and light when the same charge passes through it. What is the potential difference across the lamp?

.... **[1 mark]**

Lamps, resistors and diodes

1 Doris wants to investigate how the resistance of a wire depends on its length.

  a What two quantities should Doris measure so that she can calculate the resistance of the wire?

.... [2 marks]

  b Complete the circuit diagram below that Doris could use for her investigation.

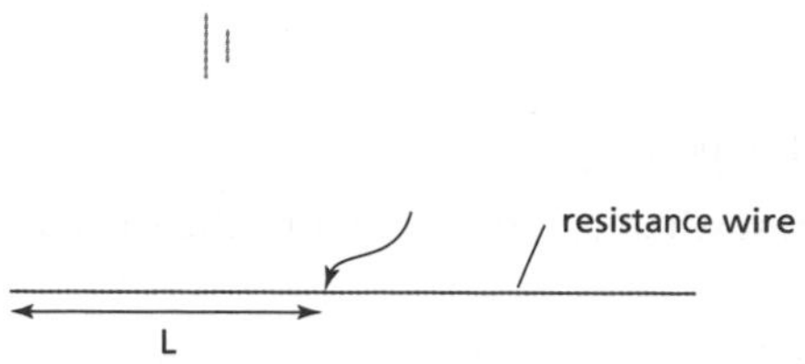

resistance wire

L

[3 marks]

  c What relationship should Doris find that relates the resistance and the length of the wire?

....

....

.... [2 marks]

G–E

2 The graph of current against voltage for a component is a straight line.
What does this tell you about the resistance of the component?

.... [1 mark]

3 The resistance of a filament lamp increases as the current increases.

  a Explain why this happens.

....

.... [2 marks]

  b What would a graph of current against voltage for a filament lamp look like?

....

.... [2 marks]

D–C

4 Explain how the resistance of a diode changes with the current through it.

....

....

.... [3 marks]

B–A*

Heating effects, LDRs and thermistors

1 Tulsey purchases a projector to use with her laptop. The instructions say that when the projector is switched off after use, the power must not be disconnected from the projector until the fan has stopped.

Explain the reason for this.

....

....

.... [3 marks]

2 An electric heater has a power of 4400 W.

a What current will flow through it when it is plugged into the mains supply of 220 V? Show your working.

....

.... [3 marks]

b What energy will be transferred by the heater in 5 minutes? Show your working.

....

.... [3 marks]

3 When current flows through a wire, electrons travel through the lattice of the metal.

Explain how this causes the wire to heat up.

....

....

....

.... [4 marks]

4 In what ways does the resistance of a thermistor depend on temperature?

....

.... [2 marks]

5 Suggest whether it would be more suitable to use an LDR or a thermistor in the following circuits.

a A central-heating controller.

.... [1 mark]

b A circuit for automatically controlling street lamps.

.... [1 mark]

c A circuit to control the cooling fan in a laptop.

.... [1 mark]

6 In most metals, increasing the temperature increases the resistance of the metal. In a thermistor, increasing the temperature decreases the resistance.

Explain why this is so.

....

....

.... [2 marks]

Scalar and vector quantities

1 Explain the difference between a scalar quantity and a vector quantity.

....

.... **[2 marks]**

2 Reggie sets out from home on his bicycle. First he goes to the shop, which is 500 m away, then he rides to his friend's house, which is a further 1200 m. He then returns home.

a What total distance has Reggie travelled when he returns home?

.... **[1 mark]**

b What is his displacement when he is at his friend's house?

.... **[1 mark]** G–E

c What is his displacement when he returns home?

.... **[1 mark]**

3 Jesse Rose walks along a footpath for 6 minutes. She then cuts across a field, which takes a further 4 minutes. She walks a total distance of 900 m.

What is her average speed?

....

.... **[2 marks]**

4 A racing car accelerates from rest to 30 m/s in 6.0 s.

a What is its average acceleration?

.... **[1 mark]**

b The car continues to accelerate at an acceleration of 3.0 m/s².

What is the car's speed after a further 2.0 s?

.... **[1 mark]** D–C

c The car then turns a corner at a constant speed.

Explain why the car is still accelerating.

....

.... **[2 marks]**

5 A light aeroplane has a deceleration of 2.5 m/s², and it is able to hit the runway at a minimum speed of 35 m/s. The runway of a remote airport is 400 m long.

a How many seconds will the plane take to stop?

.... **[1 mark]**

b Will the plane be able to land without overrunning the runway? Explain your answer. B–A*

....

....

.... **[2 marks]**

Distance–time and velocity–time graphs

G–E

1 a What does the gradient of a distance–time graph represent?

.... [1 mark]

D–C

b A graph of distance against time for a car is a curved line with a decreasing gradient.

What does this tell you about the motion of the car?

.... [1 mark]

B–A*

2 A graph representing a ball rolling down a slope from rest has an increasing gradient.
By taking a tangent to the line, Xenia finds that the gradient at 0.40 s is 1.6 m/s.

Find the acceleration of the ball. Show your working.

....

.... [2 marks]

G–E

3 On a velocity–time graph, what does a horizontal line represent?

.... [1 mark]

4 The graph below shows how the velocity of a motorbike varies with time.

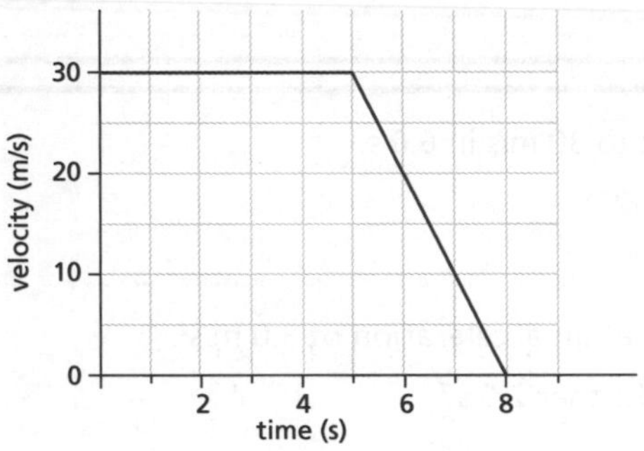

D–C

a Describe the motion of the motorbike from 0 to 8 seconds.

....

.... [2 marks]

b Calculate the deceleration of the motorbike.

....

.... [2 marks]

B–A*

c What is the distance travelled by the motorbike during the first 5 seconds?

.... [1 mark]

d What is the distance travelled by the motorbike in the last 3 seconds?

.... [1 mark]

Understanding forces

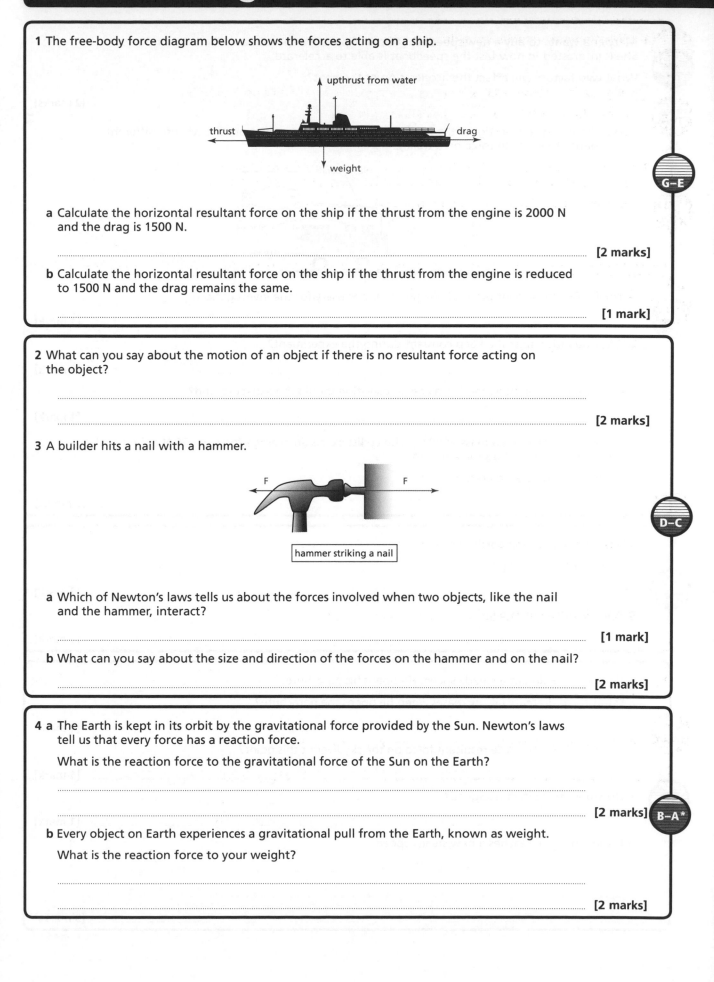

1 The free-body force diagram below shows the forces acting on a ship.

upthrust from water

thrust

drag

weight

G–E

a Calculate the horizontal resultant force on the ship if the thrust from the engine is 2000 N and the drag is 1500 N.

.... **[2 marks]**

b Calculate the horizontal resultant force on the ship if the thrust from the engine is reduced to 1500 N and the drag remains the same.

.... **[1 mark]**

2 What can you say about the motion of an object if there is no resultant force acting on the object?

....

.... **[2 marks]**

3 A builder hits a nail with a hammer.

F   F

hammer striking a nail

D–C

a Which of Newton's laws tells us about the forces involved when two objects, like the nail and the hammer, interact?

.... **[1 mark]**

b What can you say about the size and direction of the forces on the hammer and on the nail?

.... **[2 marks]**

4 a The Earth is kept in its orbit by the gravitational force provided by the Sun. Newton's laws tell us that every force has a reaction force.

What is the reaction force to the gravitational force of the Sun on the Earth?

....

.... **[2 marks]** **B–A***

b Every object on Earth experiences a gravitational pull from the Earth, known as weight.

What is the reaction force to your weight?

....

.... **[2 marks]**

Force, mass and acceleration

1 Marylena wants to buy a new speedboat to participate in a speedboat race next month. She is interested in how fast the speedboat is able to accelerate.

What **two** factors will affect the acceleration of the boat?

.... [2 marks]

2 Cassandra wants to investigate the link between force and acceleration. She will use different elastic bands to vary the force on a trolley.

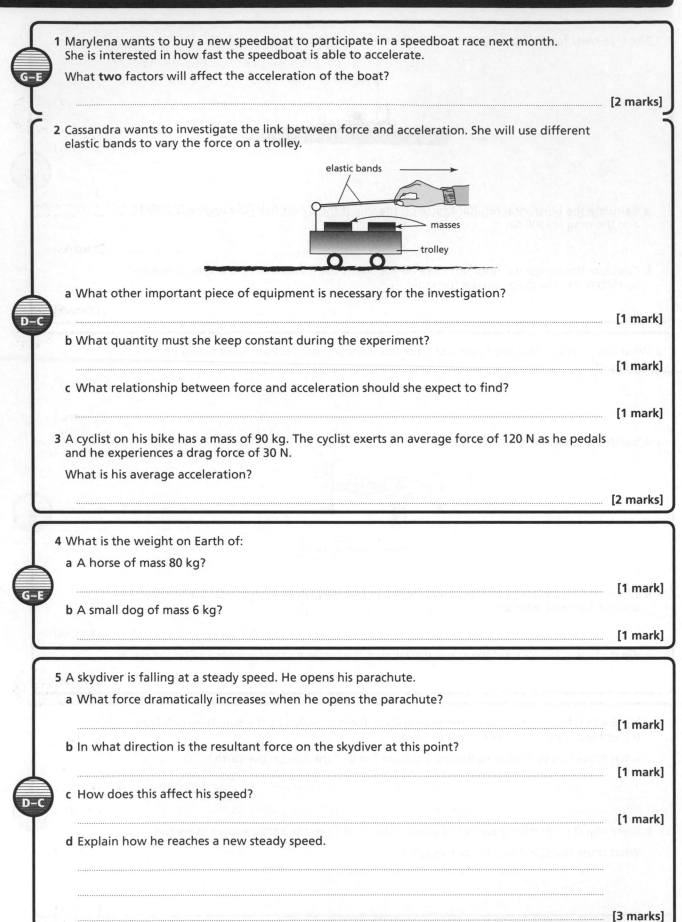

elastic bands

masses

trolley

a What other important piece of equipment is necessary for the investigation?

.... [1 mark]

b What quantity must she keep constant during the experiment?

.... [1 mark]

c What relationship between force and acceleration should she expect to find?

.... [1 mark]

3 A cyclist on his bike has a mass of 90 kg. The cyclist exerts an average force of 120 N as he pedals and he experiences a drag force of 30 N.

What is his average acceleration?

.... [2 marks]

4 What is the weight on Earth of:

a A horse of mass 80 kg?

.... [1 mark]

b A small dog of mass 6 kg?

.... [1 mark]

5 A skydiver is falling at a steady speed. He opens his parachute.

a What force dramatically increases when he opens the parachute?

.... [1 mark]

b In what direction is the resultant force on the skydiver at this point?

.... [1 mark]

c How does this affect his speed?

.... [1 mark]

d Explain how he reaches a new steady speed.

....

....

.... [3 marks]

Stopping distance

1 A car is travelling at 20 m/s. The driver sees a deer on the road and hits the brakes.

  a What is the name given to the time between the moment that she sees the deer and the moment that she hits the brakes?

    .... [1 mark]

  b Find the distance travelled during this time if it takes the driver 0.3 s to apply the brakes.

    .... [1 mark]

2 Complete the following equation:

  Stopping distance = ... + ... [2 marks]

G–E

3 Give **three** factors that could increase the braking distance of a car.

  ....

  ....

  .... [3 marks]

4 Friction opposes motion.

  a Name a situation in which friction is useful.

    .... [1 mark]

  b Name a situation when friction is a nuisance and explain why.

    .... [2 marks]

D–C

5 The graph below shows the thinking and braking distances for a car with an initial velocity of 20 m/s.

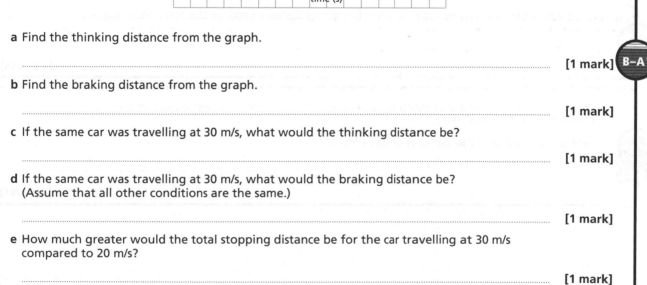

  a Find the thinking distance from the graph.

    .... [1 mark]

B–A*

  b Find the braking distance from the graph.

    .... [1 mark]

  c If the same car was travelling at 30 m/s, what would the thinking distance be?

    .... [1 mark]

  d If the same car was travelling at 30 m/s, what would the braking distance be? (Assume that all other conditions are the same.)

    .... [1 mark]

  e How much greater would the total stopping distance be for the car travelling at 30 m/s compared to 20 m/s?

    .... [1 mark]

Momentum

1 A 200-kg truck is travelling at 12 m/s.

a What is its momentum?

.... **[1 mark]**

b Momentum is a vector quantity. What does this mean?

.... **[1 mark]**

G–E

2 Denise likes to skateboard. In one particular trick, she runs and jumps onto the skateboard. Denise has a mass of 45 kg and her skateboard has a mass of 6.0 kg.

a Denise runs at a velocity of 4.0 m/s. What is her momentum?

.... **[1 mark]**

b Initially, the skateboard is stationary. What is its initial momentum?

.... **[1 mark]**

c When Denise jumps onto the skateboard, both she and the skateboard move off together. What is their new velocity? Show your working.

....

....

.... **[3 marks]**

D–C

3 Two identical railway carriages move towards each other at velocities of 2.0 m/s and 3.0 m/s respectively. They collide and couple together.

What is their combined direction and velocity after the collision?

.... **[2 marks]**

B–A*

4 a Crash barriers are deliberately designed to crumple if a car collides with them. Explain how this helps protect the passengers in the car.

....

....

.... **[2 marks]**

G–E

b Name **two** other safety features in cars that are designed to reduce the force of an impact.

....

.... **[2 marks]**

D–C

5 A toy crossbow uses an elastic strap to apply force to an arrow. It applies an average force of 40 N to the arrow for 0.4 s.

What is the change in momentum of the arrow?

....

.... **[2 marks]**

B–A*

Work, energy and power

1 a Describe what is meant by 1 joule of work.

....

....

[2 marks]

b A force of 20 N is exerted on a box as it moves through a horizontal distance of 5 m. Calculate the work done.

....

[1 mark]

c What work is done against gravity when a person of mass 60 kg rises a vertical height of 4 m in a lift?

....

[1 mark]

G–E

2 A rollercoaster car is pulled up a slope by a motor. The height of the slope is 8 m and the mass of the car is 500 kg.

a What work is done in pulling the car up the slope?

....

[1 mark]

b The motor output is 6000 J as the car is pulled up the slope. Explain why this is greater than your previous answer.

....

[1 mark]

c It takes 40 seconds for the car to reach the top of the slope. Find the power of the motor.

....

[1 mark]

D–C

3 A forklift truck lifts a pallet of bricks of weight 2400 N through a vertical height of 1 m in 40 seconds.

a Find the work done.

....

[1 mark]

b Find the power used by the forklift.

....

[1 mark]

4 A remote-controlled car moves at a constant velocity of 10 m/s. The average driving force of its motor is 2000 N.

What is the output power of the car?

....

[1 mark]

B–A*

KE, GPE and conservation of energy

1 Complete these sentences:

a Kinetic energy is measured in .... [1 mark]

b Mass is measured in .... [1 mark]

c Velocity is measured in .... [1 mark]

2 a What is the kinetic energy of a remote-controlled car of 12.0 kg travelling at 2.5 m/s?

.... [1 mark]

b Give **two** ways that the kinetic energy of the car could be increased.

....

.... [2 marks]

3 What is the gain in gravitational potential energy of a climber of mass 50 kg who climbs a crag 200 m high?

.... [1 mark]

4 Mathew is on his roller blades. He skates up a slope that is 18 m high and 40 m long. Mathew's mass is 30 kg.

a What is Matthew's gain in gravitational potential energy?

.... [1 mark]

b He then skates back down the slope. What happens to the gravitational potential energy?

.... [1 mark]

c Find Matthew's maximum velocity down the slope, if he simply holds his feet still and allows himself to slide down freely.

.... [1 mark]

d In reality, when Mathew slides down freely, his maximum velocity is less than this. What has happened to some of the energy?

.... [2 marks]

e If Mathew skates down, he can achieve a higher velocity than this. What force allows him to do this?

.... [1 mark]

5 A skydiver falls a distance of 500 m. Assuming no energy is lost, what is her gain in kinetic energy?

.... [1 mark]

6 A car has a braking distance of 12 m when it is travelling at 10 m/s.

What will happen to the braking distance if the velocity of the car is doubled? Explain your answer.

....

....

.... [3 marks]

Atomic nuclei and radioactivity

1 Scientists believe that everything in the Universe is made of atoms.

a Name **two** particles that are found inside the nucleus of an atom.

.... **[2 marks]**

b A particular atom has a neutral charge. What does this reveal about its protons and electrons?

.... **[1 mark]**

c What name is given to a particle with more electrons than protons?

.... **[1 mark]**

d What charge will this particle have?

.... **[1 mark]**

e Suggest a way in which charged atoms can be produced.

.... **[1 mark]**

G–E

2 The symbol below is used to represent an atom.

$$^A_Z X$$

a What does X represent? .... **[1 mark]**

b What does A represent? .... **[1 mark]**

c What does Z represent? .... **[1 mark]**

d How could you work out how many neutrons are inside the atom?

.... **[1 mark]**

D–C

3 What is an isotope?

....

....

.... **[4 marks]**

B–A*

4 There are three types of nuclear radiation.

a Which name is given to an electron emitted from the nucleus?

.... **[1 mark]**

b Which type of radiation is a helium nucleus?

.... **[1 mark]**

c Which type of radiation is an electromagnetic wave?

.... **[1 mark]**

G–E

5 Complete the parts **a** to **d** of the table below.

Radiation	Charge	Mass	Ionising effect	Penetration
Alpha	+2	a	Strong	b
Beta	c	0.00055	Weak	Stopped by a few millimetres of aluminium
Gamma	0	0	d	Never completely stopped, but reduced significantly by thick lead or concrete

[4 marks]

D–C

6 Explain what happens to the numbers of nucleons in an atom when it emits an alpha particle.

....

.... **[2 marks]**

B–A*

Nuclear fission

1 Nuclear reactions produce energy.

G–E

   a What word describes the nuclear reaction that is responsible for the energy generated in the Sun?

   .... **[1 mark]**

   b What word is used to describe 'splitting the nucleus'?

   .... **[1 mark]**

   c What particle causes this nuclear reaction?

   .... **[1 mark]**

D–C

2 What name is given to the process by which fast-moving particles from one reaction go on to split other uranium nuclei?

.... **[1 mark]**

G–E

3 a What **two** fuels are commonly used in nuclear power stations?

   .... **[2 marks]**

   b Explain why the disposal of waste from nuclear power stations is a major concern.

   ....

   .... **[2 marks]**

4 The diagram below shows a nuclear reactor.

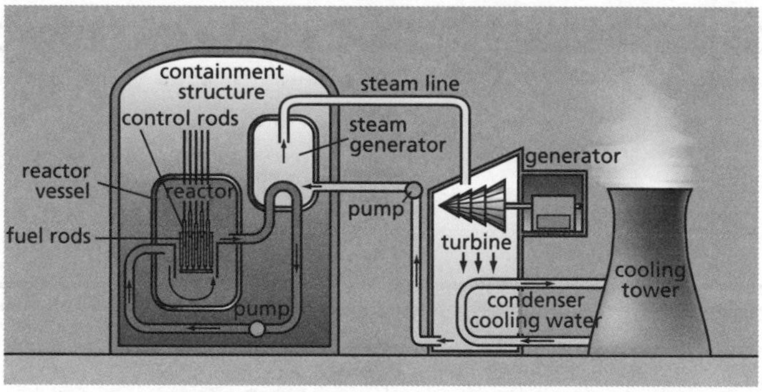

D–C

Explain the purpose of each of the following key components.

a Coolant

.... **[1 mark]**

b Moderator

....

.... **[2 marks]**

c Control rods

....

....

.... **[2 marks]**

B–A*

5 Explain the meaning of the term 'critical mass'.

....

.... **[2 marks]**

Fusion on the Earth

1 The diagram below shows the process of nuclear fusion.

Use the words below to label the diagram.

helium   **tritium**   **deuterium**   **neutron**

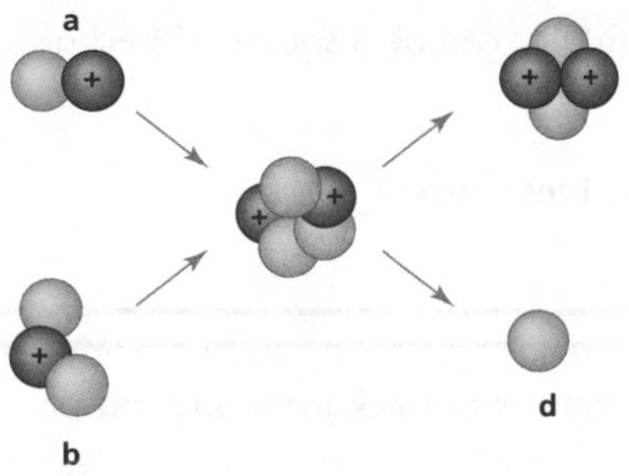

[4 marks]

2 The theory of cold fusion was proposed by scientists Stanley Pons and Martin Fleishmann in 1989.

a What is cold fusion?

.... **[1 mark]**

b Explain **two** reasons why the theory of cold fusion has been rejected.

....

....

.... **[2 marks]**

3 Fusion reactions are more difficult to start than fission reactions.

a What triggers a fission reaction?

.... **[1 mark]**

b Fusion reactions happen when two nuclei collide. Why must this collision take place at extremely high pressures and temperatures?

....

.... **[2 marks]**

c Describe how these nuclei can be controlled at such high temperatures.

....

.... **[2 marks]**

d Why is it possible for fusion to occur in stars?

....

.... **[1 mark]**

Background radiation

1 What name is given to the device that detects radiation?

.... [1 mark]

2 Background radiation comes from a variety of sources and is monitored in different regions of the UK.

  a What type of rock contains uranium that can be a source of background radiation?

.... [1 mark]

  b What radioactive gas is given off by these rocks?

.... [1 mark]

3 The pie chart below shows the main sources of background radiation.

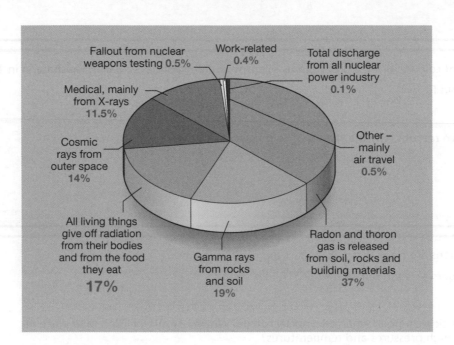

Fallout from nuclear weapons testing 0.5%
Work-related 0.4%
Total discharge from all nuclear power industry 0.1%
Medical, mainly from X-rays 11.5%
Cosmic rays from outer space 14%
Other – mainly air travel 0.5%
All living things give off radiation from their bodies and from the food they eat 17%
Gamma rays from rocks and soil 19%
Radon and thoron gas is released from soil, rocks and building materials 37%

  a List **three** artificial sources of background radiation.

....

.... [3 marks]

  b List **three** natural sources of background radiation.

....

.... [3 marks]

  c Does most background radiation come from man-made or natural sources?

.... [1 mark]

4 Radon-222 is an isotope of radon.

  a What type of radiation is emitted by radon-222?

.... [1 mark]

  b With reference to the penetration power of this type of radiation, explain how it can be a health hazard.

....

....

.... [3 marks]

Uses of radioactivity

1 Suggest **two** uses of radiation in industry.

....

....

....
[2 marks]

G–E

2 The diagram below shows a paper-making machine in a factory. The thickness of the paper is controlled by a radiation source.

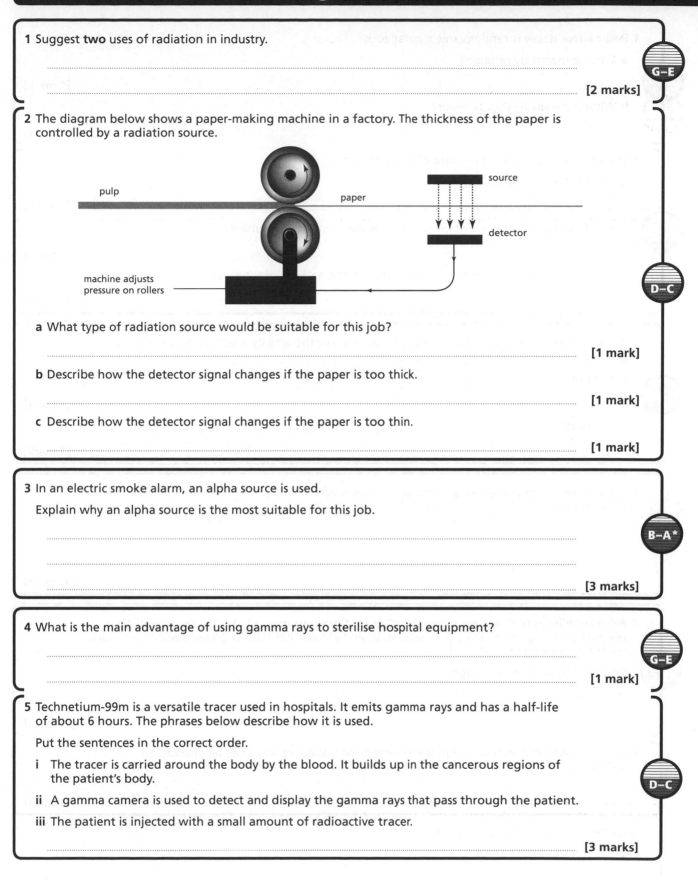

pulp

paper

source

detector

machine adjusts
pressure on rollers

a What type of radiation source would be suitable for this job?

....
[1 mark]

b Describe how the detector signal changes if the paper is too thick.

....
[1 mark]

c Describe how the detector signal changes if the paper is too thin.

....
[1 mark]

D–C

3 In an electric smoke alarm, an alpha source is used.

Explain why an alpha source is the most suitable for this job.

....

....

....
[3 marks]

B–A*

4 What is the main advantage of using gamma rays to sterilise hospital equipment?

....

....
[1 mark]

G–E

5 Technetium-99m is a versatile tracer used in hospitals. It emits gamma rays and has a half-life of about 6 hours. The phrases below describe how it is used.

Put the sentences in the correct order.

i   The tracer is carried around the body by the blood. It builds up in the cancerous regions of the patient's body.

ii A gamma camera is used to detect and display the gamma rays that pass through the patient.

iii The patient is injected with a small amount of radioactive tracer.

....
[3 marks]

D–C

Activity and half-life

1 Radioactive decay is random and spontaneous.

G–E

a What is meant by random?

.... [1 mark]

b What is meant by spontaneous?

.... [1 mark]

2 The activity of a source is the rate of decay of nuclei.

D–C

a What is the unit of activity?

.... [1 mark]

b What happens to the activity when the number of nuclei is doubled?

.... [1 mark]

c Describe the relationship between activity and the half-life of the source.

.... [1 mark]

3 A radioactive sample has a half-life of 6 hours. Initially the activity is 800 Bq. What will the activity be after:

B–A*

a 12 hours?

.... [1 mark]

b 24 hours?

.... [1 mark]

4 State **three** safety precautions that should be used when handling radioactive sources in a laboratory.

G–E

....

....

.... [3 marks]

5 Patricia believes that nuclear power is better for the environment than coal because nuclear reactors do not give off any CO_2 emissions. Brian believes that nuclear power stations are worse for the environment.

D–C

Explain why Brian may be right.

....

.... [2 marks]

6 Describe **two** ways in which high-level waste from nuclear reactors can be disposed of.

B–A*

....

.... [2 marks]

P2 Extended response question

Aran is given a radioactive source, suitable for use in a school laboratory.

Plan an experiment that Aran could use to decide whether the source emits alpha, beta or gamma radiation. You should describe any apparatus that is needed and explain how the results will determine the type of source.

The quality of written communication will be assessed in your answer to this question.

[6 marks]

B1 Grade booster checklist

I can describe how organisms are classified into groups according to the characteristics they have in common.	
I know what the term 'species' means and can explain and state the limitations of this definition.	
I can construct keys that show how species can be identified.	
I know that Charles Darwin was the first (with Alfred Russel Wallace) to explain that natural selection is the mechanism of evolution.	
I understand that variation can be caused by genes or the environment.	
I know that the cell's nucleus contains chromosomes on which genes are located, and understand how forms of genes called alleles lead to differences in inherited characteristics.	
I know that genetic disorders can be caused by changes in genes (mutations).	
I know that homeostasis means maintenance of a stable internal environment, and that examples include maintaining the right body temperature and amount of water in the body.	
I know that the central nervous system consists of the brain and spinal cord, and that nervous responses can be voluntary or involuntary.	
I know that hormones are produced in endocrine glands and transported in the blood to their target organs.	
I know that lack of insulin causes Type 1 diabetes and resistance to insulin causes Type 2.	
I know that plants use hormones to respond to stimuli such as light and gravity.	
I know what a drug is and can describe how stimulants, depressants, painkillers and hallucinogens affect the brain.	
I know the effects of some chemicals in cigarette smoke.	
I know the harmful effects of alcohol abuse.	
I know that organs can be transplanted.	
I know that infectious diseases are caused by pathogens, including those in animal vectors.	
I know that antiseptics help to prevent the spread of infection.	
I know how energy is transferred along food chains and between trophic levels.	
I can describe the sources of pollution, including sulfur dioxide, and nitrates and phosphates.	
I know that decomposers recycle dead organic matter.	
I am working at grades G/F/E	

I can describe the characteristics of the five kingdoms and the hierarchy of classification, from kingdom to species.	
I can explain why scientists use the binomial system to name and identify species.	
I know that the individuals of a species vary, and this variation can be continuous or discontinuous.	
I can describe how and why organisms are adapted to their environment, including extreme environments such as polar regions.	
I can explain Darwin's theory of natural selection in terms of variation, overproduction, competition, survival, inheritance and gradual change.	
I can explain the meanings of different genetic terms, e.g. dominant, recessive, homozygous, heterozygous, phenotype, genotype.	
I can interpret monohybrid inheritance using Punnett squares, genetic diagrams and family pedigrees.	
I can describe the symptoms of sickle cell disease and cystic fibrosis.	
I can explain how body temperature and body water content are regulated.	
I can explain how blood glucose levels are regulated by insulin.	
I can explain how insulin injections control Type 1 diabetes and that diet and exercise help to control Type 2 diabetes.	
I can explain how hormones bring about plant responses and use data to support explanations.	
I can describe how drug abuse can lead to drug addiction.	
I can explain the effects of some chemicals in cigarette smoke.	
I can evaluate some of the harmful effects of alcohol abuse.	
I can describe how the demand for organs can be supplied.	
I can describe how different pathogens are spread.	
I can explain how animals and plants are able to defend themselves against pathogens by physical and chemical means.	
I understand that antibiotics are used to treat and control infections.	
I understand why living things are interdependent, giving mutualism and parasitism as examples.	
I can explain and apply understanding to the transfer of energy along food chains and between trophic levels in a pyramid of biomass.	
I can explain how the increase in human population has an impact on the environment and how the impact can be reduced by recycling.	
I can explain the processes of eutrophication and acid rain formation.	
I can explain how scientists can use indicator species to assess the impact of these phenomena.	
I understand how carbon is recycled.	
I am working at grades D/C	

I understand the issues surrounding the classification of viruses and vertebrates and can explain that most classifications reflect the relatedness.	
I can apply understanding of binomial classification to the conservation of species.	
I can explain how variations such as hybridisation and ring species complicate classification.	
I can apply understanding of adaptations to organisms near hydrothermal vents.	
I can explain how speciation occurs as a result of geographic isolation.	
I can explain how genetics, including DNA evidence and bacterial resistance, can be used to support Darwin's theory.	
I can analyse patterns of monohybrid inheritance, and calculate and analyse the outcomes of the crosses.	
I can evaluate pedigree analysis when screening for sickle cell disease and cystic fibrosis.	
I can explain how vasoconstriction and vasodilation are used in negative feedback during thermoregulation.	
I can describe the function of dendrons and axons, and explain how different components work together to bring about a reflex action.	
I can apply understanding of the effect of insulin to the role of glucagon regulating blood glucose levels.	
I can evaluate the correlation between obesity (including BMI calculations) and Type 2 diabetes.	
I understand the usefulness of plant hormones as weedkillers and substances that improve the growth of crop plants.	
I understand that drugs that affect the nervous system mostly have their effect at the synapse.	
I can evaluate data that establishes the correlation between smoking and ill-health.	
I can evaluate some of the harmful effects of alcohol abuse.	
I can discuss the ethical issues of organ transplants.	
I understand that antibacterials produced by plants may be useful to humans.	
I can evaluate evidence that resistant strains of pathogenic bacteria (including MRSA) can arise from the misuse of antibiotics.	
I can apply understanding to include mutualism in nitrogen-fixing bacteria in legumes and chemosynthetic bacteria in tube worms.	
I can explain and use data to show how the increase in human population has an impact on the environment and how the impact can be reduced by recycling.	
I can apply understanding of indicator species.	
I understand how nitrogen is recycled.	
I am working at grades B/A/A*.	

B2 Grade booster checklist

I know why light microscopes and electron microscopes are useful when studying cells.	
I can name the main components in plant and animal cells.	
I know what a gene is and can describe the basic structure of DNA.	
I know that the work of scientists Watson, Crick, Franklin and Wilkins helped discover the structure of DNA.	
I understand what we mean by genetic engineering.	
I can state some uses of genetically modified (GM) crops and know some issues associated with them.	
I know that mitosis results in two daughter cells identical to the parent cell, and that it is used for growth, repair and asexual reproduction.	
I can describe how plants are cloned by using cuttings and how embryo transplants are used.	
I know that stem cells in the embryo can differentiate into all other types of cell.	
I understand that each protein has its own specific sequence of amino acids, resulting in different-shaped molecules and that the order of bases in a gene decides the order of amino acids in a protein.	
I can explain what a mutation is.	
I know that enzymes are biological catalysts and can state what conditions affect their activity.	
I know the word equations for aerobic and anaerobic respiration.	
I know the meanings of the words diffusion and osmosis.	
I can explain why heart and breathing rate increase with exercise	
I know the word equation for photosynthesis and the conditions that affect its rate.	
I can describe how water, glucose and mineral salts are transported through a plant via the xylem, phloem and roots.	
I can describe how fieldwork techniques are used to investigate the distribution of organisms in an ecosystem.	
I understand that fossils are evidence for evolution.	
I know that growth is an increase in size, length and mass that occurs through cell division.	
I know the components in blood.	
I can name the four chambers of the heart and the four major blood vessels associated with it.	
I can name the parts of the digestive system and know the function of digestive enzymes.	
I know what a functional food is.	
I am working at grades G/F/E	

I can carry out simple magnification calculations.	
I can describe the functions of each cell component.	
I can describe the structure of DNA.	
I can evaluate the roles of Watson, Crick, Franklin and Wilkins in the discovery of the structure of DNA.	
I can describe the stages of the genetic-engineering technique.	
I can discuss the advantages and disadvantages of golden rice.	
I can describe simply how mitosis results in two diploid daughter cells and how meiosis results in four haploid daughter cells.	
I can describe how plants are cloned by using cuttings.	
I can discuss the advantages, disadvantages and risks of cloning mammals.	
I can describe how stem-cell therapy can be used.	
I can describe why the order of the bases in a gene decides the shape and function of a protein.	
I can describe how a mutation can be harmful, beneficial or neither.	
I can use the 'lock and key' hypothesis to describe how enzymes work.	
I can explain why a change in conditions will affect the activity of an enzyme.	
I can describe why muscles may start to respire anaerobically and state what EPOC is.	
I can describe how the movement of oxygen and glucose in the body is facilitated by diffusion.	
I can explain why heart and breathing rate increase with exercise.	
I can describe how the structure of a leaf is adapted for photosynthesis.	
I can describe how root hair cells are adapted to take up water by osmosis and explain why transpiration is important in the movement of water and mineral salts.	
I can describe simply the process of active transport.	
I can describe how fieldwork techniques are used to investigate the distribution of organisms in an ecosystem.	
I can explain why there are gaps in the fossil record.	
I can describe how growth and development happens in both plants and animals.	
I can describe the function of the components in blood.	
I can describe how the circulatory system transports substances around the body.	
I can describe how food is moved along the alimentary canal by peristalsis and evaluate Visking tubing as a model of the small intestine.	
I can describe some examples of functional foods.	
I am working at grades D/C	

I can explain why the development of modern microscopes has enabled us to see cells with more clarity and detail.	
I can explain the role of mitochondria and chloroplasts.	
I can explain what we mean by complementary base pairing.	
I can evaluate the implications of the Human Genome Project.	
I can evaluate the advantages and disadvantages of using bacteria to produce human insulin.	
I can evaluate the use of herbicide-resistant crops.	
I can explain in detail the process of mitosis and meiosis to include what happens to the chromosomes.	
I can describe how plants are cloned by using cuttings.	
I can explain each stage of the production of a cloned mammal.	
I can evaluate the use of stem-cell therapy.	
I can explain what happens in each stage of protein synthesis.	
I can explain the role of enzymes in DNA replication and protein synthesis.	
I can explain why a change in conditions will affect the activity of an enzyme.	
I can apply the term 'concentration gradient' to explain diffusion.	
I can calculate heart rate, stroke volume and cardiac output.	
I can explain how limiting factors affect the growth of a plant.	
I can explain why transpiration is important in the movement of water and mineral salts.	
I can explain how active transport is used in the absorption of mineral salts through the roots.	
I can explain how to reduce errors when using sampling techniques.	
I can explain how the pentadactyl limb provides evidence for evolution.	
I can interpret growth by using percentile charts.	
I can explain how the heart pumps blood.	
I can explain the role of the gall bladder in digestion and how the structure of villi allows efficient absorption of products from digestion into the blood.	
I can evaluate Visking tubing as a model of the small intestine.	
I can evaluate the evidence for the claimed benefits of the use of functional foods.	
I am working at grades B/A/A*.	

B2 Grade booster checklist 257

I can describe how the Earth's early atmosphere and oceans were formed.	
I can describe the composition of today's atmosphere and the early atmosphere.	
I can describe how carbonate rocks formed from carbon dioxide.	
I can describe the formation of igneous, metamorphic and sedimentary rock and give an example of each.	
I can describe the chemical name for limestone, chalk and marble as calcium carbonate.	
I can list some uses of limestone, and some advantages and disadvantages of quarrying.	
I can write a word equation for the thermal decomposition of calcium carbonate.	
I know that atoms are rearranged in chemical reactions, not created or destroyed.	
I can explain that hydrochloric acid is naturally present in the stomach and why indigestion remedies are used.	
I can explain that acids are neutralised by bases, which are metal oxides, hydroxides or carbonates.	
I understand that a salt is produced in a neutralisation reaction.	
I can describe electrolysis as using electricity to split up compounds.	
I can describe the tests for hydrogen, chlorine and oxygen and can list some uses of chlorine.	
I can explain that most metals are extracted from metal compounds that are in rocks, but that a few metals are found as elements.	
I can explain that most metals are extracted by heating with carbon or by electrolysis.	
I can describe oxidation as gain of oxygen and reduction as loss of oxygen.	
I can list some reasons for recycling metals.	
I can describe some properties and uses of aluminium, copper, gold and steel.	
I understand that metals can be mixed to make alloys to improve their properties.	
I can describe hydrocarbons as molecules that contain hydrogen and carbon only.	
I can describe how crude oil is split up by fractional distillation.	
I can write a word equation for the combustion of hydrocarbons, and name the products of incomplete combustion.	
I can describe what a greenhouse gas does.	
I understand that fossil fuels are a non-renewable resource but biofuels are renewable.	
I can describe hydrogen as a good fuel and explain what makes it a difficult fuel to use.	
I can name the first three alkanes.	
I can describe an alkene as a molecule that is unsaturated, decolourises bromine water and that is used to make polymers.	
I can describe how cracking converts long chain alkanes into shorter alkanes and alkenes.	
I can explain what polymerisation means and list some uses for particular polymers.	
I can explain why disposing of polymers is a problem and list some solutions.	
I am working at grades G/F/E	

I can explain how rocks provide a record of the atmosphere from years ago.	
I can explain how photosynthesis decreases the amount of carbon dioxide and increases the amount of oxygen in the atmosphere.	
I can use experimental results to calculate the amount of oxygen in the atmosphere.	
I can explain why there are small changes in the composition of the modern atmosphere.	
I can describe the difference between intrusive and extrusive igneous rock.	
I can explain why sedimentary rock may have fossils and be easily eroded.	
I can describe an experiment to compare how easily different metal carbonates decompose to metal oxides.	
I can describe the use of limestone products to neutralise soil and acidic gases from power stations.	
I know that the products of a reaction have different properties to the reactants, but that the mass of reactants and products is the same.	
I can describe how to compare the effectiveness of different indigestion remedies.	
I can describe chlorine as a toxic gas that can be obtained by electrolysis of seawater.	
I know that the electrolysis of water produces oxygen and hydrogen and that the hydrolysis of dilute hydrochloric acid produces chlorine and hydrogen.	
I can describe the extraction of a metal from its oxide as reduction and corrosion as oxidation.	
I can explain why some metals are extracted by electrolysis.	
I can draw a diagram to explain why alloying increases the strength of a metal.	
I can explain some reasons for turning iron into steel alloys.	
I can describe some properties and some uses for each fraction of crude oil.	
I can describe some causes of and some effects of acid rain.	
I can describe the relationship between atmospheric carbon dioxide levels and global temperatures.	
I can list some advantages and disadvantages of biofuels.	
I can describe an experiment to compare the energy released from different fuels.	
I can draw a labelled diagram of how to crack liquid paraffin.	
I can draw the structure of methane, ethane, propane, ethene and propene.	
I can list the properties of poly(ethene), poly(chloroethene), poly(propene) and poly(tetrafluoroethene)	
I can explain why a particular polymer might be chosen for a particular use.	
I can evaluate some solutions to the problem of disposing of polymers.	
I am working at grades D/C	

I can explain how the evolution of life changed the atmosphere.	
I can describe and explain an experiment to find the amount of oxygen in the atmosphere.	
I can explain how the crystal size of igneous rocks shows where the rock formed.	
I can suggest ways in which the negative effects of quarrying can be minimised, and write a balanced argument for and against limestone quarrying.	
I can explain the use of calcium carbonate to prevent the release of acidic gases from power stations.	
I can write equations for the thermal decomposition of calcium carbonate, the reaction between water and calcium oxide and the addition of carbon dioxide to limewater.	
I can identify when a formula equation is not balanced.	
I can predict the mass of a product given the mass of all other reactants and products.	
I can name the salt that would be produced by reacting an acid and a base.	
I can describe an experiment to electrolyse hydrochloric acid.	
I can relate the reactivity of a metal to the method chosen to extract it and to the ease with which it corrodes.	
I can discuss why chemists develop new materials and explain how shape memory alloys are suited to particular functions.	
I can explain why different fractions of crude oil have different boiling points.	
I can explain why the products of incomplete combustion are dangerous.	
I can explain how greenhouse gases warm the Earth, and evaluate the evidence supporting the theory of climate change.	
I can explain some methods of reducing acid rain and atmospheric carbon dioxide levels.	
I can write the reaction equation for the fermentation of glucose into ethanol.	
I can write arguments for and against replacing fossil fuels with biofuels.	
I can discuss the difficulties of using hydrogen as a fuel and describe a hydrogen fuel cell.	
I can explain why cracking is necessary.	
I can write an equation for the polymerisation of ethene.	
I can describe some recent developments in developing biodegradable polymers.	
I am working at grades B/A/A*	

C2 Grade booster checklist

I can describe the structure of an atom showing the position of protons, neutrons and electrons.	
I can explain that the number of electrons is the same as the number of protons.	
I can explain the meaning of atomic mass and atomic number.	
I can describe an element as metal or non-metal by looking at the periodic table.	
I can describe how Mendeleev organised his periodic table and how the modern periodic table is arranged.	
I can describe how positive and negative ions are formed.	
I can explain that ionic bonding is electrostatic attraction between oppositely charged ions.	
I can write the formula for an ion, and say which ions are present in an ionic compound from its name.	
I know that ionic substances are crystalline and have high melting and boiling points.	
I can use the solubility rules to decide if an ionic substance is soluble.	
I can list the flame colours of sodium, potassium, calcium and copper.	
I can describe a covalent bond as a shared pair of electrons and name some covalent molecules.	
I can draw a dot and cross diagram for hydrogen, hydrogen chloride, water and methane.	
I can list some differences in properties between ionic and covalent substances.	
I can describe how to separate two immiscible liquids.	
I can describe an experiment to separate colours using paper chromatography.	
I can list the physical properties of ionic, covalent and metallic substances.	
I can draw a diagram of metallic structure as a regular arrangement of ions surrounded by a sea of delocalised electrons.	
I can describe the properties of alkali metals as soft and reactive.	
I can explain that the reactivity of the alkali metals gets higher going down the group.	
I can describe the colour and physical state of the halogens.	
I can describe the properties of the noble gases.	
I can name a use for helium and for argon, and describe the properties that make it good for this use.	
I can describe the difference between an exothermic and endothermic reaction.	
I can use experimental results to decide if a reaction is exothermic or endothermic.	
I can list the factors that affect the rate of a reaction.	
I can explain what is meant by a catalyst, and describe why catalytic convertors are used in vehicles.	
I can calculate the relative formula mass of a compound or element.	
I can explain why the yield of a reaction is never 100% of what it could be.	
I can explain why industry must dispose of waste products carefully.	
I am working at grades G/F/E	

I can list the mass and charge of protons, neutrons and electrons.	
I can explain the meaning of isotope.	
I can describe how Mendeleev predicted the existence and properties of unknown elements by looking at the properties of known elements.	
I can write the electron arrangement of an atom using the periodic table and the 2.8.8 system.	
I can write the electronic structure of an ion and describe the properties of ionic substances.	
I can use the name of an ionic compound to write its formula.	
I can write a formula equation for a precipitation reaction using the correct state symbols.	
I can describe an experiment to make an insoluble salt.	
I can describe tests and results for the presence of carbonates, sulfates and chloride ions.	
I can explain why covalent bonds form and why carbon usually makes four bonds and hydrogen makes one bond.	
I can explain why simple covalent substances have a low melting point but giant covalent substances have a high melting point.	
I can explain that fractional distillation can be used to separate miscible liquids.	
I can analyse the results from a paper chromatogram to decide what is present in a mixture.	
I can use physical properties of an unknown substance to decide what structure it has.	
I can use the structure of metals to explain why they are malleable and can conduct electricity.	
I can describe the reaction of alkali metals with water and write the formula equations.	
I can write equations for the reactions of halogens with hydrogen and metals.	
I can describe the observations, hypothesis and experiments that led to the discovery of the noble gases.	
I can describe an experiment to measure the energy change in a reaction.	
I can describe experiments to show how the rate of a reaction is affected by concentration, surface area, temperature and catalysts.	
I can draw graphs to show the expected results from experiments about rates of reaction.	
I can explain how the design of catalytic converters helps them to work more effectively.	
I can write word equations for the reactions catalysed by catalytic converters.	
I can calculate the percentage by mass of an element in a compound.	
I can calculate the percentage yield of a reaction from the actual yield and the theoretical yield.	
I am working at grades D/C	

I can state the number of protons and neutrons in an atom given the periodic table.	
I can calculate the relative atomic mass of an element given the abundance of each isotope.	
I can predict the properties of an element given the properties of another element with the same number of electrons in the outer shell.	
I can predict whether an element will form a positive or negative ion and how many charges it will carry.	
I can write a balanced formula equation from a word equation.	
I can explain why ionic substances conduct electricity when dissolved or molten.	
I can suggest suitable reactants to make an insoluble salt.	
I can explain the use of barium sulfate in X-rays	
I can explain how scientists use emission spectroscopy.	
I can predict the formula of a simple covalent molecule by looking at how many unpaired electrons each atom has.	
I can draw dot and cross diagrams for covalent molecules containing double bonds.	
I can draw the structure of diamond and graphite and use it to explain their different properties.	
I can explain why the components separate during fractional distillation of air.	
I can calculate Rf values from a paper chromatogram.	
I can explain why ionic substances are only able to conduct electricity when dissolved or molten.	
I can predict the outcome of a halogen displacement reaction.	
I can use the trend in properties to predict the properties of an unknown noble gas.	
I can draw an energy level diagram for an exothermic and endothermic reaction.	
I can explain why reactions are endothermic or exothermic using ideas about bond breaking and making.	
I can use to collision theory to explain why concentration, surface area and temperature affect the rate of a reaction.	
I can use reacting masses to work out the empirical formula of a compound.	
I can use a symbol equation to calculate the masses of each reactant and product.	
I can discuss some considerations that industrial chemists must weigh up when deciding about the economics of reactions.	
I am working at grades B/A/A*	

I know that ideas and models about the Solar System and the Universe have changed throughout history.	
I understand that waves are reflected and refracted at boundaries between different materials.	
I know that convex lenses are used in telescopes to magnify distant objects.	
I understand that waves transmit energy and information without transferring matter.	
I understand the terms frequency, wavelength and amplitude.	
I can describe the two types of wave – longitudinal and transverse.	
I can recall the electromagnetic waves in order of increasing wavelength and/or frequency.	
I understand the hazards of electromagnetic waves.	
I know uses for all the types of electromagnetic waves.	
I know that alpha, beta and gamma are types of ionising radiation, which come from radioactive materials.	
I can compare the relative sizes of and distances between the Earth, the Moon, the planets, the Solar System, galaxies and the Universe.	
I know that the Big Bang theory explains the origins of the Universe.	
I know what ultrasound and infrasound are.	
I know that there are two main types of seismic wave produced by earthquakes: P waves and S waves.	
I know that seismometers are used to detect earthquakes, but that earthquakes are almost impossible to predict.	
I can recall the different layers in the Earth's structure.	
I know that electric current (measured in amps) is a flow of charge and that the voltage pushes the current around a circuit.	
I know that power is the rate of using energy and is measured in watts.	
I can name some renewable energy sources and some non-renewable energy sources.	
I know that electricity is generated when a wire or a coil moves relative to a magnetic field.	
I know that mains electricity is a.c. and electricity from a battery is d.c.	
I know that electricity is transmitted at very high voltages and that it is dangerous.	
I know that the unit of electricity supply is the kilowatt hour.	
I understand the advantages of using low-energy appliances in the home.	
I know the different types of energy and know that energy is transferred from one type to another.	
I know that energy is often wasted in the form of heat.	
I know that the amount of heat (infrared) radiated by objects depends on the temperature and the surface.	
I am working at grades G/F/E	

I understand the differences and similarities between the geocentric model and the heliocentric model of the Universe.	
I understand how scientists use different waves to find out about the Universe.	
I know how to measure the focal length of a convex lens.	
I can explain how a reflecting telescope works.	
I can calculate the speed of a wave.	
I can describe uses for all types of electromagnetic waves.	
I understand ionisation and its dangers.	
I can describe methods used to find evidence of life beyond Earth.	
I understand how modern methods of astronomy have developed our understanding of the Universe.	
I can describe the evolution of stars.	
I can describe the evidence that supports both the Big Bang and the Steady State theories.	
I know that the Doppler shift occurs when a wave source is moving relative to an observer.	
I know that the red-shift of light from stars gives evidence of the Universe expanding.	
I can describe uses of both ultrasound and infrasound.	
I know that P waves are longitudinal and that S waves are transverse, and that they travel through the Earth at different speeds and are reflected and refracted at boundaries.	
I know that the Earth's crust is made up of tectonic plates, which move due to convection currents in the mantle.	
I understand that earthquakes are caused by the relative motion of the tectonic plates at their boundaries.	
I understand that electric current is the rate of flow of electric charge, and that the voltage or potential difference is a measure of the energy carried by the charge.	
I can calculate electrical power.	
I understand the advantages and disadvantages of using renewable and non-renewable sources of energy.	
I understand that ultimately the Sun is the source of all our energy on Earth.	
I understand how a generator produces electricity.	
I know that transformers are used to change the voltage of a.c. electricity.	
I understand that electricity is transmitted at very high voltages to reduce the amount of energy wasted.	
I can calculate electricity consumption in kilowatt hours and work out the cost.	
I understand the concept of payback time.	
I understand the principle of conservation of energy.	
I can calculate efficiency of energy transfers from data or from Sankey diagrams.	
I know that for an object to remain at constant temperature, the amount of energy absorbed must equal the amount of energy emitted.	
I am working at grade D/C	

I understand that refraction happens because the wave changes speed and direction at boundaries between different materials.	
I can interpret data about modern astronomical observations.	
I understand the evolution of stars.	
I understand how the wavelength and frequency change due to the Doppler effect.	
I understand that the red-shift of light from stars supports both the Big Bang theory and the Steady State theory, and the presence of cosmic background radiation only supports the Big Bang theory.	
I understand how P and S waves travel through the Earth.	
I can use the turns ratio equation for transformers to calculate relative potential differences across transformers.	
I can draw scaled Sankey diagrams to show the efficiency of energy transfers.	
I am working at grade B/A/A*	

I can recall that like charges repel and unlike charges attract.	
I can demonstrate an understanding of attraction by electrostatic induction and the dangers of electrostatic charges.	
I can recall that current is the rate of flow of charge.	
I can recall that cells and batteries supply direct current.	
I can describe how an ammeter is connected in series and measures current in amperes.	
I can understand that the current in a circuit can be changed using a variable resistor.	
I know the advantages and disadvantages of heating effect of an electric current.	
I can demonstrate an understanding of displacement and velocity.	
I can recall that velocity is speed in a stated direction and is a vector quantity.	
I can interpret distance–time and velocity–time graphs.	
I can draw and interpret free-body force diagrams.	
I can calculate resultant forces.	
I can use the equation $W = mg$.	
I can recall that stopping distance = thinking distance + braking distance.	
I can describe the factors that affect stopping distance.	
I can recall that momentum is a vector quantity.	
I can use the equation work done = force × distance moved in the direction of force.	
I can use the equation KE = $\frac{1}{2} mv^2$.	
I can state how atoms may gain or lose electrons to form ions.	
I can recall that alpha and beta particles and gamma rays are ionising radiations.	
I can describe nuclear fusion and fission reactions as a source of energy.	
I can describe how nuclear power stations get their energy from fission reactions and convert this into electricity.	
I can describe the process of nuclear fusion and recognise it as the energy source for stars.	
I can describe background radiation and the regional variation of radon gas.	
I can describe uses of radioactivity including irradiating food and sterilising equipment.	
I can recall how activity of a source decreases over a period of time.	
I can state that half-life is the time taken for half the undecayed nuclei to decay.	
I understand that ionising radiations damage tissues and can cause mutations.	
I can describe how scientists have changed their awareness of the hazards of radioactivity over time.	
I am working at grades G/F/E	

I can explain how an insulator can be charged by friction.	
I can explain electrostatic charging in terms of transfers of electrons.	
I understand common electrostatic phenomena in terms of movement of electrons.	
I can explain the uses of electrostatic charges, including paint and insecticide sprayers.	
I can explain how earthing safely removes excess charges.	
I can use the equation $Q = It$.	
I can explain that current is conserved at a junction.	
I can describe how a voltmeter is connected in parallel and measures potential difference in volts.	
I can use the equation $V = IR$.	
I understand how current varies with potential difference for filament lamps, diodes and fixed resistors.	
I can use the equations $P = VI$ and $E = IVt$.	
I can describe how the resistance of an LDR changes with light intensity and how the resistance of a thermistor changes with temperature.	
I can use the equation speed = distance / time.	
I can use the equation $a = (v-u) / t$.	
I can calculate acceleration from the gradient of a velocity–time graph.	
I can demonstrate an understanding of action and reaction forces.	
I understand how a body accelerates in the direction of a resultant force.	
I can use the equation $F = ma$.	
I can describe how bodies fall at the same acceleration in vacuum.	
I understand the relationship between weight and mass.	
I know how thinking and braking distance affect stopping distance.	
I can use the equation momentum = mass x velocity.	
I can describe the rate of change of momentum when applied to seat belts, crumple zones and air bags.	
I can describe power as the rate of doing work and can recall it is measured in watts.	
I can use the equation GPE = mgh.	
I can describe the idea of conservation of energy in energy transfers.	
I can describe an alpha particle as a helium nucleus, beta particle as an electron and gamma ray as electromagnetic radiation.	
I can compare alpha, beta and gamma radiations in terms of ionisation and abilities to penetrate.	
I can explain fission of uranium-235 and the principles of a controlled chain reaction.	
I can explain how a chain reaction is controlled in a nuclear reactor.	
I can describe how heat energy in a reactor is turned into electrical energy.	
I can explain the difference between nuclear fission and nuclear fusion.	
I can explain the origins of background radiation on the Earth.	
I can describe uses of radioactivity including smoke alarms, tracing and gauging thicknesses, and diagnosis and treatment of cancer.	
I can discuss the long-term disposal and storage of nuclear waste.	
I can evaluate the advantages and disadvantages of nuclear power for generating electricity.	
I am working at grades D/C	

I can explain that potential difference is energy transferred per unit charge.	
I can explain the transfer of heat in components in terms of collision between electrons and atoms in the lattice.	
I can calculate distance travelled from the area under a velocity–time graph.	
I can explain how resultant forces affect the motion of objects.	
I can describe the motion of objects falling in air.	
I can describe conservation of linear momentum.	
I can use and apply the equation force = rate of change of momentum.	
I can use the equation power = work done / time.	
I can use calculations to show that braking distance is directly proportional to the square of speed.	
I can describe the nuclei of isotopes.	
I can explain how atoms form ions.	
I can explain why nuclear fusion cannot occur at low pressures and temperatures and the difficulties of using it for power generation.	
I can use the concept of half-life to carry out calculations of the decay of nuclei.	
I am working at grades B/A/A*	

Key

1	relative atomic mass
H	**atomic symbol**
hydrogen	name
1	atomic (proton) number

Group 1	Group 2											Group 3	Group 4	Group 5	Group 6	Group 7	Group 8
																	4 **He** helium 2
7 **Li** lithium 3	9 **Be** beryllium 4											11 **B** boron 5	12 **C** carbon 6	14 **N** nitrogen 7	16 **O** oxygen 8	19 **F** fluorine 9	20 **Ne** neon 10
23 **Na** sodium 11	24 **Mg** magnesium 12											27 **Al** aluminium 13	28 **Si** silicon 14	31 **P** phosphorus 15	32 **S** sulfur 16	35.5 **Cl** chlorine 17	40 **Ar** argon 18
39 **K** potassium 19	40 **Ca** calcium 20	45 **Sc** scandium 21	48 **Ti** titanium 22	51 **V** vanadium 23	52 **Cr** chromium 24	55 **Mn** manganese 25	56 **Fe** iron 26	59 **Co** cobalt 27	59 **Ni** nickel 28	63.5 **Cu** copper 29	65 **Zn** zinc 30	70 **Ga** gallium 31	73 **Ge** germanium 32	75 **As** arsenic 33	79 **Se** selenium 34	80 **Br** bromine 35	84 **Kr** krypton 36
85 **Rb** rubidium 37	88 **Sr** strontium 38	89 **Y** yttrium 39	91 **Zr** zirconium 40	93 **Nb** niobium 41	96 **Mo** molybdenum 42	[98] **Tc** technetium 43	101 **Ru** ruthenium 44	103 **Rh** rhodium 45	106 **Pd** palladium 46	108 **Ag** silver 47	112 **Cd** cadmium 48	115 **In** indium 49	119 **Sn** tin 50	122 **Sb** antimony 51	128 **Te** tellurium 52	127 **I** iodine 53	131 **Xe** xenon 54
133 **Cs** caesium 55	137 **Ba** barium 56	139 **La*** lanthanum 57	178 **Hf** hafnium 72	181 **Ta** tantalum 73	184 **W** tungsten 74	186 **Re** rhenium 75	190 **Os** osmium 76	192 **Ir** iridium 77	195 **Pt** platinum 78	197 **Au** gold 79	201 **Hg** mercury 80	204 **Tl** thallium 81	207 **Pb** lead 82	209 **Bi** bismuth 83	[209] **Po** polonium 84	[210] **At** astatine 85	[222] **Rn** radon 86
[223] **Fr** francium 87	[226] **Ra** radium 88	[227] **Ac*** actinium 89	[261] **Rf** rutherfordium 104	[262] **Db** dubnium 105	[266] **Sg** seaborgium 106	[264] **Bh** bohrium 107	[277] **Hs** hassium 108	[268] **Mt** meitnerium 109	[271] **Ds** darmstadtium 110	[272] **Rg** roentgenium 111							

Elements with atomic numbers 112–116 have been reported but not fully authenticated.

* The Lanthanides (atomic numbers 58–71) and the Actinides (atomic numbers 90–103) have been omitted.

Cu and Cl have not been rounded to the nearest whole number.

Answers

B1 Influences on life

Page 146 Classification and naming species

1 a i amphibian ii mammal iii fish iv bird v reptile

 b It is not a vertebrate/it is an invertebrate (it has no backbone)

2 They are cold-blooded (poikilotherms); so need the warmth of the Sun to keep their internal body at the right temperature

3 a They are able to make their food (by photosynthesis)

 b Chlorophyll; a cell wall made of cellulose

4 They are made up of both fungi and protoctists; so cannot be classified in one kingdom

5 It has a backbone so would be a vertebrate; but would be difficult to classify into a group because it has characteristics of both birds; and reptiles

6 a Organisms that are capable of breeding together to produce fertile offspring

 b Hybrids

7 a *Rattus*

 b *Rattus exulans*

8 They need to be sure that the spiders are of different species; as the biodiversity can only be reported as high if the area contains a number of different species

Page 147 Identification, variation and adaptation

1 To identify the types of organism they find

2 a Any two from: height; colour of leaves; length of roots; number of branches, etc.

 b i continuous

 ii discontinuous

 iii continuous

 c A line graph to display the hand-span data; because it is continuous; a bar chart to display the eye-colour data; because it is discontinuous

3 The birds live in a geographically close area; the birds in the ring will be able to interbreed but the two species at either end cannot

4 a Able to survive for long periods without food or water

 b Stops them sinking into the sand

 c Stops sand entering the nostrils

 d They have a small surface area relative to their mass; so they do not lose very much body heat from their surface

 e The hairs are full of air; which is a poor conductor of heat (good insulator); so prevents the heat energy from the caribou escaping

5 They would share some characteristics because they belong to the same phylum; but they would differ because they live in very different places; and would have different adaptations to survive in their environment

Page 148 Evolution

1 a Because they are adapted to live in different environments

 b i Food / shelter / mate

 ii They have died out / are no longer living

 iii They were not able to catch prey / they were killed by predators / died of disease

 c Those with dark skin are not well camouflaged; so they will not survive; because they will be spotted by their prey

 d Snakes showed variation – some had rattles and some didn't; the snakes that had rattles were able to scare off predators and not get eaten (better adapted) so they survived and reproduced; the genes for the rattle were passed on to their offspring; the snakes that did not have a rattle were eaten and so did not reproduce

2 a Any two from: there may be predators on Anguilla which they have no survival instincts against; there may not be suitable food on Anguilla; the climate may not be suitable

 b Variation will exist in the population; those with the advantageous characteristics will survive and reproduce; over time the changes in the iguana will be so great that they will become different enough to those on the Virgin Islands to be classed as a new species

Page 149 Genes and Variation

1 a Inherited

 b Environmental

 c Inherited

2 a DNA

 b Genes

 c Chromosomes

3 He has probably inherited the tallness gene; from both parents

4 As their DNA is so similar; they must have a common ancestor; which shows that they both evolved over time from this animal

5 a She inherited one from her mother and one from her father

 b Heterozygous

6 a Grey fur

 b Grey

 c Because the babies have inherited alleles from each parent, each baby must carry the allele for white fur; but its effect is hidden by the allele for grey

7 a They are brother and sister

 b The father (Chris) inherited the allele from his father (Robert); Holly inherited the allele from Chris; as the allele is hidden in Chris it is probably recessive

Page 150 Monohybrid inheritance

1 a The allele for red flowers

 b A mixture of red and white flowers

2 He was not a scientist (he was a priest) / he did not publish his findings

3 a Any two upper-case letters, e.g. AA

 b Any two lower-case letters, e.g. aa

 c Any upper-case letter and its corresponding lower-case letter, e.g. Aa

Answers

4

Cross: Rr × rr		
Parental gametes	R	r
r	Rr	rr
r	Rr	rr

5 a 1 in 2 (50%)

  b 1 in 2 (50%)

  c 1 in 2 (50%)

  d 0 (0%)

  e 1 in 2 (50%)

6 a

Cross: Rr × Rr		
Parental gametes	R	r
R	RR	Rr
r	Rr	rr

  b 1 in 4 (25%)

Page 151 Genetic disorders

1 a A change in the DNA/a gene

  b The mutation will be passed to the children during fertilisation (mutated sperm fuses with an egg)

2 a An illness that is inherited

  b Any two from: blocked airways of the lungs, which makes breathing difficult; lung infections because of bacteria becoming trapped in the mucus; problems digesting food, which can lead to malnutrition

  c The sickle-shaped cells cause a blockage in a blood vessel; that leads to an organ; which starves the organ of oxygen (and nutrients)

3 a To predict the risk of someone inheriting a particular disorder

  b I-1, I-2, II-1, II-2

  c

Cross: Aa × Aa		
Parental gametes	A	a
A	AA	Aa
a	Aa	aa

  d 1 in 2 (50%)

Page 152 Homeostasis and body temperature

1 Any two from: temperature; water content of blood; glucose content of blood; ion (salt) content in blood

2 a There will be an increased amount of water in the blood

  b They will produce more urine

  c Her brain detects an increase in water in blood and sends a message via hormones to the kidneys; the kidneys produce more urine

  d Because the body does it automatically

  e The kidneys would have produced less urine; her blood water content would have increased

3 a 37 °C

  b Muscles under the skin start moving (contracting and relaxing) and produce heat

  c The thermoregulatory centre in the brain; detects when his temperature has risen back up to 37 °C; and sends responses to the muscles to stop contracting and relaxing

  d Any four from: the thermoregulatory centre in the brain; detects when his temperature has fallen below 37 °C; sends responses to the walls of blood vessels under the skin; stimulating contraction of the muscles so they narrow (constrict); this results in the flow of blood through the skin being reduced; therefore less heat is lost through the surface of the skin

Page 153 Senses and the nervous system

1 a Brain; spinal cord

  b Any two from: skin; eye; ear; tongue; nose

  c As electrical impulses; along nerves (neurons)

2 a i Skin ii Eye iii Ear iv Nose / tongue

  b They contain more touch receptors than any other part

3 a Retina

  b Impulses; travel along the optic nerve

4 a Diagram A **b** Motor neurone

Page 154 Responses and coordination

1 a Involuntary **b** Involuntary **c** Voluntary

2 Blinking stops the eye being damaged / pulling hand away from hot pan stops skin being burnt

3 Chemicals (neurotransmitters) are released into the gap; they travel across it and stimulate an impulse in the next neurone

4 a Nerve impulses have less distance to travel

  b To bring about a quick response; and minimise damage to the body

5 Receptors in the skin detect pain and an impulse is sent along a sensory neurone to the spinal cord; the impulse travels across the synapse to the relay neurone; the impulse travels across the synapse to the motor neurone; the impulse travels along the motor neurone to the arm muscle; the muscle contracts, pulling hand away

Page 155 Hormones and diabetes

1 a Testosterone

  b Testes

  c In the bloodstream

2 a Insulin

  b To bring the amount of glucose in the blood back down to normal

3 a 30 minutes / 360 minutes

  b Between 210 minutes; and 300 minutes

  c The contracting muscles took glucose from the blood; to provide energy

  d Insulin; was released by the pancreas into the bloodstream; which converted glucose into glycogen

  e Glucagon

Answers

4 a Insulin

  b Type 1 diabetes

  c Daily insulin injections

  d Her blood glucose levels would become high; which can result in serious health problems

  e She inherited a mutated gene; from one of her parents

Page 156 Plant hormones

1 a Positive

  b Negative

2 a Maximises the rate of photosynthesis in the leaves

  b To firmly anchor the plant into the ground so it doesn't fall over / to enable roots to absorb water

3 a Side A

  b Phototropism

  c Side B

  d The high concentration of auxin on side B; will cause the cells here to elongate more than the cells on side A; which makes the shoot bend towards the light

4 a To stimulate the growth of roots on the cuttings

  b To produce large numbers of new plants quickly; because the new plants are identical to the parent plants, growers can make sure that desirable qualities of the parent are kept in the new plants

5 a It affects the growth of plants

  b The higher the concentration, the taller the plant

Page 157 Drugs, smoking and alcohol abuse

1 a Depressant

  b Stimulant

  c Painkiller

  d Hallucinogen

  e Stimulant

2 a They crave the short-term effects of the drug (a sense of well-being); their body becomes used to the changes so they become dependent on the drug

  b Heroin is an illegal drug so is only available through drug dealers, who charge a lot of money for the drug; addicts may turn to crime to get money to pay for it

3 a They got quicker

  b Caffeine is a stimulant; so it increases the activity of the brain by enhancing the release of neurotransmitters

4 a Tar

  b Nicotine

5 They have tar in their lungs; they are coughing to try and remove it

6 a Any one from: lowered inhibitions; slowed reaction times; blurred vision; difficulty controlling the arms and legs

  b Eight units

  c The safe daily amount of alcohol to drink per day is four units; Paul has drunk more than this and so if he does it often he is at risk of developing an illness (e.g. liver cirrhosis, brain damage, heart disease, cancer, raised blood pressure)

7 a Any one from: treatment became better; living conditions improved; disease transmission slowed; better diagnosis

  b More people were smoking tobacco

  c They could have been another reason why cases of lung cancer rose (e.g. more pollution from cars) so the data does not directly link smoking with lung cancer

Page 158 Ethics of transplants

1 a She needs a kidney from a donor to be put into her body

  b As she is related to Maria, there is less chance of Maria rejecting her kidney

2 Because lungs can only be donated by someone who is dead

3 Any two from: there is an increased risk of transplanting diseased organs into recipients because poor donors do not receive regular health care; the deal exploits poor people and violates their human rights; it encourages people to enter into an illegal activity

4 a They are about the same size

  b The animal organ will probably be rejected by the human body

  c Genetic modification

  d Animal cruelty

5 There is a lack of organs available for transplant

6 Answer should include a discussion of the following: the age of the recipient; the cause of their illness; how healthy their lifestyle is

Page 159 Infectious diseases

1 a Pathogen

  b It can be spread from person to person

  c The lungs have ideal conditions for bacteria growth / lungs are warm and moist

2 a Cold / flu

  b *Salmonella*

  c Cholera

3 The bacteria produce endotoxins (poisons); which stimulate the small intestine wall to contract violently and more frequently than normal; absorption of digested food and water is prevented; as a result, the victim quickly dehydrates

4 a An animal that spreads pathogens

  b An (*Anopheles*) mosquito bites the infected person and sucks up the pathogen that causes malaria; before biting another person and transferring the pathogen into their blood; which then causes them to develop malaria

5 The cream will stop any microbes around the cut from multiplying; to stop the cut becoming infected

6 Colds are caused by viruses; antibiotics only control infection by bacteria

7 a MRSA cannot be treated with most antibiotics; so the patients could get an infection which cannot be treated and this could lead to health problems

  b Some bacteria contain resistance genes; these individuals survive antibiotic treatments; they reproduce and spread quickly

Answers

Page 160 Defences and interdependency

1 To stop blood loss; to prevent pathogens from entering the body

2 a Eyes (tears)

  b (Glands in the) skin

  c (Glands in the) stomach

3 They do not have as many white blood cells to destroy pathogens; so the pathogens will be able to multiply in the body

4 Plants may be found that have antibacterial properties; we need new antibacterial chemicals because many of the ones we are using today are no longer effective against bacteria

5 a Parasitism; because the aphids are removing the plants' food, which damages them and the plant gets nothing in return

  b Mutualism; because both the ants and aphids get something useful out of the relationship

6 a The nitrogen-fixing bacteria live in its root nodules and convert nitrogen in the air into nitrogen-containing compounds; that the bean plant uses to make proteins; the plant produces sugars, which the bacteria need as a food source

  b The bean plant uses the corn plant to grow up so it can reach the light; the leaves of the bean plant are prickly, which deters animals from eating the bean and the corn plant / the squash leaves cover the ground to stop weeds growing around the bean/corn plant and competing for water/minerals

Page 161 Energy, biomass and population pressures

1 a i Seaweed

  ii Limpet / octopus / seal

  b It only eats plants

  c Predator; octopus

  d Scavenger

2 a Consumers usually eat more than one type of food; only food webs represent this, as in a food chain consumers just eat one type of food

  b Energy is lost at each stage in the food chain, so eventually the food energy available dwindles to zero; no energy available means no further links in the food chain

3 Pyramid should have levels labelled with the name of the organism, starting with the seaweed at the bottom; each level should get progressively smaller

4 a There are more adults than children

  b It will decrease; because there are fewer children to replace the adults that will die

5 a Demand will increase; because the population is increasing

  b Oil is non-renewable so it will run out if we keep using it to make plastics

6 a Landfill; incinerator

  b Recycling saves energy as it removes the need to make a new items from raw materials; this process uses up vast quantities of energy – more than is required to collect and sort the items in the recycling process

Page 162 Water and air pollution

1 a Chemicals that are harmful to human health and to wildlife

  b Nitrates / phosphates

  c The pollutants are not absorbed by the plants; so they enter groundwater, which is a source of drinking water, in the run-off from the fields

2 a The pollutants cause plants and algae to grow at a fast rate; there is an increase in dead plant material; as bacteria decompose the plant material, the amount of bacteria rises

  b The bacteria use up the oxygen in the water; and create toxic chemicals

3 The oxygen concentration is low; which means that the water is highly polluted

4 a Sulfur in the fossil fuel reacts with oxygen to release sulfur dioxide; this dissolves in water vapour, which falls as acid rain or snow

  b **Either**: acid rain falls into water and causes the gills of fish to overproduce mucus; this clogs the gills and the fish die of oxygen starvation; **or**: acid rain washes substances important for healthy tree growth out of the soil; poisons are released from the soil and the trees die

5 a As the distance increases, so does the number of different species of lichen

  b They show that as you get further away from the city the level of pollution decreases; because many species of lichen are very sensitive to sulfur dioxide; so the higher the levels, the less variety of species of lichen there are growing

6 Car engines burn petrol and release sulfur dioxide; the sulfur dioxide gets carried in the wind to Sweden; where it dissolves in water vapour and falls as acid rain

Page 163 Recycling carbon and nitrogen

1 a Bacteria / fungi

  b Decomposers break down the dead plants and animals; to form nutrients that the plants need to build new tissue

2 a Combustion/burning

  b Photosynthesis

3 a From eating plants or other animals

  b The animal will die; its tissue will be decomposed by decomposers; the decomposers carry out respiration, releasing carbon into the atmosphere

4 a They convert the ammonia from the decomposition of dead organic matter; to nitrates which plants can absorb

  b They convert nitrates into nitrogen gas

  c The water has forced oxygen out of the soil so the oxygen levels are low; this means that nitrifying bacteria die so the levels of nitrate in the soil decrease; the grass cannot make protein so cannot grow; also, as oxygen levels are low denitrifying bacteria can thrive; removing nitrates further from the soil

Answers

Page 164 B1 Extended response question

0 marks
Insufficient or irrelevant science. Answer not worthy of credit.

1–2 marks
Answer may be simplistic. There may be limited use of specialist terms. Errors of grammar, punctuation and spelling prevent communication of the science.

3–4 marks
For the most part the information is relevant and presented in a structured and coherent format. Specialist terms are used for the most part appropriately. There are occasional errors in grammar, punctuation and spelling.

5–6 marks
All information in answer is relevant, clear, organised and presented in a structured and coherent format. Specialist terms are used appropriately. Few, if any, errors in grammar, punctuation and spelling.

B2 The components of life

Page 165 Seeing cells and cell components

1 a The cells are too small to be seen without a microscope/ with the naked eye

   b × 200

2 Higher magnification / more detail; better resolving power / clearer image

3 a 1 = nucleus; 2 = vacuole

   b A plant cell; because it has a cell wall / chloroplasts / vacuole

4 a Mitochondria produce energy; sperm cells need energy to propel them towards the egg

   b The cell wall strengthens the cell; stem cells need to be strong to support the rest of the plant

   c Chloroplasts are used for photosynthesis; roots are underground so cannot photosynthesise as there is no light; leaf cells do receive light and carry out photosynthesis so require many chloroplasts

5 a Nucleus

   b Cell wall

   c No; bacterial cells do not have a nucleus

Page 166 DNA

1 a It is two strands; twisted into a spiral shape

   b TTACGAAT

   c The order of the bases determines the type of protein made; so changing the order will change the protein

2 a To break the cell membrane; and release the chromosomes

   b To break down the protein core of the chromosomes; and release the DNA

   c It causes the DNA to come out of solution; so the strands can be removed

3 So the two strands of DNA can separate easily; to enable the cell to make protein/divide

4 They used X-ray crystallography; to discover where the atoms were on the strand

5 Genetic engineering involves transferring a gene into the cells of another organism; these cells have to be able to recognise the coding; in order to manufacture the protein coded by the foreign gene

Page 167 Genetic engineering and GM organisms

1 a They have genes from another organism

   b The bacteria grow best at this temperature; so insulin is produced more quickly

2 A = restriction enzyme; B = restriction enzyme; C = ligase enzyme

3 Benefit – any one from: it is cheap to produce and available in large quantities; it is human insulin so there are no allergic reaction or intolerances in users; it does not use animal products and so is not a problem for religious groups of vegetarians; drawback – GM organisms might have unknown and unforeseen effects on other organisms, including humans

Answers

4 a They will not have to use pesticides; they will get higher yields of cotton as less are destroyed by pests and thus will make more profit

b Any one from: the toxins may affect wildlife; may disrupt food chains; the Bt gene may be transferred to other plants

5 a Golden rice contains beta-carotene; which is used to make vitamin A in the body

b Any one from: aiming for a balanced diet is a better solution to vitamin A deficiency; trying to deal with the vitamin A problem with a single GM solution is too limiting; golden rice has never undergone feeding trials to check it is safe for people to eat

6 Farmers can use herbicides on their fields but only the weeds die, not the wheat; this means that the wheat has more space/water/light to grow; and so the farmer has more to harvest and sell

Page 168 Mitosis and meiosis

1 a The parent cell divides in two

b They have the same chromosomes

c To replace old skin cells that die / to repair cuts in the skin

2 a Mitosis   b Meiosis

3 a 1 = 30; 2 and 3 = 30

b They are genetically identical to each other / they have the same chromosomes

c They have two sets of chromosomes

4 a It is the process by which sex cells (gametes) are formed; which join together during fertilisation

b 1 = 30; 2 and 3= 30; 4, 5, 6 and 7 = 15

c The daughter cells (4–7); they have half the number of chromosomes; so during fertilisation they will form a zygote with a full set of chromosomes

d Chromatids can swap pieces of DNA; different chromosomes end up in each daughter cell

Page 169 Cloning plants and animals

1 a Living things that have the same genetic material as each other

b So she has many copies of her award-winning plant

c Either: use tissue culture; take a piece of the plant and grow it in a nutrient liquid or gel; which is sterile; or: take a cutting by cutting a small piece of the stem; dip in hormone rooting powder; and place in soil

2 a Each cell will grow into a baby

b Because they all came from one fertilised egg; so they have the same genetic material

c So he can produce many babies; which have the desired characteristics of a single female cow

d He could make an exact copy of a cow with the desired characteristics; fertilisation would result in offspring not identical to the parent

3 a They could be born with defects, which would result in a short and perhaps painful life

b Take eggs from a goat of a similar species and remove the nucleus; take a preserved cell from the ibex and remove the nucleus; insert the nucleus from the ibex cell into the egg; put the formed embryo into a donor mother goat

Page 170 Stem cells and the human genome

1 a Unspecialised

b Differentiation

2 a Adult stem cells can differentiate into more types of cell than adult stem cells; so a wider range of tissues can be repaired

b The embryo is destroyed when the stem cells are removed; some people feel that an embryo is a human life and so killing it is wrong

3 Any three from: their practice is probably unregulated and so potentially dangerous or will not work; they may use embryonic stem cells, which some people find morally wrong; the stem cells may be rejected; the stem cells they use may carry genetic mutations for disease; the treatment may trigger side effects

4 They are unspecialised; so they have the ability to differentiate into any type of cell and can therefore be used for a range of treatments; there are no ethical concerns relating to their use because they are adult stem cells taken from the person being treated

5 a To sequence the DNA in human cells

b Any two from: the workload could be shared to get the project done more quickly; different areas of expertise working together; costs were shared between different countries

c Any one from: to help us work out if a person is vulnerable to certain disease; to produce personalised drugs

Page 171 Protein synthesis

1 a Amino acids

b It contains five amino acids / it only contains a few amino acids; proteins contain over 50

2 a RNA is single stranded while DNA is double stranded; RNA contains the base U instead of the T in DNA

b To carry the protein-making code from the DNA into the cytoplasm; where the protein is made

3 a CCA CGA UGC AUC GGA UUA

b To copy protein-coding sequence onto mRNA; which is able to leave the nucleus; and enter the cytoplasm where the protein is made

c 2; because it has a codon which is complementary to the first codon on the mRNA

d It will cause a change in the mRNA; which would result in a different amino acid being placed in the chain; this may affect the shape of the protein; and its function which will affect the activity of the cell

Page 172 Mutations

1 a Beneficial; because the bacteria can no longer be killed by penicillin

b The offspring will have identical DNA to the parent cell

c The mutant bacteria are not killed by the penicillin; so they can reproduce to pass on the mutation; these offspring also survive; the non-resistant bacteria will die and so cannot reproduce

2 a i Mutation 2

ii Mutation 1

iii Mutation 3

b The order of the amino acids may be changed; which will affect the shape of the protein

Answers

3 a The mutation has changed the second G into an A; this codon still codes for lysine; so the structure of the protein will be the same

b Silent mutation

Page 173 Enzymes

1 a They speed up the rate of reaction; inside a living thing

b To break down food; into soluble substances that can enter the blood

c Activity will be low; because pH10 is alkaline; pepsin works best in acid

2 a Protein **b** Amino acids

c An enzyme will only fit to one substrate; because the substrate fits into the active site of the enzyme

3 The enzyme speeds up the rate at which amino acids join in the newly formed protein; it is needed so proteins can be made quickly

4 a It is increasing

b 5 °C

c Cold; because the enzyme works best at cold temperatures

5 a There is more substrate to bind with enzymes; therefore more enzymes will be active; and the reaction will be quicker

b All the active sites on the enzyme are full

6 The amalyse will denature; which means its active site will change shape; starch will not be able to fit into the active site and bind to form an enzyme substrate complex; so it will not be broken down into glucose

Page 174 Respiring cells and diffusion

1 a It is higher in exhaled air

b Oxygen gets used up; in respiration

c It does not get used up or made in processes inside the body

2 a Aerobic

b Anaerobic

c Lactic acid was building up in them

d Anaerobic respiration releases less energy from glucose molecules

e Oxygen is reaching her muscle cells; and breaking down lactic acid into carbon dioxide and water

3 The coffee molecules are in constant motion; there will be a net movement of molecules to where there is a high concentration to a lower one; so the coffee molecules will spread away from the pot into the kitchen and from there into the rest of the house

4 a Glucose needs to be able to enter the blood so it can be transported to cells around the body; as they need it for respiration; to provide energy for life processes

b In the small intestine; because glucose will diffuse from where its concentration is high to where it is low; and glucose needs to enter the blood

5 a The blood is constantly flowing through the capillary; so the glucose is moved away, decreasing the concentration of glucose; so it remains lower than the concentration in the small intestine

b To keep diffusion at a high rate

Page 175 Effects of exercise

1 a So it can be transported to muscles; which need it for respiration; to release energy for movement

b To remove carbon dioxide from the blood

c Heart

2 a Breathe into a volume of limewater before and after exercise; measure how long it takes for the limewater to go cloudy; limewater will go cloudy more quickly after exercise

b Measure the number of pulses per minute before and after exercise; the pulse rate will be quicker after exercise

3 a i 220 dm^3 **ii** 18 **iii** 0.3 dm^3

b i Vishram **ii** Tom **iii** Chelsea

4 $0.07 \times 72 = 5.04$ l/min (1 mark for each side of the equation and 1 for the correct units)

Page 176 Photosynthesis

1 a It makes glucose; which is needed for growth

b No; because the cells do not contain chlorophyll

2 Any one from: they have a large surface area; which maximises the absorption of light; cells on the upper surface are packed with chloroplasts containing chlorophyll; for a maximum rate of photosynthesis; air spaces enable reactants for photosynthesis; to reach cells; stomata enable oxygen from the atmosphere; to reach the leaf cells for photosynthesis

3 a The plants would be taller / the tomatoes would be bigger / the tomatoes would taste sweeter

b i Allows a high light intensity

ii Increases the temperature

iii Makes sure the plants always have enough water for photosynthesis

c A source/increased concentration of carbon dioxide

4 a Take a glowing splint and put it into a tube of collected gas; if it is oxygen, the splint will relight

b As the light intensity increases, the rate of photosynthesis increases

c At 15 cm the light intensity has reached its optimum; so even though light intensity increases further, the rate of photosynthesis does not

d The heat from the lamp has increased the temperature of the water-weed; so the enzymes involved in photosynthesis have denatured; so photosynthesis happens at a slower rate

Page 177 Transport in plants, osmosis and fieldwork

1 a Tube A = xylem; tube B = phloem

b Mineral salts

c Glucose

2 a Hot; because the rate of evaporation would be faster

b To stop too much water being lost

3 Out of the cell; the water travelled from where it was most concentrated to where it was least concentrated; by osmosis

Answers

4 a The mass did not change so no water has entered or been lost from the cells

b The concentration of water (salt) was the same in the solution and in the potato cells; so osmosis did not occur

5 Oxygen is needed for respiration; which produces energy which is needed for active transport; active transport is the mechanism by which mineral salts are absorbed from the soil

6 a $4 \times 16 = 64$

b Make sure the quadrat is thrown randomly (e.g. close your eyes when you throw); use the quadrat several times and calculate an average

Page 178 Fossil record and growth

1 a The remains or impression made by a dead organism

b Millions of years ago dead organisms fell to the bottom of the sea; and were covered in sediment; which formed sedimentary rock

2 a There are fossils missing from a continuous series of change; between ancestors and their descendants

b Any one from: fossils do not always form; they get destroyed in geological cycles; many fossils are yet to be found

3 a They have five 'fingers' (pentadactyl limb)

b Because each animal uses its limbs differently / because they have adapted in different ways

c It shows that modern invertebrates all descended from a common ancestor; but through natural selection each one has changed to adapt to different environments and modes of life

4 They will divide and elongate so its mass will increase

5 Similarity: growth involves cell division / differentiation; Difference: in plants growth also involves cell elongation / in plants growth happens throughout life, in animals growth stops at adulthood / in plants growth only occurs in one type of tissue (meristems), in animals growth occurs in all tissues

6 a 20 kg/m^2

b 25th percentile

c He has a BMI equal to or greater than 25% of boys of the same age

Page 179 Cells, tissues, organs and blood

1 a Tissue   **b** Cell

c Cell   **d** Organ

e Organ

2 a Ovary – 4 reproductive system

b Stomach – 1 digestive system

c Heart – 3 circulatory system

d Brain – 2 nervous system

3 Their structure has changed; to enable them to carry out a function

4 a Contains haemoglobin; to carry oxygen

b Has a tail; so it can swim to the egg for fertilisation

5 Embryonic cells are all identical; genes are switched on and off; which causes the cells to be different

6 a White blood cell

b Red blood cell

c Plasma / platelets

7 Red blood cells contain haemoglobin; in the lungs, oxygen binds with haemoglobin to produce oxyhaemoglobin; this is transported to the respiring tissues, where the oxygen leaves the haemoglobin

8 a Platelets

b It plugs the wound to stop blood loss; it restricts bacteria and viruses from entering the body

Page 180 The heart and circulatory system

1 a Carries blood from the heart to the lungs

b Makes sure the blood flows in the right direction

c Brings blood to the heart from around the body

2 Vena cava ➝ right **atrium** ➝ **right** ventricle ➝ **pulmonary** artery

3 It has to pump blood to all areas of the body; so requires thick muscle to produce enough force; the right ventricle only has to pump blood to the lungs (which is a shorter distance)

4 For 1 mark: each side pumps blood along a different route; for 2 marks: the left side pumps blood all around the body; the right side pumps blood to the lungs

5 a atria walls contract; the increase in pressure opens the valves; blood is forced into the ventricles

b The valves close

6 Vein; artery; capillary

7 a Vein **b** Vein **c** Artery **d** Capillary

8 To prevent the blood going in the wrong direction (backflow); because the blood in the veins is at a low pressure

9 To provide a large surface area; for the efficient exchange of materials; between the blood and the tissues

Page 181 The digestive system

1 a Oesophagus

b Stomach

c Small intestine

2 a To break it into smaller pieces so it can be swallowed / give it a larger surface area for the action of saliva

b To start the breakdown of food / to make it slippery and easier to swallow

3 To move food along the alimentary canal

4 a To carry bile from the gall bladder; into the small intestine

b Bile will not be able to enter the small intestine; so she will not be able to digest fats properly; so she may lose weight (or other sensible outcome of this)

5 They speed up the breakdown of food; into soluble products; that can enter the bloodstream

6 a No; because it is a large molecule that would not be able to get through the membrane of the Visking tubing

b Yes; because the protease would have broken down the protein; and the amino acids are small enough to pass through the membrane

c No; because there was no enzyme to break down the starch

7 At least two features plus an explanation of how they relate to its function: villus wall is only one cell thick; so digested food does not have far to diffuse into the blood / a good blood supply; means that the digested food is quickly taken away from the villus so more can diffuse across to replace it / the membrane of the villus is permeable to food molecules; so they can reach the blood

Page 182 Functional foods

1 a It has health-promoting properties

   b A probiotic

2 a They have added sugars (oligosaccharides); which are a food supply for the 'good' bacteria; which increases their numbers

   b It decreases the number of 'bad' bacteria; which can lead to diseases of the alimentary canal

3 a It lowers cholesterol levels in the blood

   b They are very believable; because scientific studies have been carried out that prove them correct

4 Answer should state that the claim is invalid, because the study was not scientific. Give at least two reasons why: it was only carried out on a small number of children; data was not collected; it is just the opinion of the teacher; the improvement could be due to other factors, which have not been ruled out

Page 183 B2 Extended response question

0 marks
Insufficient or irrelevant science. Answer not worthy of credit.

1–2 marks
Answer may be simplistic. There may be limited use of specialist terms. Errors of grammar, punctuation and spelling prevent communication of the science.

3–4 marks
For the most part the information is relevant and presented in a structured and coherent format. Specialist terms are used for the most part appropriately. There are occasional errors in grammar, punctuation and spelling.

5–6 marks
All information in answer is relevant, clear, organised and presented in a structured and coherent format. Specialist terms are used appropriately. Few, if any, errors in grammar, punctuation and spelling.

C1 Chemistry in our world

Page 184 Early Earth

1 iii Oxygen

2 The early atmosphere was made by volcanoes; if scientists know which gases come from volcanoes they can tell what was in the early atmosphere

3 a Oxidation

   b Iron sulfide rocks form when there is little oxygen in the atmosphere; iron oxide rocks form when there is a lot of oxygen in the atmosphere; the amount of oxygen in the atmosphere has increased between 4 billion and 1.8 billion years ago

4 a There was much more carbon dioxide in the early atmosphere / less carbon dioxide in the modern atmosphere

   b Any two from: carbon dioxide dissolved into the oceans; it was used by marine animals to make shells; it was used by early plants for photosynthesis

5 a $100 - (75 + 5 + 2); = 18\%$

   b It is (3%) lower than the modern atmosphere

   c The equation shows the reaction in photosynthesis; when plants evolved they took in carbon dioxide so the level of carbon dioxide in the atmosphere dropped; they gave out oxygen so the level of oxygen in the atmosphere rose; carbon dioxide gas from the atmosphere dissolved into the oceans (aq)

Page 185 Today's atmosphere

1 a iii Digital thermometer

   b Burning fossil fuels; cutting down forests/trees

   c So they can detect any changes.

2 a iii Combustion of hydrocarbons has released carbon dioxide

   b i It increases carbon dioxide; and decreases oxygen levels

   ii Trees do lots of photosynthesising so fewer trees means less photosynthesis; less carbon dioxide is taken up and less oxygen is produced

3 a $CH_4 + 2O_2 \longrightarrow CO_2 + 2H_2O$ (1 mark for each 2 in the correct place; 1 mark for including **no** balancing number in front of **both** CH_4 and CO_2)

   b Show state symbols on the equation as liquid (l) and gas (g)

   c No carbon is oxidised and no carbon dioxide is produced, only water; $2H_2 + O_2 \longrightarrow 2H_2O$ (1 mark for the explanation, 1 mark for the correct formula, 1 mark for the correct balancing numbers)

Page 186 Types of rock

1 a iii It forms when rock comes under heat and pressure

   b The size of the crystals reveals whether the rock cooled under the ground or on the surface; small crystals mean it cooled on the surface and large crystals mean it cooled underground

   c The plankton died and the skeletons settled on the bottom of the ocean; many more layers fell on top, causing pressure; the layers hardened into rock

Answers

2 a Tube 1; because has the biggest difference in temperature between tube and surroundings

b Tube 2

c Agree (1 mark); any three explanations from: tube 2 cooled slowly; because it was insulated; cooling magma underground is insulated by rock; cotton wool is like the rock underground; it formed large crystals like granite; granite is intrusive igneous rock, which cools slowly

3 a $0.78 + 0.92 + 1.20 + 0.93 + 1.1 + 0.86 + 0.76 + 0.61 = 7.16$
$7.16 \div 8 = 0.90$ (0.895);
$2.15 + 1.76 + 0.89 + 1.24 + 0.98 + 1.35 + 1.29 + 1.17 = 10.83$
$10.83 \div 8 = 1.35$ (1.354);
$1.35 - 0.90 = 0.45$ mm / $1.354 - 0.985 = 0.369$ mm

b Any four from: slow cooling gives large crystals; because particles have time to move and join on to a few crystals; cooling underground is slow because surrounding rocks provide insulation; Hell Roaring Mountain rock is intrusive because crystals are large; Tower Falls rock is extrusive because crystals are smaller; particles form many crystals at the same time; Obsidian Cliffs cooled very fast so there was not enough time for crystals to form

Page 187 Sedimentary rock and quarrying

1 a Cementation = Concentrated solutions of salts stick rock fragments together
Weathering = Rocks break into small fragments.
Sedimentation = Small fragments of rock are deposited on the sea bed
Erosion = Rock fragments are removed from the surface of the rock

b Weathering, erosion; sedimentation, cementation

c Heat; and pressure; below the ground

2 Organisms fall into sediments; they are covered and leave an impression/caste; the caste fills with minerals; evidence is that they are only found in sedimentary rock / limestone was under the sea and contains sea shells

3 a Layers can be seen, which mean it has eroded so must be soft

b Answer must include the idea of different hardness of layers; and different rates of erosion

4 Any three from: dust; noise; damage to roads by heavy lorries; increased traffic; destruction of local wildlife habitats

5

Ingredient	Glass	Cement	Concrete
Limestone	✓	✓	✓
Sand	✓		✓
Clay		✓	✓
Gravel			✓
Water			✓
Sodium carbonate	✓		

(Each column completely correct for 1 mark)

6 Answer should include suggestions for: noise reduction; dust reduction; improving traffic flow; reclaiming the land

Page 188 Atoms and reactions

1 ii Atoms from one element change to another element during a chemical reaction

2 a The grey metal changed to white powder so its properties changed; a bright light was produced indicating energy change

b 2.0 g

3 Physical: green / solid / melting point of 200 °C;
Chemical: decomposes/reacts with acid to form salt

4 a 395.65 g

b KNO_3 is soluble and PbI_2 is insoluble / KNO_3 is colourless and PbI_2 is yellow

5 Other measurements – the mass at the end of the reaction; calculation – starting mass minus end mass; CO_2 is a gas so will escape from the beaker; other products remain in the beaker; since mass is conserved, any decrease in mass is due to carbon dioxide

Page 189 Thermal decomposition and calcium

1 a Calcium carbonate ($CaCO_3$)

b By heating (strongly)

c Carbon dioxide

d Thermal decomposition

2 Magnesium carbonate; because metal carbonates decompose; to give metal oxides

3 a $Ca(OH)_2$ (1 mark for correct symbols for Ca and OH, 1 mark for $(OH)_2$)

b Limewater detects carbon dioxide (CO_2); calcium oxide reacts with carbon dioxide to make calcium carbonate ($CaCO_3$); which is insoluble; so is cloudy

4 a Zinc carbonate ⟶ zinc oxide; + carbon dioxide

b Put the *same mass* of zinc and copper carbonate in separate boiling tubes; heat using the *same flame* in the apparatus shown; *time* how long it takes for the limewater to become cloudy

c Copper carbonate will decompose first; because it decomposes at a lower temperature than zinc carbonate

5 a To increase the pH of the soil

b Calcium oxide is fast-acting; less is needed

c Three times more limestone would be needed so transportation costs would be higher; calcium oxide is difficult to handle/harmful

Page 190 Acids, neutralisation and their salts

1 a ii Sodium chloride

b acid + base; ⟶ salt + water

c Alkali

d The acid is neutralised when the pH reaches 7; the more drops of alkali needed to reach pH7, the more acid the vinegar contains

2 a Measure a known volume of acid (any sensible volume); weigh the antacid tablet or use the number of tablets recommended as one dose for that brand; measure the pH of the acid before and after mixing with the tablet

b The higher the rise in pH, the more acid was neutralised by the tablet and therefore the more effective the tablet

c **Either** Yes, because this is similar to taking antacid tablets in real life: you would keep adding until the acid was neutralised; by crushing up the tablet you can add smaller amounts each time, which is more accurate; **or** no, because in real life you would take whole tablets not parts of tablets so spatulas of crushed powder are not an accurate measure; you don't neutralise all the stomach acid in real life

3 a Hydrochloric acid

b $CaCO_3 + 2HCl \longrightarrow CaCl_2 + CO_2 + H_2O$ (1 mark for the correct formula of hydrochloric acid, 1 mark for the remaining formulae and 1 mark for correct balancing)

Answers

Page 191 Electrolysis and chemical tests

1 a iv Electrolysis

b Chlorine is toxic (poisonous); any two from: breathing apparatus; any named protective clothing or safety goggles; regular health checks; regular checks of chlorine levels in the working atmosphere

c Poly(chloroethene) / polyvinyl chloride (PVC)

2 a i Chlorine can be collected from the cathode

b The electrodes have different charges – the anode is positive and the cathode is negative; the particles in solution are charged; and are attracted to the oppositely charged electrode

3 a Anode = 3.3 cm³; cathode = 7.3 cm³

b It contains more of one element than the other (approximately twice as much)

c Oxygen at the anode; hydrogen at the cathode; oxygen will relight a glowing splint; hydrogen will give a squeaky pop with a lighted splint

d They continued the electrolysis for different lengths of time

e Group 5 – they failed to capture all the gas at the anode / they allowed air to enter the tube at the cathode

Page 192 Metals – sources, oxidation and reduction

1 a ii A naturally occurring compound

b Gold is very unreactive

2 a The abundance of the ore; the percentage of metal in the ore; the cost of extraction

b Electrolysis is passing electricity through (molten) ore; it is used to extract some metals from their ores

3 a To reduce; the metal oxide

b Orange specks; in black powder

c It would work with lead or tin (allow iron and zinc); it would not work with aluminium, magnesium, calcium, sodium, potassium

4 a Al is oxidised; Fe_2O_3 is reduced

b It is more expensive than reduction with carbon

c Reduction with carbon; $2Fe_2O_3$ (s) + 3C (s); → 4Fe (s) + $3CO_2$ (g) / iron oxide + carbon; → iron + carbon dioxide; heating to high temperature / heating in a blast furnace

5 a Steel corrodes when it comes into contact with oxygen and water; coating forms a barrier

b Any two from: zinc is more reactive than iron; it provides sacrificial protection; it corrodes first even when damaged

Page 193 Metals – uses and recycling

1
Gold	Does not corrode	Electronic connectors
Aluminium	Low density	Aircraft
Copper	Malleable	Plumbing

2 a An alloy

b It makes the metal stronger; diagram should show layers of metal atoms of the same size, able to slide over each other; and layers of atoms of different sizes

c Chromium; it would make it resistant to corrosion

3 By alloying with a white metal; which makes the gold stronger/harder; and less expensive

4 Any three from: use recycled metal; recycle the scrap metal produced in the plant; recycling reduces the energy used in extracting/transporting/mining; only 5% of energy is needed to recycle

5 Answer should include references to the following: metal ores as a finite resource; mining and extraction are damaging to the environment; the high energy cost of extraction by electrolysis; the percentage energy saving

Page 194 Hydrocarbons and combustion

1 a Crude oil contains many different molecules; made from carbon and hydrogen only; which are not chemically bonded / can be easily separated

b Different boiling points

2 It must be divided into fractions (with similar boiling point); uses should include two from: refinery gases for heating and cooking; petrol for transport fuel; kerosene for aircraft fuel; fuel oil for ships and power stations; diesel for lorries, some cars and trains; bitumen for roads and roofs

3 Any four from: crude oil is a mixture of molecules of different sizes; the larger the molecule the stronger the intermolecular forces holding the molecules as a liquid; increasing the temperature increases kinetic energy of the molecules; smaller molecules require less kinetic energy to break away as gases / larger molecules require more energy to become gases; different-sized molecules become gases at different temperatures / have different boiling points so can be separated by gradually increasing the temperature and collecting the gases /gradually cooling the vaporised oil

4 a i It is a reduction reaction

b They are both (invisible) gases; carbon dioxide and water

5 a The wax combusted/was oxidised/turned to products

b The oxygen had all reacted

c Carbon dioxide; and water

d Carbon monoxide

e Carbon dioxide: limewater test (becomes cloudy); cobalt chloride (paper): changes from pink to blue

6 a When there is not enough oxygen; it is toxic

b Yellow flame and soot; this happens because there is not enough oxygen; it is called incomplete combustion

7 Incomplete combustion of hydrocarbons; which absorb/ reflect light from the sun

Page 195 Acid rain and climate change

1 a iv It reduces the yield of some crops

b The sulfur dioxide; is carried away in the atmosphere (to northern Norway)

2 a Hydrocarbon fuels contain sulfur; which reacts with oxygen during combustion; to form sulfur dioxide / acid rain

b Damages leaves; makes soil acidic

3 a Clams and snails

b Perch depend on mayfly species as a food source

4 Any three from: minimise energy use; choose low-sulfur fuels; remove sulfur from fuels before burning; capture sulfur dioxide by scrubbing waste gases

5 iii, i, iv, ii (iii before i = 1 mark; i before iv = 1 mark; all correct = 3 marks)

6 a The temperature has risen

b There is a correlation between the levels of carbon dioxide and the temperature; burning fossil fuels has increased the level of carbon dioxide

Answers

7 Any four from: seeding oceans with iron will increase growth of algae; because iron is a limiting factor for algal growth; algae take carbon dioxide from the atmosphere for photosynthesis and for making carbonate shells; carbon dioxide from the atmosphere will be stored within algae; carbon dioxide is a greenhouse gas; reducing carbon dioxide in the atmosphere will reduce the amount of energy trapped on the planet

Page 196 Biofuels and fuel cells

1 a Any three from: fossils fuels are non-renewable; supplies are running out; fossil fuels produce a lot of pollution; fossils fuels are becoming more expensive

   b Ethanol

   c Growing fuel crops will use up land needed to grow food

2 Any three from: high energy release; low carbon dioxide production; low polluting emissions/sulfur dioxide/nitrogen oxides; low ash/smoke; ease of transport; ease of storage; cost; cost of technology needed to use it

3 Any four from: wood residue requires too much land use to be sustainable as a fuel source; algae use little land; high farmland use for sugar cane and rapeseed is likely to have a significant effect on food production; although algae uses little land the energy cost are high and therefore the benefit of using biofuel in terms of reducing greenhouse gas emissions is reduced; use a mixture of sources, e.g. where there is surplus farmland use sugar cane or rapeseed; use waste wood residue from other industries; research less energy intensive ways of producing biodiesel from algae

4 It is a renewable resource / can be made from water; it does not produce polluting gases

5 a They used the water from the first experiment that was already heated

   b Fuel 1; it gave the highest temperature rise

   c Any one of: use a heat shield; take the mass of the fuel before and after burning; control the size of the flame; stir the water

6 Producing the hydrogen has an energy cost; most electricity is generated using fossil fuels; which produce carbon dioxide

Page 197 Alkanes and alkenes

1 a C

   b A and C (methane and ethane)

   c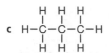

   d Each carbon has 4 bonds; to 4 different groups

2 a

Substance	Mass (g)	% of total
Natural gas	80 g	100
Methane	64	80
Ethane	8	10
Propane	2.4	3
Butane	5.6	7

   b Methane has only 1 carbon; propane has 3 carbons (and 4 more hydrogens)
   c It is a gas at room temperature; increasing the pressure converts it to a liquid, which takes up less space

3 C_2H_6; the atoms are held together by covalent bonds; which are shared pairs of electrons.

4

Propene...	True	False
Has four carbon atoms		✓
Is an unsaturated molecule	✓	
Contains a double bond	✓	
Does not react with bromine water		✓

5 Any three from: burning fuel produces polluting gases; oil is a finite resource so we should preserve it; oil is a raw material for the chemical industry; it is used to make (any named product); there is no other energy source as effective as crude oil

6 a iii Butene   b Ethene; C_2H_4

Page 198 Cracking and polymers

1 a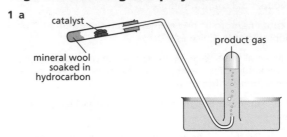

   b Heat; the catalyst/aluminium oxide

   c Alkanes; alkenes

2 a To reduce energy costs / to reduce the temperature of reactions

   b Catalysts are not used up in the reaction; so the platinum can be reused

3 Any four from: cracking as breaking long chain alkanes into shorter alkanes and alkenes; describe the petrol/gases fraction that is the most valuable; the supply is less than the demand and why; cracking can increase the supply; fuel oil/bitumen supplies are far greater than demand

4 a Poly(propene)

   b Poly(ethane)

   c Poly(tetrafluoroethene)

   d Poly(chloroethene)

5 iii, ii, iv, i (1 mark each for putting ii before iv and i after iv)

6 Any two from: polymers are made from oil and we will run out of oil; recycling polymers does not use oil; polymers can be burned but this produces toxic gases, or buried in landfill sites but we are running out of sites; polymers are not biodegradable so they stay in the ground for a very long time; recycling polymers saves energy, which is good for the environment

7 a Starch; cellulose

   b Plant polymers are biodegradable

   c Preserves oil supplies / renewable resource used

Answers

Page 199 C1 Extended Response question

0 marks
Insufficient or irrelevant science. Answer not worthy of credit.

1–2 marks
Answer may be simplistic. There may be limited use of specialist terms. Errors of grammar, punctuation and spelling prevent communication of the science.

3–4 marks
For the most part the information is relevant and presented in a structured and coherent format. Specialist terms are used for the most part appropriately. There are occasional errors in grammar, punctuation and spelling.

5–6 marks
All information in answer is relevant, clear, organised and presented in a structured and coherent format. Specialist terms are used appropriately. Few, if any, errors in grammar, punctuation and spelling.

C2 Discovering chemistry

Page 200 Atomic structure and the periodic table

1 a **iii** It is where the neutrons are found

  b Same: both contain protons and electrons / both have a nucleus with electrons surrounding it; different: helium has neutrons and hydrogen does not / helium has two protons and electrons and hydrogen has one

  c Atoms contain subatomic particles / protons, neutrons and electrons; most of an atom is space / atoms are not solid

2 a Protons have a positive charge; neutrons have no charge

  b 7

3 a Mg

  b **i** 9 **ii** 12 **iii** 4

4 a They have different numbers of neutrons

  b They have the same number of protons / they have the same atomic number; the number of protons decides the element

5 There are different isotopes of magnesium; which have different numbers of neutrons (different atomic masses); relative atomic mass is the average mass of an atom of the element; taking into account the abundance of all isotopes

Page 201 Electrons

1 a

Element symbol	Electronic configuration
F	2.7
Al	**2.8.3**
S	2.8.6

  b **i** Electron **ii** 11 **iii** Electron shells

2 a Magnesium; 2.8.2

  b Group 7; number of protons is the same as the number of electrons so there would be 17 electrons; 2 electrons in the first shell and 8 in the second means 7 in the outer shell; the group number is the number of electrons in the outer shell

  c Ne = 2.8; Ar = 2.8.8

  d The electronic configuration shows that they each have 8 electrons in their outer shell; they are in the same group in the periodic table; elements in the same group have similar properties

  e Magnesium

  f Period 4

3 a Mendeleev knew about the chemical properties of the elements; he grouped elements according to their properties; he knew the atomic masses and listed elements in order of atomic mass; the chemical properties depend on the electronic structure

  b Mendeleev realised that not all elements had been discovered; he left gaps to allow for undiscovered elements; he predicted the properties of the undiscovered elements; when the elements were discovered, his predictions were found to be correct

Page 202 Ionic bonds and naming ionic compounds

1 a Electrostatic attraction = force that holds ions together; negative ion = atom that has gained electrons; positive ion = atom that has lost an electron; ionic compound = made from positive and negative ions

Answers

2 a The 2 electrons; in the outer shell; of magnesium are transferred to; the outer shell of oxygen; 1 mark for correct diagram

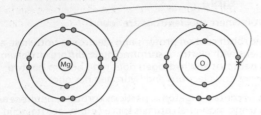

b $F = F^-$; $Na = Na^+$; $S = S^{2-}$; $Ca = Ca^{2+}$

c Potassium has 1 electron in its outer shell (and loses it to form a 1+ ion); chlorine has 7 electrons in its outer shell (and gains one more to form a 1– ion)

3 a Cation = Ca^{2+}; anion = CO_3^{2-}

b Diagram should show 2 electrons in the inner shell and 8 in the outer shell; charge should be shown as negative

c A sodium ion has a full outer shell of electrons; a sodium atom only has 1 outer electron

4 a

Name	Ions	Elements
Sodium chloride	Sodium and chloride	Sodium and chlorine
Potassium fluoride	**Potassium and fluoride**	Potassium and fluorine
Calcium oxide	**Calcium and oxide**	Calcium and oxygen
Magnesium nitrate	Magnesium and nitrate	**Magnesium, nitrogen, oxygen**

b Calcium hydroxide; ammonium chloride; magnesium bromide; sodium carbonate

c Potassium sulfide contains potassium and sulfur only; potassium sulfate contains potassium, sulfur and oxygen

Page 203 Writing chemical formulae

1 a

Name of ion	Charge	Formula
Sulfide	2–	S^{2-}
Fluoride	1–	F^-
Hydrogen	1+	H^+
Ammonium	1+	NH_4^+
Potassium	1+	K^+
Sulfate	2–	SO_4^{2-}
Carbonate	2–	CO_3^{2-}
Hydroxide	1–	OH^-
Nitrate	1–	NO_3^-
Tin	2+	Sn^{2+}
Copper	2+	Cu^{2+}
Silver	1+	Ag^+
Iron	3+	Fe^{2+}
Iron	2+	Fe^{2+}

b Barium sulfate is an ionic compound that contains barium and <u>sulfur</u> ions. The barium ion is a <u>polyatomic</u> cation and the <u>sulfate</u> ion is a monoatomic <u>cation</u>. (1 mark for underlining sulfur **or** sulfate)

2 a $CaCl_2$ / James; Cl^- has one negative charge; Ca^{2+} has two positive charges; two negative ions are required for each positive ion to make a neutral compound

b Potassium fluoride = KF; calcium nitrate = $Ca(NO_3)_2$; aluminium oxide = Al_2O_3; ammonium sulfate = $(NH_3)_2SO_4$

3 a Two ammonia molecules are used in the equation

b Two hydrogen atoms are in the formula

c The brackets show that the whole ammonium ion is multiplied by two

4 a i Sodium chloride + silver nitrate $\longrightarrow$ silver chloride + sodium nitrate (1 mark for each correct reactant; 1 mark for the correct product)

ii Potassium hydroxide + hydrogen chloride/ hydrochloric acid $\longrightarrow$ potassium chloride + water (1 mark for each correct reactant; 1 mark for the correct product)

b i $Na2(SO4) + BaCl2 \longrightarrow BaSO_4 + 2NaCl$ (3 correct formulae, 3 marks; 2 in front of NaCl, 1 mark)

ii $Mg(NO_3)_2 + 2NaOH \longrightarrow Mg(OH)_2 + 2NaNO_3$ (3 correct formulae, 3 marks; 2 in front of either NaOH or $Mg(OH)_2$, 1 mark)

Page 204 Ionic properties and solubility

1 a iii Forms crystals, high melting point

b Diagram should show alternating positive and negative ions; ions must be labelled as specific substances

2 No it is not an ionic compound; because it melted at a low temperature/in a Bunsen flame

3 a Aluminium oxide has a giant lattice structure; large amounts of energy are needed to break strong ionic bonds; it must be melted because it does not conduct electricity as a solid

b Calcium oxide; it is the only compound where both ions have a double charge; this makes a stronger ionic bond so more energy is needed to break the bonds

4 Soluble – sodium chloride; magnesium nitrate; ammonium carbonate; insoluble – calcium carbonate; silver chloride

5 iii $AgNO_3$ (aq) + NaCl (aq) $\longrightarrow$ AgCl (s) + $NaNO_3$ (aq)

6 ii Silver nitrate and sodium chloride

Page 205 Preparation of ionic compounds

1 a Choose the two **soluble** compounds. Then dissolve them and mix them together. The insoluble salt that you want will be **precipitated as a solid**.

b Precipitation reaction

2

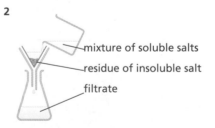

mixture of soluble salts

residue of insoluble salt

filtrate

3 Barium sulfate is X-ray opaque; it leaves a silhouette of the bowel on the X-ray so doctors can spot any abnormalities; it is not absorbed as it is insoluble

4 Calcium = brick red; sodium = yellow/orange; potassium = lilac; copper = green/blue

5 a Dip nichrome wire into acid then the solution; put in flame; yellow/orange flame means sodium; add silver nitrate and nitric acid; white precipitate means chloride is present

b The white powder is a carbonate; test the gas with limewater; cloudy limewater confirms that the gas is carbon dioxide

Answers

6 Helium has a unique spectrum; which did not match the spectrum of any known element; so it must be an unknown element

Page 206 Covalent bonds

1 a iv iron oxide

  b i H_2O

  ii 2

2 a The atoms are held together by a shared; pair of electrons

  b To become more stable

3 a To become stable by having a full outer shell of electrons

  b Hydrogen has 1 electron in its outer shell; it needs 2 to fill the outer shell; there is no space for more than 2 electrons so it can only bond with one atom

4

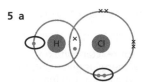

5 a

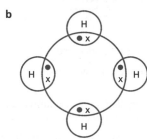

  b

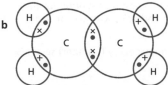

  c i 8

  ii 2

  iii Carbon needs 8 electrons to have a full outer shell; but hydrogen only needs 2 electrons to have a full outer shell

6 a Two shared pairs of electrons; between the same two atoms

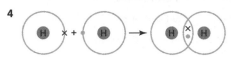

  b

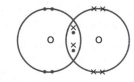

Page 207 Properties of elements and compounds

1 a ii It is white

  b Diagram showing a circuit; including a power source; light bulb or ammeter; two separated electrodes in a beaker of solution

  c Sucrose

  d Simple – water/ammonia; methane/any named alkane/ glucose/carbon dioxide; giant – diamond/graphite/silicon dioxide/silicon/silicon carbide

2 a 1 simple covalent; 2 ionic; 3 giant covalent (structure and bonding must both be correct for 1 and 3)

  b Test whether the solution conducts electricity; if it is ionic it will conduct electricity; test whether it melts in a Bunsen flame/at a low temperature; ionic will not melt

  c Ionic substances contain charged particles; which can move when molten or dissolved; covalent substances have no charged particles

3 a Graphite has a layered structure; with weak bonds between layers; the layers easily slide over each other; **or** a clearly labelled diagram of the structure of graphite showing weak bonds between layers

  b Silicon dioxide has a giant structure; because it has a high melting point (since it is solid at room temperature); carbon dioxide has a simple structure; because it has a low boiling point/is a gas at room temperature; giant covalent structures have high melting points/require large amounts of energy to break bonds between particles, simple covalent structures have low melting points/require little energy to break forces between particles

Page 208 Separating solutions

1 Put into a separating funnel; allow to settle into two layers; open the tap and allow the lower layer to run into collecting vessel

2 iii, ii, iv, i, v (completely correct = 3 marks; **iii** before **ii** = 1 mark; **ii** before **i** = 1 mark)

3 Cool air to below −219 °C; remove solid carbon dioxide and allow temperature to rise; nitrogen turns to gas first (at −196 °C); argon turns to gas next (at −186 °C); oxygen can be collected as a liquid

4 Diagram should show a pencil line or cross near the bottom; with two spots higher up

5 Glucose; fructose

6 a Substance A

  b Substance C

  c Run a chromatogram using a pure sample of phenylalanine; measure the Rf value / see how far the spot travels; if it matches one of A, B or C then that confirms the identity

Page 209 Classifying elements

1 a i Metallic **ii** Ionic lattice

  b Simple molecular covalent

2 a Giant lattice; of alternating sodium and chloride ions / positive and negative ions (or a clearly labelled diagram)

  b Lots of energy; is needed to break strong ionic bonds

3

	Melting point	Solubility	Electrical conductivity	Structure
Silicon carbide	2730 °C	**Insoluble**	Non-conductor	**Giant molecular covalent**
Boron trifluoride	−127 °C	Very soluble	Non-conductor	**Simple molecular covalent**
Copper(II) oxide	1201 °C	Insoluble	**Only conducts when molten**	Ionic
Cerrosafe	74 °C	**Insoluble**	Conducts as solid	**Metallic**

Answers

4 a Alkali metals

b Group 0

c i It has a low melting point

5 a Labelled diagram showing: a regular arrangement; positive ions; sea of electrons

b Metals are malleable because layers of ions slide over each other; in the sea of electrons; metals conduct electricity because electrons can move; when a voltage is put across them

6 Two pairs from below (the reason must match the explanation):

Mobile sea of electrons; means it conducts electricity

Layers of ions slide over each other in the sea of electrons; therefore ductile/easily made into wires

Regular lattice of positive ions has strong metallic bonds; therefore has a high melting point so does not melt when electricity is put through it

Page 210 Alkali metals

1 a Line D

b Any two from: low melting point; soft; react with water

2 a i Fizzing

ii Lighted splint; gives a squeeky pop

iii Either: test the pH; lithium hydroxide would have a high value; **or:** add indicator; the water is alkaline

iv $2Li\ (s) + 2H_2O\ (l) \longrightarrow 2LiOH\ (aq) + H_2\ (g)$ (1 mark each for: formula for LiOH; formula for H_2; balanced correctly; left-hand side state symbols; right-hand side state symbols)

b (1 mark each for: correct number of shells; correct number of electrons; loss of one outer electron leaves a 1+ ion; which leaves a full/ stable shell of electrons)

3 a Reactivity increases down the group

b Potassium has two more shells of electrons; the outer electron is further from the nucleus in potassium; the inner shells of electrons shield the outer electron from the positive pull of the nucleus; the outer electron is more easily lost (in potassium)

c Rubidium would have a very vigorous reaction/more vigorous than lithium, sodium or potassium; produces hydrogen/rubidium hydroxide; $2Rb + 2H_2O \longrightarrow 2RbOH + H_2$

4 a Atomic mass increases; as reactivity increases

b At conferences; in scientific journals

Page 211 Halogens and noble gases

1 Fluorine = yellow gas; chlorine = green gas; bromine = brown liquid; iodine = grey solid (1 mark each for three correct lines in left column and three in right)

2 a iii $H_2\ (g) + Br_2\ (l) \longrightarrow 2HBr\ (g)$

b Low pH (any number below 4)

3 a Chlorine water + potassium bromide, a colour change from green/colourless to brown; $Cl_2 + 2KBr \longrightarrow Br_2 + 2KCl$; iodine water and potassium would have no reaction

b Chlorine is more reactive than bromine, so it displaces bromine from potassium bromide; iodine is less reactive than bromine since it does not displace bromine from potassium bromide; therefore the order of reactivity is Cl>Br>I

4 He, Ne, Ar, Xe (1 mark each for any two in the correct order)

5 a Neon 2.8; argon 2.8.8

b Both neon and argon have 8 electrons in their outer shell/ have a full outer shell; they do not need to share, lose or gain electrons; to increase stability

6 a Any value between 1.5 and 4.5 g/dm^3

b Use the periodic table to find the atomic number/ mass of radon (in group 0); plot a graph of the atomic number/mass; against the boiling point; extend the line/ extrapolate and read off the boiling point at the atomic number/mass of radon

Page 212 Endothermic and exothermic reactions

1 a iv The reaction takes in energy overall

b i Endothermic **ii** Exothermic

iii Endothermic **iv** Exothermic

2 a Diagram should show an insulated container; and a thermometer

b Temperature before and after mixing the reactants

c The temperature would rise

3 Reactions start by breaking bonds (of reactants)/taking in energy; next, new bonds are made/energy is given out; for exothermic reactions more energy is given out in making bonds; than is taken in when breaking bonds

4 a Exothermic

b $(22 + 24) \div 2 = 23$; $43 - 23 = 20$; °C; increase

c Silver nitrate + sodium chloride; $\longrightarrow$ silver chloride + sodium nitrate

d Reactants and products are labelled; energy is labelled at the y-axis; energy of products is below energy of reactants

5 The energy needed to break bonds; in oxygen and methane; is less than the energy needed to make bonds; in carbon dioxide and water

Page 213 Reaction rates and catalysts

1 ii Cut the carrots into small pieces.

2 A lower concentration of acid in the water; reduces the rate of reaction

3 a Diagram should show **either**: flask/test tube containing magnesium and acid; sealed connection; to inverted measuring cylinder under water/gas syringe; **or** flask containing magnesium and acid; with cotton wool in the neck; on a balance

b The change in volume/mass; every (number) seconds/ minute

c As a line graph

d The higher the concentration; the steeper the (initial) gradient (or a graph showing this)

4 Only collisions with sufficient energy; result in a reaction

5 a A catalyst increases the rate of a reaction; without being used up by the reaction

b To reduce; polluting emissions/carbon monoxide and unburned hydrocarbons

c Hydrocarbon + oxygen $\longrightarrow$ carbon dioxide and water

6 a Open structure allows gases to pass through; and increases the surface area of the catalyst

b $2CO + O_2 \longrightarrow 2CO_2$ (1 mark for correct formula; 1 mark for correct balancing)

Answers

Page 214 Mass and formulae

1 a 36.5

b 160

c 85

d 74

2 a $(12 \div 44) \times 100\%$; 27.3%

b $(48 \div 167) \times 100\%$; 28.7%

c $(28 \div 148) \times 100\%$; 18.9%

3 a Mass of copper = 4.95 g; mass of copper oxide = 6.20 g

b 6.20 − 4.95 = 1.25

c $4.95 \div 63.5 = 0.078$; $1.25 \div 16 = 0.075$; formula CuO

4 $(43 \div 65) \times 100\%$; = 66.2%

5 Actual yield from 10 cm³ = (50% of 100) 50cm³ oxygen; volume required to make 100cm³ = **20cm³** hydrogen peroxide

6 Mr CuSO$_4$ = 160; Mr CuO = 80; $(160 \div 80) \times 5 = 10$ g

7 H_2O is not damaging to the environment / no problems disposing compared to HCl; faster process means more profit

Page 215 C2 Extended response question

0 marks
Insufficient or irrelevant science. Answer not worthy of credit.

1–2 marks
Answer may be simplistic. There may be limited use of specialist terms. Errors of grammar, punctuation and spelling prevent communication of the science.

3–4 marks
For the most part the information is relevant and presented in a structured and coherent format. Specialist terms are used for the most part appropriately. There are occasional errors in grammar, punctuation and spelling.

5–6 marks
All information in answer is relevant, clear, organised and presented in a structured and coherent format. Specialist terms are used appropriately. Few, if any, errors in grammar, punctuation and spelling.

P1 Universal physics

Page 216 The Solar System

1 a Geocentric means that Earth is in the centre; and all the other objects orbit the Earth

b Mercury, Venus, Mars, Jupiter and Saturn (1 mark for two or three correct, 2 marks for four, 3 marks for all five)

c The ancient Greeks could only see five planets with the naked eye; nowadays we use telescopes to see further into space so can see more planets

d The retrograde motion of Mars / sometimes Mars appeared to be going backwards in space; the geocentric model explained this by using epicycles; the heliocentric model predicted this motion

2 Any three from: the Catholic Church believed that God had created the Universe; all the heavenly bodies were perfectly spherical and orbited Earth; if there were mountains on the Moon this meant that it wasn't a perfect sphere; if there were moons orbiting Jupiter then not everything could be orbiting Earth

3 Any two from: telescopes give brighter images; images are more magnified in a telescope; using a telescope enables astronomers to see more distant objects

4 The asteroid belt is made up of lumps of rock; orbiting the Sun between Mars and Jupiter

5 Any two from: the Hubble space telescope has a much larger aperture than ground-based telescopes; so it can see much dimmer / more distant objects; light doesn't have to pass through the atmosphere; so there is not as much distortion

Page 217 Reflection, refraction and lenses

1 Incidence; angle; reflected; normal

2
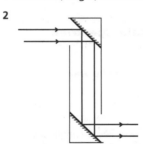
1 mark for right-angle turn at top mirror; 1 mark for right-angle turn at bottom mirror; 1 mark for arrows in correct direction (1 mark deducted if only one ray complete)

3 Diagram C is correct

4 Air; because it is the least dense material

5 a The focal point is the point at which parallel rays of light entering the lens; will converge and meet

b Lens A

Page 218 Lenses in telescopes

1 Obtain a focused image of a distant object; on a screen; and measure the distance from the image to the lens

2
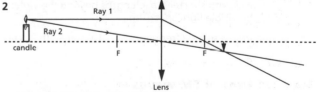

a Straight line; drawn through F

b Straight line; drawn through centre of the lens

c Image drawn where the rays cross

d Magnified

e Magnification = 0.5/1 = 0.5

f A real image can be projected; onto a screen and a virtual image cannot

Answers

3

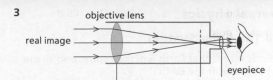

4 A diminished image of the distant object; is formed at the focal point of the objective lens; the eyepiece magnifies the image

5 It is a reflecting telescope; reflecting telescopes are lighter than refracting ones so it was easier to launch into space; curved mirrors can be much larger than lenses so it can collect much more light to see distant objects

Page 219 Waves

1 Energy; matter; oscillations

2 a B **b** E or A

3 Frequency of wave = 15 waves/60 seconds = 0.25 Hz; speed = wavelength × frequency = 8 × 0.25 = 4; m/s

4 a Reaction-time errors; because it is a very short time

b Average time is 0.578 s; speed = distance/time so 200/0.578; = 346 m/s

5 Wavelength = speed/frequency; 1500/500 = 3 m

6 Longitudinal – sound wave; transfer energy; transverse – radio wave; vibrations at right angles to the direction of travel; light waves; transfer energy

7 a A wave caused by earthquakes or explosions; that travels through the Earth

b It detects the vibrations of the earth caused by seismic waves; the movement of the earth is detected relative to a heavy pendulum

Page 220 The electromagnetic spectrum

1 a Transverse; vacuum

b Gamma rays; ultraviolet; infrared; radio waves

2 a Radio waves

b Gamma rays

c Any two from: apply suntan lotion; wear sunglasses; cover our skin with clothes

3 The spectrum of white light is all the colours light is made up of / the colours of the rainbow (red, orange, yellow, green, blue, indigo, violet); you can produce it by shining a ray of light through a glass prism

4 a Herschel found that the temperature was highest; beyond the red end of the spectrum where there was no visible colour; this meant that there were waves with wavelengths longer than visible light

b Ritter found that the reaction was fastest; beyond the area of violet light in the spectrum; this meant that there were waves with shorter wavelengths beyond the violet end of the visible light spectrum

5 a Wavelength = speed / frequency; $3 \times 10^8 / 60 \times 10^6 = 5$ m

b Radio wave

Page 221 Uses of EM waves

1 Microwaves = mobile-phone communication; X-rays = airport-security scanner; infrared = TV remote control; ultraviolet = water sterilisation

2 The microwaves travel in a straight line; from a transmitter/ aerial to a satellite; the satellite retransmits the signal to a receiver/aerial in the USA

3 Any three from: take an infrared photograph of some homes; the more infrared emitted, the lighter the image will be; the better the insulation of the home, the less infrared emitted (or vice versa); the better-insulated homes will be darker on the infrared photograph

4 a i Ray of light refracted; away from the normal

ii Ray of light reflected; angle *I* = angle *r*

b Diagram should show reflections off inside of fibre; emerging from far end; reflected angles should be roughly equal to incident angles

Page 222 Gamma rays, X-rays, ionising radiation

1 X-rays can be used to investigate broken bones; the benefits outweigh the risks

2 Similarity – they are both highly ionising / both can travel through solids; difference – gamma rays come from the nucleus of radioactive atoms while X-rays come from metals bombarded with electrons

3 High-energy electrons from a heated cathode; are accelerated through a very high-voltage tungsten anode; electrons collide with the tungsten atoms and emit X-rays

4 Alpha; beta; gamma

5 a Michael is correct; radioactivity is not affected by physical conditions such as temperature; it only depends on the amount of radioactive material in the sample

b An alpha/beta particle enters the GM tube; and ionises the gas inside the tube; allowing a pulse of electric current to flow; each count represents a single particle

Page 223 The Universe

1 Comet, Moon, Mercury, Jupiter, Sun, Milky Way (3 marks if all correct, 1 mark deducted for each object incorrectly placed)

2 a 11.20/0.38 = 29.5 times larger

b 5.20/1.52 = 3.4 times further away

c 1.00/0.27 = 3.7 times larger

3 a Time in seconds = 8 × 60 = 480; distance = speed × time = $3 \times 10^8 \times 480$; = 144×10^9 m

b Light from distant stars takes several years to reach Earth; so we see the stars as they were when the light left them

4 Stars and galaxies give out different types of EM radiation; some objects do not emit any visible light so can only be seen using other parts of the spectrum

5 a Any two from: content of soil; content of rocks; detect signs of water; detect microorganisms such as bacteria

b SETI is a project that is searching for intelligent life forms elsewhere in the Universe; a radio signal is sent into space; in the hope that another form of life will try to communicate

Page 224 Analysing light

1 a A device to split up light into its component colours

b Use an old CD or DVD as a reflector with a box with a slit in it; let light enter the box via the slit; to reflect off the disc to see the spectrum

2 a Different stars are different temperatures; so they emit different-coloured light

b The chemical composition of the star; the temperature of the star

3 Oxygen and ozone absorb mostly ultraviolet rays; UV rays cause skin cancer; so if there are holes in the ozone, more UV will reach Earth

Answers

4 a 3 mm

  b It has reduced

  c The pitch would be higher

5 Red-shift is when a light source moves away from you; the wavelength appears to lengthen; because red light has a longer wavelength than blue light

Page 225 The life of stars

1 E, C, D, B (3 marks for all correct, 1 mark deducted for each letter in the incorrect position)

2 a The force of gravity is balanced; by outward pressure from the nuclear reactions

  b The force of gravity will become larger than the outward force; and the Sun will shrink to become a white dwarf

3 a They are hotter than average-sized stars like the Sun, so they emit more blue light

  b Any four from: it will expand to become a super red giant; some heavier elements will start to be produced in the core; nuclear fusion reactions eventually stop; the star collapses due to gravity; an explosion called a supernova occurs; leaving a neutron star/black hole

4 White dwarf, main sequence star, red giant, supernova (3 marks for all correct, 1 mark deducted for each object in the incorrect position)

5 Black holes have such an enormous force of gravity; that no EM radiation is emitted from them so they cannot be observed

Page 226 Theories of the Universe

1 Big Bang theory to ideas 2, 3 and 4; Steady State theory to ideas 1, 2 and 5 (1 mark for each correct connection)

2 a The red-shift; of light from stars

  b The left-over radiation; from the Big Bang

  c The Steady State theory of the Universe cannot account for the existence of cosmic background radiation; so the only theory that supports it is the Big Bang theory

3 a The detection of waves; of a different wavelength to those you are expecting to record

  b To get rid of all known sources; so if the background radiation was still detected, it must come from space

  c Because the radiation from the Big Bang; would have been emitted equally in all directions

  d The original radiation from the Big Bang would have been in the gamma spectrum; because of the Doppler effect / expansion of the Universe; the wavelength would have increased into the microwave region

Page 227 Ultrasound and infrasound

1 a 50 Hz; 2000 Hz

  b Higher

  c Infrasound is sound with a frequency too low for humans to hear / below 20 Hz; ultrasound is sound with a frequency too high for humans to hear / above 2000 Hz

2 a Pulses of ultrasound are passed through the womb; they reflect from different surfaces; which allows an image to built up of the unborn child

  b Ultrasound does not harm the baby inside the womb; it allows doctors to check for healthy development

3 Man-made – nuclear explosion / drilling for oil; natural – earthquake / volcanic eruption / animal movement / meteorite strikes

4 a 6 milliseconds / 0.006 seconds

  b Depth = speed × time/2; 1500 × 0.006/2; = 4.5 m

  c The reflected pulse will move; to the right

  d Emit the pulses less frequently

Page 228 Earthquakes and seismic waves

1 a Crust; mantle; outer core; inner core

  b Semi-solid rock beneath Earth's crust

2 a Tectonic plates

  b Tectonic plates move; due to convection currents in the mantle; earthquakes occur at plate boundaries when the plates suddenly slip

  c The coastlines of Africa and South America look like they fit together; similar rock types have been found in Africa and South America; fossils of the same species have been found on both continents

3 a P waves

  b P waves

  c Both P and S waves

4 Scientists cannot measure the stress/pressure of the rocks; the fault lines are too deep below the ground

5 a A is the P wave and B is the S wave

6 a The P waves are refracted through the core; so there is a region that no P waves can reach

  b Waves should start at the epicentre; some should be shown stopping at the core; others should be shown as curved paths reflected off the outer core and stopping at the crust

Page 229 Electrical circuits

1 a

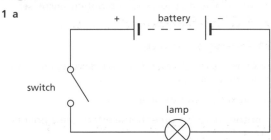

  b Any three from: they carry energy from the cell to the bulb; the charged particles in this circuit are electrons; the electrons/charged particles are given energy in the battery; the electrons/charged particles flow all the way round the circuit; at the bulb the energy carried by the electrons/charged particles is converted to heat and light

2 X = 0.3 A; Y = 0.3 A; Z = 0.6 A

3 Bulb B

4 Current; voltage; voltage; voltage; voltage (3 marks for all correct, 1 mark deducted for each incorrect answer)

5 Potential difference is the scientific term for voltage/ measured in volts; it is a measure of how much energy is given to/converted from the charged particles; one volt means that one joule of energy is given to/converted for each unit of charge

Answers

Page 230 Electrical power

1 a Kettle

b Iron

c Iron

d

Appliance	Energy used in joules	Energy used in kW h
Kettle	1 200 000 J	0.33 kW h
Iron	2 160 000 J	0.6 kW h
Vacuum cleaner	1 200 000 J	33 kW h

e The watt is too small a unit / the number of watts used by domestic appliances is too large a number

f Kettle – current = 2000/230 = 8.7 A; 13 A fuse; iron – current = 1200/230 = 5.2 A; 13 A fuse; vacuum cleaner – current= 1000/230 = 4.3 A; 5 A fuse

2 a Any two from: less pollution; lowers energy bills; longer-lasting; saves energy resources

b Initial cost of bulbs has to be paid; but the savings on energy bills can be used to offset the cost; payback time is the amount of time taken for the total savings to add up to the initial cost

3 a

Device	Initial cost (£)	Annual saving (£)	Payback time (years)
Double glazing	7000	350	20
Loft insulation	450	75	6
Draught excluders	40	5	8
Cavity-wall insulation	550	110	5

b Double glazing; it increases sound-proofing / increases the value of the property

Page 231 Energy resources

1 E, D, B, C (3 marks for all correct, 1 mark deducted for each sentence in the wrong place)

2 a Carbon dioxide; sulfur dioxide

b They contribute to the greenhouse effect; cause pollution / acid rain

c Any two from: radioactive waste is produced; the consequences of an accident can be disastrous; the start-up time is very long; building/decommissioning a nuclear power station is very expensive

3 Infrared radiation emitted from the surface of the Earth; is absorbed by gases in the atmosphere, making the temperature rise; causing it to re-radiate infrared radiation

4 Renewable – wind, biomass, hydroelectric, solar, tidal; non-renewable – coal, oil, nuclear (half a mark for each correct answer)

5 a Wind turns the turbine blades; which are connected to a generator

b Any two from: they cause noise pollution; they are unsightly in the countryside; they need a large area of land for several turbines; they only produce electricity when the wind is blowing

6 a 2.3 MW (a range of 2.2–2.4 MW acceptable)

b 15 m/s

c No electricity is produced at speeds lower than about 4 m/s; if the wind is too strong, the turbine has to be switched off to avoid damage

Page 232 Generating and transmitting electricity

1 a Current

b Moves to the right

c Stays still

2 a The output voltage would be higher

b The output voltage would be higher

c Graph should show a negative sine wave; with the same amplitude

3 a Step-down; step-up; step-down; neither

b V_1, V_2 V_3, V_4

4 a To reduce energy loss in the cables; because lower currents are needed

b Cables are either buried underground; or suspended high up using pylons

5 The output from a battery is direct current; which will create a constant magnetic field; for a voltage to be induced on a secondary coil; there must be a varying magnetic field

Page 233 Energy and efficiency

1 Roller coaster = gravitational potential energy; rubber band = elastic potential energy; battery = chemical energy; cup of tea = heat energy

2

Device	Energy input	Useful energy output	Wasted energy output
Electric fan	Electricity	Kinetic	Heat and sound
Television	Electricity	Light and sound	Heat
Catapult	Elastic potential	Kinetic	Heat
Gas ring on a cooker	Chemical	Heat	Light and sound

3 Total energy remains constant; energy can only be transferred; from one type to another / from one object to another

4 a Chemical; heat; kinetic; electrical

b Efficiency = 144/360; = 0.4/40%

c It is wasted as heat; in the atmosphere

5 Sankey diagram showing chemical energy in coal at start, with four arrows detailing: heat energy wasted from burning coal; heat energy wasted in turbine; heat energy wasted in generator; useful electrical energy from generator

Page 234 Radiated and absorbed energy

1 a ii 200 ml of water at 40 °C

b i A cup of tea in a black cup

c i A black water bottle

2 a 20 °C

b 20 °C

c Some of the energy was used to heat up the ice cube; some energy was used to heat up the water; some energy was used to heat up the air

Answers

3 a Correctly plotted all points on a suitably scaled axis (3 marks); line of best fit drawn (1 mark)

 b Any two from: use the same amount of water in each beaker; make sure the beakers are the same size; use the same starting temperature; make sure the beakers are at the same room temperature

 c Cooling curve; below the plotted line

 d The black surface will emit more; infrared/heat radiation; so the water will cool more quickly

Page 235 P1 Extended response question

0 marks
Insufficient or irrelevant science. Answer not worthy of credit.

1–2 marks
Answer may be simplistic. There may be limited use of specialist terms. Errors of grammar, punctuation and spelling prevent communication of the science.

3–4 marks
For the most part the information is relevant and presented in a structured and coherent format. Specialist terms are used for the most part appropriately. There are occasional errors in grammar, punctuation and spelling.

5–6 marks
All information in answer is relevant, clear, organised and presented in a structured and coherent format. Specialist terms are used appropriately. Few, if any, errors in grammar, punctuation and spelling.

P2 Physics for your future

Page 236 Electrostatics

1 a Protons; neutrons

 b The protons have a positive charge; the neutrons have no charge

2 a There is friction between the jumper and Brendan's body

 b Electrons

 c All the hair has the same charge; and like charges repel each other

3 a The charged balloon repels some of the electrons away from the surface of the wall; leaving the wall with a positive charge; the opposite charges attract

 b Induction / induced charges

4 a They are the same

 b Ion

 c It loses electrons

5 The charged rod repels/attracts electrons down/up the metal rod of the gold-leaf electroscope away from/towards the cap; the rod and the gold leaf are charged with a negative/positive charge; they repel one another, causing the leaf to rise

Page 237 Uses and dangers of electrostatics

1 Less insecticide can be used; the plant gains an even coating of insecticide; every part of the plant's leaves attracts the insecticide, even the underside

2 a Field lines

 b A stronger field

3 He will experience a small electric shock; because the electrons on the surface will jump the tiny distance of air; and travel through him to the earth

4 a The fuel rubs against the pipe, causing static; the charges can build up and create spark, igniting the fuel

 b The cable will channel the charges safely to earth; where they are dispersed

5 a Embedded in the ground

 b A positive charge

Page 238 Current, voltage and resistance

1 The electrons travel in one direction only

2 a 2.0 A

 b 2.0 A

3 a Parallel

 b Series

4 a In parallel

 b An ammeter

 c In series

5 7.0 V

6 a 0.15 J

 b 30 V

Answers

Page 239 Lamps, resistors and diodes

1 a Potential difference; current

b

resistance wire

c The resistance of the wire is directly proportional to its length; when the length doubles, the resistance doubles

2 It is constant

3 a As the current increases the filament gets hotter; which increases resistance

b The graph would curve; with a decreasing gradient

4 The diode only conducts in one direction; it has an infinite resistance when the current is zero; the diode has low resistance when it conducts

Page 240 Heating effects, LDRs and thermistors

1 The current in the lamp; causes it to heat up; excessive heat could cause a fire

2 a Current (A) = power (W) / potential difference (V);
= 4400 / 220; = 20 A

b Energy (J) = power (W) × time (s);
= 4400 × 300; = 1 320 000 J

3 The heating is a result of collisions between electrons; and the ions in the lattice; the collisions cause increased vibrations (around fixed positions) of the ions; which is what we mean by heat energy

4 The resistance of a thermistor decreases as its temperature increases; greater temperature means less resistance

5 a Thermistor

b LDR

c Thermistor

6 In thermistors an increase in temperature frees up more electrons from the atoms; and this causes the resistance to decrease

Page 241 Scalar and vector quantities

1 A scalar quantity only has magnitude; a vector quantity has magnitude and direction

2 a 3400 m

b 1700 m

c 0 m

3 1.5; m/s

4 a 0.5 m/s^2

b 36 m/s

c Its direction is changing; therefore its velocity is changing so it is accelerating

5 a 14 s

b Yes, the plane will be able to land; the distance required at an average speed of 17.5 m/s is 245 m, which is shorter than the runway at 400 m

Page 242 Distance–time and velocity–time graphs

1 a Speed

b Its speed is decreasing

2 1.6 / 0.4; = 4.0 m/s^2

3 Constant velocity

4 a It travels with a constant velocity of 30 m/s for 5 seconds; then it decelerates to 0 m/s in a further 3 seconds

b 10; m/s^2

c 150 m

d 45 m

Page 243 Understanding forces

1 a 500 N; forwards

b Zero

2 It will either remain stationary; or will continue moving with a constant velocity

3 a Newton's third law

b They are equal; and opposite

4 a The gravitational force; of the Earth on the Sun

b The gravitational force; of you pulling back on the Earth

Page 244 Force, mass and acceleration

1 Driving force; mass

2 a A motion sensor

b The mass of the trolley

c It will be proportional

3 120 − 30 = 90 N so 90 / 90; = 1 m/s^2

4 a 800 N

b 60 N

5 a Air resistance / drag

b Upwards

c He will slow down

d As he slows down the force of air resistance/drag will reduce; until it balances his weight; the resultant force will then be zero and he will continue at a constant speed

Page 245 Stopping distance

1 a Reaction time

b 6 m

2 Thinking distance; braking distance

3 The mass of the car increases; the speed of the car increases; there is reduced friction between the tyres and the road because of a wet or icy surface

4 a When walking / braking

b Between moving parts of a car; because it wastes energy

5 a 10 m

b 20 m

c 15 m

d 30 m

e 15 m

Answers

Page 246 Momentum

1 a 2400 kg m/s

  b It has both magnitude and direction

2 a 180 kg m/s

  b Zero

  c (45.0 × 2.0) + (6.0 × 0) = (51 × v) / 51 × v = 1800 kg m/s; v = 1800 / 51; = 3.5 m/s

3 0.5 m/s; in the direction of the 3.0 m/s carriage

4 a It increases the time it takes for the car (and its passengers) to stop; reducing the average force of the impact

  b Safety belts; air bags

5 40 N × 0.4 s; = 16 kg m/s

Page 247 Work, energy and power

1 a The work done when a force of 1 newton moves through a distance of 1 metre; in the direction of the force

  b 100 J

  c 240 J

2 a 4000 J

  b Friction causes energy to be wasted as heat

  c 100 W

3 a 2400 J

  b 60 W

4 20 000 W

Page 248 KE, GPE and conservation of energy

1 a Joules (J)

  b Kilograms (kg)

  c Metres per second (m/s)

2 a 37.5 J

  b Increase the mass; increase the velocity

3 100 000 J

4 a 5400 J

  b It is transferred to kinetic energy

  c 19 m/s

  d It has been transferred to heat; and sound

  e Friction (between the roller blades and the ground)

5 50 m/s

6 It increases by 4; for a given car, the braking force and mass are constants; therefore braking distance is directly proportional to velocity squared

Page 249 Atomic nuclei and radioactivity

1 a Protons; neutrons

  b It has the same number of each

  c Ion

  d Negative

  e Rubbing insulators together

2 a The element's chemical symbol

  b Mass number / nucleon number

  c Atomic number / proton number

  d By working out the difference between A and Z

3 A nuclei of an element with the same number of protons; but a different number of neutrons; it has the same chemical properties; because it has the same number of electrons

4 a Beta particle

  b Alpha particle

  c Gamma ray

5 a 4

  b Stopped by paper, skin or about 6 cm of air

  c −1

  d Very weak

6 Two protons; and two neutrons are removed from the nucleus

Page 250 Nuclear fission

1 a Fusion

  b Fission

  c Neutron

2 Chain reaction

3 a Uranium; plutonium

  b Because nuclear waste can remain radioactive for thousands of years; so it is difficult to dispose of safely

4 a Removes thermal energy produced in the core

  b Slows down the fast-moving neutrons; giving them a greater chance of reacting with other uranium nuclei

  c Absorb the neutrons; to control the chain reaction

5 The minimum mass of a fissile material; required to sustain a chain reaction

Page 251 Fusion on the Earth

1 a Deuterium

  b Tritium

  c Helium

  d Neutron

2 a Nuclear fusion carried out at room temperature

  b Pons and Fleischmann failed to produce sufficient details of their experiment; so their experiment could not be reproduced and therefore validated by other scientists

3 a Uncharged neutrons

  b Because the positively charged hydrogen nuclei repel each other; so they must move rapidly to overcome the electrostatic forces

  c By strong electromagnetic fields; produced by electromagnets

  d Because there are such high temperatures and pressures in stars

Page 252 Background radiation

1 Geiger counter

2 a Granite

  b Radon

3 a Any three from: nuclear power industry; nuclear weapons tests; air travel; medical; work-related

  b Cosmic rays; rocks; food

  c Natural sources

Answers

4 a Alpha radiation

b Alpha radiation cannot pass through the skin, but radon gas can be inhaled; so the alpha radiation enters the lungs; where it can cause cancer

Page 253 Uses of radioactivity

1 Any two from: irradiating food; detecting cracks in metal; detecting leaks in underground pipes; checking water quality

2 a Beta radiation

b It would get weaker

c It would get stronger

3 It is the most ionising radiation; so it creates more ions in the air than any other radiation; which makes the smoke alarm most effective

4 Equipment can be sterilised when sealed in a package

5 ii, iii, i

Page 254 Activity and half-life

1 a It cannot be predicted

b It is not affected by external conditions

2 a Bequerel (Bq)

b It doubles

c It is inversely proportional

3 a 200 Bq

b 50 Bq

4 Any three from: point the source away from other people; use special tools or gloves to handle the source; only remove the source from its lead-lined container when you need it; wash your hands after using the source

5 Nuclear power stations produce waste that remains radioactive for thousands of years; so disposing of it safely is a problem

6 It can be stored deep underground; or in special tunnels made under mountains.

Page 255 P2 Extended response question

0 marks
Insufficient or irrelevant science. Answer not worthy of credit.

1–2 marks
Answer may be simplistic. There may be limited use of specialist terms. Errors of grammar, punctuation and spelling prevent communication of the science.

3–4 marks
For the most part the information is relevant and presented in a structured and coherent format. Specialist terms are used for the most part appropriately. There are occasional errors in grammar, punctuation and spelling.

5–6 marks
All information in answer is relevant, clear, organised and presented in a structured and coherent format. Specialist terms are used appropriately. Few, if any, errors in grammar, punctuation and spelling.